AF389613

InSAR Signal and Data Processing

InSAR Signal and Data Processing

Editors

Mengdao Xing
Zhong Lu
Hanwen Yu

MDPI • Basel • Beijing • Wuhan • Barcelona • Belgrade • Manchester • Tokyo • Cluj • Tianjin

Editors

Mengdao Xing
Xidian University
China

Zhong Lu
Southern Methodist University
USA

Hanwen Yu
University of Houston
USA

Editorial Office
MDPI
St. Alban-Anlage 66
4052 Basel, Switzerland

This is a reprint of articles from the Special Issue published online in the open access journal *Sensors* (ISSN 1424-8220) (available at: https://www.mdpi.com/journal/sensors/special_issues/InSARSDP).

For citation purposes, cite each article independently as indicated on the article page online and as indicated below:

LastName, A.A.; LastName, B.B.; LastName, C.C. Article Title. *Journal Name* **Year**, *Article Number*, Page Range.

ISBN 978-3-03936-984-3 (Hbk)
ISBN 978-3-03936-985-0 (PDF)

Contents

About the Editors

Mengdao Xing (Professor) received B.S. and Ph.D. degrees from Xidian University, China, in 1997 and 2002, respectively. He is currently a professor with the National Laboratory of Radar Signal Processing, Xidian University. He holds the appointment of Associate Dean of the Academy of Advanced Interdisciplinary Research. His current research interests include synthetic aperture radar (SAR), inversed synthetic aperture radar (ISAR), sparse signal processing, and microwave remote sensing. He has written or co-written more than 200 refereed scientific journal papers. He has also authored or co-authored two books about SAR signal processing. His research has been cited more than 9350 times. He was rated one of the Most Cited Chinese Researchers by Elsevier. He has over 40 authorized Chinese patents. Dr. Xing's research has been supported by various funding programs, such as the National Science Fund for Distinguished Young Scholars. He is an IEEE Fellow. Currently, he serves as an Associate Editos for *Radar Remote Sensing* of IEEE, *Transactions on Geoscience and Remote Sensing*, and *Sensors*. He was a Guest Editor of the Special Issue "InSAR in Remote Sensing" of the IEEE Geoscience and Remote Sensing Magazine from 2018 to 2020, and a Guest Editor of the Special Issue InSAR Signal and Data Processing of *Sensors*.

Zhong Lu received B.S. and M.S. degrees from Peking University, Beijing, China, in 1989 and 1992, respectively, and a Ph.D. from the University of Alaska Fairbanks, USA, in 1996. He was a physical scientist with the United States Geological Survey during 1997–2013, and is now a professor and endowed Shuler-Foscue chair at Roy M. Huffington Department of Earth Sciences, Southern Methodist University, USA (www.smu.edu/dedman/lu). His research interests include technique developments of interferometric synthetic aperture radar (InSAR) processing and their applications to the study of volcano, landslide, and human-induced geohazards. He has published more than 200 peer-reviewed journal articles and book chapters focused on InSAR techniques and applications, and a book entitled "InSAR Imaging of Aleutian Volcanoes: Monitoring a Volcanic Arc from Space" (Springer, 2014). He is a member of NASA-India SAR (NISAR) Science Team (2012–), Senior Associate Editor of *Remote Sensing* and *Frontier in Earth Sciences*, and a member of the editorial boards of the *International Journal of Image and Data Fusion* and *Geomatics, Natural Hazards and Risk*.

Hanwen Yu (Doctor) was born in Xi'an, Shaanxi, China, in 1985. He received B.S. and Ph.D. degrees in electronic engineering from the Xidian University, Xi'an, in 2007 and 2012, respectively. He is currently a Post-Doctoral Research Fellow with the Department of Civil and Environmental Engineering, National Center for Airborne Laser Mapping, University of Houston, USA. He has authored or coauthored more than 25 research articles in high-impact peer-reviewed journals, such as the IEEE *Transactions on Geoscience and Remote Sensing* and the IEEE *Transactions on Image Processing*, and *Remote Sensing of Environment*. His research interests include phase unwrapping, machine learning, and synthetic aperture radar (SAR) interferometry (InSAR) signal processing and applications. Dr. Yu was a recipient of the Recognition of Best Reviewer of IEEE *Transactions on Geoscience and Remote Sensing* in 2019. Dr. Yu serves as a Topical Associate Editor of SAR remote sensing in IEEE *Transactions on Geoscience and Remote Sensing*. He was a Guest Editor of the Special Issue "InSAR in Remote Sensing" of the IEEE Geoscience and Remote Sensing Magazine from 2018 to 2020 and a Guest Editor of the Special Issue "InSAR Signal and Data Processing" of Sensors from 2019 to 2020.

Editorial

InSAR Signal and Data Processing

Mengdao Xing [1,*], Zhong Lu [2] and Hanwen Yu [3,4]

[1]   National Laboratory of Radar Signal Processing, Xidian University, Xi'an 710071, China
[2]   Roy M. Huffington Department of Earth Sciences, Southern Methodist University, Dallas, TX 75205, USA; zhonglu@mail.smu.edu
[3]   Department of Civil and Environmental Engineering, University of Houston, Houston, TX 77004, USA; yuhanwenxd@gmail.com
[4]   National Center for Airborne Laser Mapping, University of Houston, Houston, TX 77004, USA
*   Correspondence: xmd@xidian.edu.cn

Received: 30 June 2020; Accepted: 2 July 2020; Published: 7 July 2020

Abstract: We present here the recent advances in exploring new techniques related to interferometric synthetic aperture radar (InSAR) signal and data processing and applications.

1. Introduction

This Special Issue "InSAR Signal and Data Processing" of Sensors collects eleven articles from several InSAR researchers over several countries. The selected articles cover both InSAR signal processing techniques and their practical applications in Earth sciences. Readers of all levels will be able to gain a better understanding of InSAR as well as the when, the how, and the why of applying this technology.

2. Special Issue Contents

The first paper [1], "Polarimetric Stationarity Omnibus Test (PSOT) for Selecting Persistent Scatterer Candidates with Quad-Polarimetric SAR Datasets", proposes the polarimetric stationarity omnibus test method for improving the spatial density and the phase quality of persistent scatterer (PS) points. The experimental results show that the proposed method can achieve the polarimetric optimization of the interferometric phase of the PS, suppress the sidelobe of the strong scatterer effectively, and hence better reveal the details of the ground object. The second article [2], "Coherent Markov Random Field-Based Unreliable DSM Areas Segmentation and Hierarchical Adaptive Surface Fitting for InSAR DEM Reconstruction", proposes a novel InSAR digital elevation model reconstruction method using a digital surface model generated by an InSAR system with a coherent Markov random field technique. The comparison results shown in the experimental section indicate the superiority of the proposed algorithm. The third paper [3], "Multibaseline Interferometric Phase Denoising Based on Kurtosis in the NSST Domain", and the fourth article [4], "Extended Phase Unwrapping Max-Flow/Min-Cut Algorithm for Multibaseline SAR Interferograms Using a Two-Stage Programming Approach", focus on phase denoising and phase unwrapping techniques of multibaseline InSAR, respectively. The fifth paper [5], "Mining-Induced Time-Series Deformation Investigation Based on SBAS-InSAR Technique: A Case Study of Drilling Water Solution Rock Salt Mine", shows an InSAR case study concerning salt extraction based on solution mining. The study applies the SBAS-InSAR technique to obtain the spatial–temporal characteristics of the ground subsidence caused by solution mining activities. The sixth paper [6], "Phase Difference Measurement of Under-Sampled Sinusoidal Signals for InSAR System Phase Error Calibration", discusses the issue related to phase error calibration in spaceborne single-pass InSAR. The proposed method of the phase difference measurement of the high-frequency internal calibration signal of the InSAR system is suitable for the phase error calibration. As the highest elevation permafrost region in the world, the Qinghai-Tibet Plateau (QTP) permafrost is quickly degrading due to global warming, climate change, and human activities. The seventh article [7], "Permafrost Deformation Monitoring Along the Qinghai-Tibet Plateau Engineering Corridor Using InSAR Observations with Multi-Sensor SAR Datasets from 1997–2018", presents an application using a time-series InSAR technique with multiple SAR datasets to monitor the permafrost ground deformation along the QTEP from 1997 to 2018. GaoFen-3 is a new Chinese InSAR remote sensing satellite. The eighth article [8], "ScanSAR Interferometry of the Gaofen-3 Satellite with Unsynchronized Repeat-Pass Images", discusses interferometric analysis and processing methods for GaoFen-3 images in ScanSAR mode. The ninth paper [9], "A Highly Efficient Heterogeneous Processor for SAR Imaging", concerns the hardware design of a SAR signal processor consisting of two 18 × 16 heterogeneous arrays that provide 115.2 GOPS throughput. In the tenth paper [10], "Monitoring the Land Subsidence Area in a Coastal Urban Area with InSAR and GNSS", 34 scenes of Sentinel-1A SAR images are used for SBAS and PS processing to obtain the surface deformation field of a large region spanning the Shenzhen, China, and Hong Kong Special Administrative Regions. The last article [11], "Safe Helicopter Landing on Unprepared Terrain Using Onboard Interferometric Radar", proposes an interferometric radar survey system for the generation of ground surface topography for helicopter landing sites. The system generates high-quality three-dimensional terrain surface topography data and estimates the slope of the site with the required accuracy.

Acknowledgments: The guest editors would like to thank the authors' contribution to this Special Issue and all reviewers for providing valuable and constructive comments. In particular, we would like to thank the in-house editor of the Sensors journal for the administrative support.

Conflicts of Interest: The authors declare no conflict of interest.

References

1. Luo, X.; Wang, C.; Shen, P. Polarimetric Stationarity Omnibus Test (PSOT) for Selecting Persistent Scatterer Candidates with Quad-Polarimetric SAR Datasets. *Sensors* **2020**, *20*, 1555. [CrossRef] [PubMed]
2. Qian, Q.; Wang, B.; Hu, X.; Xiang, M. Coherent Markov Random Field-Based Unreliable DSM Areas Segmentation and Hierarchical Adaptive Surface Fitting for InSAR DEM Reconstruction. *Sensors* **2020**, *20*, 1414. [CrossRef] [PubMed]
3. Liu, Y.; Li, S.; Zhang, H. Multibaseline Interferometric Phase Denoising Based on Kurtosis in the NSST Domain. *Sensors* **2020**, *20*, 551. [CrossRef] [PubMed]
4. Zhou, L.; Lan, Y.; Xia, Y.; Gong, S. Extended Phase Unwrapping Max-Flow/Min-Cut Algorithm for Multibaseline SAR Interferograms Using a Two-Stage Programming Approach. *Sensors* **2020**, *20*, 375. [CrossRef] [PubMed]
5. Liu, X.; Xing, X.; Wen, D.; Chen, L.; Yuan, Z.; Liu, B.; Tan, J. Mining-Induced Time-Series Deformation Investigation Based on SBAS-InSAR Technique: A Case Study of Drilling Water Solution Rock Salt Mine. *Sensors* **2019**, *19*, 5511. [CrossRef] [PubMed]
6. Yuan, Z.; Gu, Y.; Xing, X.; Chen, L. Phase Difference Measurement of Under-Sampled Sinusoidal Signals for InSAR System Phase Error Calibration. *Sensors* **2019**, *19*, 5328. [CrossRef] [PubMed]
7. Zhang, Z.; Wang, M.; Wu, Z.; Liu, X. Permafrost Deformation Monitoring Along the Qinghai-Tibet Plateau Engineering Corridor Using InSAR Observations with Multi-Sensor SAR Datasets from 1997–2018. *Sensors* **2019**, *19*, 5306. [CrossRef] [PubMed]
8. Sun, Z.; Yu, A.; Dong, Z.; Luo, H. ScanSAR Interferometry of the Gaofen-3 Satellite with Unsynchronized Repeat-Pass Images. *Sensors* **2019**, *19*, 4689. [CrossRef] [PubMed]
9. Wang, S.; Zhang, S.; Huang, X.; An, J.; Chang, L. A Highly Efficient Heterogeneous Processor for SAR Imaging. *Sensors* **2019**, *19*, 3409. [CrossRef] [PubMed]

10. Hu, B.; Chen, J.; Zhang, X. Monitoring the Land Subsidence Area in a Coastal Urban Area with InSAR and GNSS. *Sensors* **2019**, *19*, 3181. [CrossRef] [PubMed]
11. Shimkin, P.E.; Baskakov, A.I.; Komarov, A.A.; Ka, M.-H. Safe Helicopter Landing on Unprepared Terrain Using Onboard Interferometric Radar. *Sensors* **2020**, *20*, 2422. [CrossRef] [PubMed]

 *sensors*

Article

Extended Phase Unwrapping Max-Flow/Min-Cut Algorithm for Multibaseline SAR Interferograms Using a Two-Stage Programming Approach

Lifan Zhou [1,*]**, Yang Lan** [2,3]**, Yu Xia** [1] **and Shengrong Gong** [1]

[1]   School of Computer Science and Engineering, Changshu Institute of Technology, Changshu 215500, China; cslgxiayu@163.com (Y.X.); shrgong@cslg.edu.cn (S.G.)
[2]   National Laboratory of Radar Signal Processing, Xidian University, Xi'an 710071, China; lanyangxd@hotmail.com
[3]   Collaborative Innovation Center of Information Sensing and Understanding, Xidian University, Xi'an 710071, China
*   Correspondence: zhoulifan_rs@163.com

Received: 21 November 2019; Accepted: 7 January 2020; Published: 9 January 2020

Abstract: Multi-baseline (MB) phase unwrapping (PU) is a key step of MB synthetic aperture radar (SAR) interferometry (InSAR). Compared with the traditional single-baseline (SB) PU, MB PU is applicable to the area where topography varies violently without obeying the phase continuity assumption. A two-stage programming MB PU approach (TSPA) proposed by H. Yu. builds the link between SB and MB PUs, so many existing classical SB PU methods can be transplanted into the MB domain. In this paper, an extended PU max-flow/min-cut (PUMA) algorithm for MB InSAR using the TSPA, referred to as TSPA-PUMA, is proposed, consisting of a two-stage programming procedure. In stage 1, phase gradients are estimated based on Chinese remainder theorem (CRT). In stage 2, a Markov random field (MRF) model of PUMA is designed for modeling local contextual dependence based on the phase gradients obtained by stage 1. Subsequently, the energy of the MRF model is minimized by graph cuts techniques. The experiment results illustrate that the TSPA-PUMA method can drastically enhance the accuracy of the original PUMA method in the rugged area, and is more efficient than the original TSPA method. In addition, the noise robustness of TSPA-PUMA can be improved through adding more interferograms with different baseline lengths.

Keywords: phase unwrapping (PU); multi-baseline (MB); two-stage programming approach (TSPA); phase unwrapping max-flow/min-cut (PUMA)

---

1. Introduction

Interferometric synthetic aperture radar (InSAR) is a powerful tool to reconstruct the digital elevation model (DEM) or surface deformation of the Earth's surface [1]. Phase unwrapping (PU), as a key processing step of InSAR, is the procedure of retrieving the absolute phase through the wrapped phase. Unfortunately, the traditional single-baseline (SB) PU is an ill-posed problem, i.e., there are infinite solutions to it, if no extra information is added. In fact, a phase continuity assumption (also known as Itoh condition) employed by most SB PU methods is that the absolute value of phase differences between neighboring pixels is less than π [2]. Unfortunately, violent terrain changes and high system noise frequently fail to observe the phase continuity assumption in reality, so it is still difficult for SB PU to generate the correct PU result. However, the multi-baseline (MB) PU problem is well-posed rather than ill-posed, which makes use of the baseline diversity to significantly increase the ambiguity intervals of interferometric phases. To be specific, MB PU can completely eliminate the phase-continuity assumption.

In recent decades, the MB PU has been widely investigated. Yu et al. [3] provided a good review article of MB PU methods, which described that there are mainly two groups of methods: parametric-based and non-parametric-based methods. The main ideas of these two groups of MB PU methods both come from machine-learning technology [3]. The methods in the first group utilize the InSAR probability density function to build a statistical framework based on maximum likelihood (ML) [4–6] or maximum a posteriori (MAP) criteria [7,8] to find the MB PU result, [9] provided a good review of the ML- and MAP-based methods, and [10] gave a comparative study of the PU accuracy between the ML- and MAP-based methods. The methods in the second group translate the MB PU problem into an unsupervised learning problem. [11] presented a fast cluster-analysis (CA)-based MB PU method, and [12] further improved it. Besides these two groups of methods, three basic MB PU methods, i.e., the Chinese remainder theorem (CRT)-based method, projection method, and linear combination method, were put forward in [13]. [14] proposed the $L\infty$-norm programming criterion applied to the MB PU. To improve the robustness to noise, [15] presented a closed-form robust CRT method, and [16] put forward a MB PU method based on the mix-integer optimization model. More than that, [17] proposed a Kalman filtering-based MB PU method, and a wavelet approach-based MB PU method was presented in [18]. It should be noted that the major difference between SB and MB PUs lies in their different processing steps. For the detailed implementation of the SB and MB PU methods, the readers can refer to [3].

However, most of the aforementioned MB PU methods suffer from poor noise robustness, and the reason for the noise robustness problem is caused by system noise, surface deformation, or atmospheric effect [3]. In addition, the ML-, MAP-, and CA-based MB PU methods are all based on machine-learning techniques, so they usually need to determine some parameters through some extra information because they do not have clear PU meanings. Under these conditions, these MB PU methods are quite limited in real application. To solve these problems, Yu and Lan [19] proposed a two-stage programming-based MB PU method, abbreviated as TSPA, that formulates a connection between SB and MB PUs, which is also known as TSPA-InSAR technology. In stage 1, TSPA estimates the ambiguity number difference between neighboring pixels using multiple interferograms with different baseline lengths based on the CRT formulation. In stage 2, TSPA obtains the final PU result through using the L^1-norm SB PU method, i.e., minimum-cost flow (MCF) PU method [20]. It is noted that there are several strongly polynomial algorithms that can be applied to solve the MCF model (e.g., minimum mean cycle-canceling algorithm and network simplex algorithm [21]). More than that, some studies indicate that the divide-and-conquer criterion can be used to further reduce the computational and peak memory consumption of the MCF model [22,23]. To further improve the noise robustness of stage 1 of TSPA, [24] proposed a local phase model, which assumes terrain height surface in the neighborhood pixels can be approximated by a plane. Furthermore, [25] used the unscented Kalman filter (UKF) to improve the performance of the stage 2 of TSPA reducing the effect of the noise gradient on the PU results. Furthermore, [26] proposed a technique for applying TSPA to the large-scale MB InSAR data set based on the MB envelope-sparsity theorem. Compared with most of the aforementioned existing MB PU methods, the two main contributions of the TSPA method are listed as follows. First, as a MB PU method, TSPA does not obey the phase continuity assumption by taking advantage of MB diversity. Second, since TSPA makes the link between SB and MB PUs, many existing classical SB PU methods can be transplanted into MB domain.

A SB PU algorithm based on graph cuts, referred to as phase unwrapping max-flow/min-cut (PUMA), was proposed by Bioucas-Dias and Valadao [27]. This algorithm uses a new energy minimization framework, which is based on the Markov random field (MRF). Under this condition, the problem of ambiguity number estimation can be translated into computing a sequence of binary optimizations (i.e., {0, 1}-cut), which can be solved by graph cuts techniques. The reason why this algorithm is so popular is that the MRF model allows a large family of potential functions (i.e., consisting of convex potential and non-convex potential), which gives flexibility to handle effectively both continuous and discontinuous phase features. For convex potentials, the PUMA algorithm exactly solves the classical minimum L^p norm PU problem with $p \geq 1$. For non-convex potentials, the potentials with exponent less than one with $0 < p < 1$ have been employed to allow discontinuity preservation [27]. However, as a SB PU algorithm, PUMA is still limited to the phase continuity assumption, so it is potentially hard for the PUMA algorithm to obtain the correct PU result in the discontinuous region. Some researchers have already noticed this issue. [28] extended the PUMA algorithm into MB domain to further increase the discontinuity preserving ability of PUMA, but it is only less influenced by the phase continuity assumption rather than violating the phase continuity assumption. Contrarily, as described earlier, TSPA does not need to satisfy the phase continuity assumption through using two-stage programming. In this case, there is a straightforward idea to transplant the PUMA algorithm into the MB domain using the TSPA approach.

In this paper, an extended PUMA algorithm for MB InSAR using the TSPA approach, abbreviated as the TSPA-PUMA method, is proposed, which consists of a two-stage programming procedure. In stage 1 of TSPA-PUMA, stage 1 of the original TSPA is utilized to estimate the phase gradients based on CRT without obeying the phase continuity assumption. In stage 2 of TSPA-PUMA, an MRF model of PUMA with different types of clique potentials is designed for modeling local contextual dependence based on the phase gradients obtained by stage 1. Subsequently, the energy of MRF model for SB PU is minimized by computing a sequence of binary optimizations solved by graph cuts techniques. This paper uses three simulated InSAR data experiments and two real InSAR data experiments to validate the proposed approach. The results show that the TSPA-PUMA method can significantly improve the PU accuracy of the original PUMA algorithm in the rugged and mountainous area, and the noise robustness of TSPA-PUMA can be improved if employing more interferograms with different baseline lengths.

The rest of this paper is organized as follows. Section 2 reviews the original PUMA method and analyzes its disadvantages of dealing with steep terrain. In Section 3, the TSPA-PUMA method is introduced in detail. Besides that, the noise robustness, time complexity, and parameter selection of TSPA-PUMA are also analyzed. Then, in Section 4, the TSPA-PUMA method is verified by a set of simulated and real MB InSAR datasets and the corresponding experimental results are discussed in detail. Finally, Section 5 concludes this paper.

2. Review and Analysis of SB PUMA

2.1. Basic Principle of PUMA

In this section, we will review the original PUMA algorithm in SB case. SB PU can be regarded as estimating the unknown integral multiple of 2π to be added at each pixel of the wrapped phase image to restore the absolute phase, given by:

$$\varphi(s) = \psi(s) - 2k(s)\pi, \tag{1}$$

where $\varphi(s)$ is the wrapped phase of the sth pixel, $\psi(s)$ is the unknown absolute phase of the sth pixel, and $k(s)$ is the unknown ambiguity number of the sth pixel, which is also known as the wrap count. From (1), we can see that directly solving (1) is an ill-posed inverse problem, because there are two unknowns in one equation, i.e., there is no unique solution to (1). Similar to other SB PU methods, the PUMA algorithm also uses the phase continuity assumption to solve this problem. The energy minimization function for PUMA is given by:

$$\arg \min_{k(s)} \sum_{(s,s-1)} w(s,s-1) \cdot V\big(\Delta_\psi(s,s-1)\big), \tag{2}$$

where the indexes s and $s-1$ denote two neighboring pixels and $w(s,s-1)$ is the weighted coefficient, which can be derived from any kind of quality map in InSAR [29]. $V(\cdot)$ is clique potential, defined by $V(\cdot) = (\cdot)^p$, and p is the potential exponent, which determines how the phase of the neighboring pixels in the clique interact. Note that changes of the MRF model of PUMA depend primarily on choosing different clique potential $V(\cdot)$. When the corresponding clique potentials are convex (i.e., $p \geq 1$), PUMA exactly solves the classical L^p minimum norm PU problem. In the case $p = 2$, PUMA will become the least square method. A drawback of the L^2-norm clique potential is that it tends to smooth discontinuities. L^1-norm clique potential ($p = 1$) performs better than L^2-norm clique potential in preserving discontinuities. The major advantage of PUMA lies its non-convex clique potential with $0 < p < 1$, which allows an increased probability of sharp transitions. $\Delta_\psi(s, s-1)$ is the absolute phase gradients, i.e., the absolute phase difference of the neighboring pixels, which is defined by:

$$\Delta_\psi(s, s-1) = \Delta_\varphi(s, s-1) + 2\pi \cdot (k(s) - k(s-1)), \tag{3}$$

where $\Delta_\varphi(s, s-1)$ is the wrapped phase differences of the neighboring pixels. The PUMA algorithm aims to estimate the wrap count $k(s)$ that minimizes the phase gradients $\Delta_\psi(s, s-1)$ obtained by Equation (2), which can be regarded as a binary optimization problem. Initially, the labels of all pixels are set to zero, i.e., $k^{t=0}(s) = 0$. At each iteration step, every pixel's label would either be 1 or 0, i.e., $k^{t+1}(s) = k^t(s) + \delta^{t+1}(s)$, in which the t denotes iteration and $\delta^{t+1}(s) \in \{0, 1\}$, meaning that every pixel's label either increases by 1 (phase plus 2π) or 0 (phase remains unchanged). Every iteration aims to decrease the value of the energy function of Equation (2) as much as possible. After each iteration, the unwrapped phase is updated, i.e., $\psi^{t+1}(s) = \varphi(s) + 2\pi \cdot k^{t+1}(s)$, and the energy function of Equation (2) is recalculated. When the energy ceases to decrease, the iteration is terminated, where the unwrapped phase is estimated, i.e., $\psi^{t=end}(s) = \varphi(s) + 2\pi \cdot k^{t=end}(s)$. The binary optimization problem in the above referred sequence can be solved by graph cuts from [30], which are computed efficiently using max-flow/min-cut algorithms. For the convex clique potential ($p \geq 1$), because it satisfies the regularity condition, this binary optimization problem can be solved exactly using the standard graph cuts algorithm. With respect to the non-convex clique potential ($0 < p < 1$), because it does not obey the regularity condition, it is impossible to minimize the energy function of Equation (2) via the standard graph cuts algorithm. To solve this issue, an approximate version of the graph cuts algorithm is devised by applying majorize-minimize (MM) approximation, which can cope with the local minima arising from non-convex potentials. For the detailed implementation of graph cuts-based optimization of the energy function of Equation (2), the readers can refer to [27].

2.2. Problem Analysis

As described above, the PUMA algorithm aims to estimate the wrap count $k(s)$ that minimizes the phase gradients $\Delta_\psi(s, s-1)$ obtained by Equation (2) according to the phase continuity assumption. From Equation (2), we can see that the credibility of the PU result of PUMA is directly related to the correctness of $\Delta_\psi(s, s-1)$. Unfortunately, violent topographic changes and high system noise frequently make the phase continuity assumption does not work well. Under this condition, it is difficult to obtain the correct $\Delta_\psi(s, s-1)$ from the phase continuity assumption. Therefore, if the accuracy of $\Delta_\psi(s, s-1)$ is too low, no matter what kind of clique potential $V(\cdot)$ is employed, it could be impossible for the PUMA algorithm to obtain the full correct PU solution. For example, Figure 1a,b show the reference unwrapped phases with two different baselines, which come from the mountainous area around the Isolation Peak region of Colorado [31]. Figure 1c,d show two simulated noise-free interferograms of Figure 1a,b. Table 1 illustrate the major parameters of the simulated system. Figure 1e,f show the PU results of Figure 1c obtained by the PUMA methods with clique potential exponent 1 and 0.5, respectively. Figure 1g,h are the errors between Figures 1a and 1e,f, respectively. Figure 1i,j show the PU results of Figure 1d obtained by the PUMA methods with clique potential exponent 1 and 0.5, and the corresponding errors between Figures 1b and 1i,j are shown in Figure 1g,h, respectively. To fairly evaluate the PU results, the same reference point and range of the color bar are used in the PU results obtained by the two PUMA methods of the same interferogram, respectively (similarly hereinafter in experiments 1, 2 3, 4, and 5). Because the pattern of the fringes in Figure 1c is simple, we can see that the two PUMA methods with clique potential exponent 1 and 0.5 both obtain the correct PU results. However, when the pattern of the fringes in Figure 1d becomes very complicated which is difficult for the PU process, the PU accuracy of these two methods will significantly decrease. The reason is that the pattern of the fringes in Figure 1d changes fiercely, which makes the failure of the phase continuity assumption, i.e., the absolute phase differences between neighboring pixels are larger than π. Under this condition, even if PUMA with non-convex potential is better than that with convex potential due to its discontinuity preserving ability, it is still difficult enough for PUMA with non-convex potential to perform correctly. Therefore, it can be seen that the PUMA method can find the correct PU result in the area where topography is comparative flat, but in the area where topography jumps more drastically, PUMA cannot find the correct PU solution anymore, no matter what kind of clique potential is chosen.

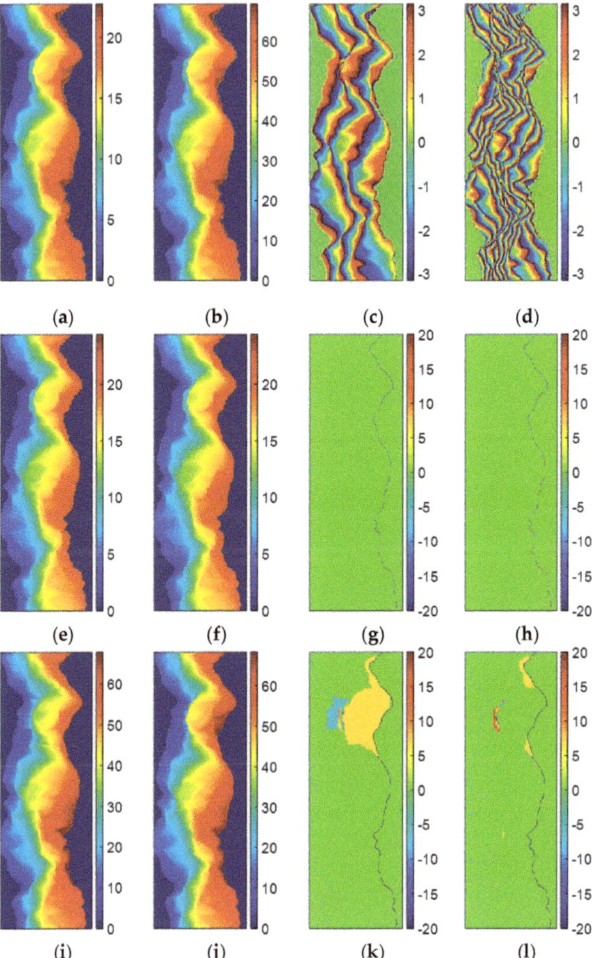

Figure 1. (**a**,**b**) Reference unwrapped phases ((**a**) short and (**b**) long baseline length). (**c**,**d**) Simulated wrapped phases of (**a**,**b**). (**e**,**f**) PU results of (**c**) obtained by (**e**) PUMA (clique potential exponent is 1), and (**f**) PUMA (clique potential exponent is 0.5). (**g**,**h**) Errors between (**a**) and PU results (**e**,**f**). (**i**,**j**) PU results of (**d**) obtained by (**i**) PUMA (clique potential exponent is 1), and (**j**) PUMA (clique potential exponent is 0.5). (**k**,**l**) Errors between (**b**) and PU results (**i**,**j**).

Table 1. Major parameters of simulated InSAR system and Interferograms.

Orbit Altitude	Incidence Angle	Wavelength
6885 km	46°	0.031 m
Interferogram	**Baseline Length**	
Figure 1c	150 m	
Figure 1d	330 m	

3. TSPA-PUMA Methodfor MB PU

According to the discussion in Section 2, we conclude that the traditional PUMA algorithm is limited to the phase continuity assumption. In this Section, we will introduce the proposed TSPA-PUMA method which can break through the limitation of the phase continuity assumption. In this Section, we only consider the dual-baseline (DB) case for simplicity, and the MB case can be extended easily. A schematic representation of the proposed TSPA-PUMA is illustrated in Figure 2. In the following, we will introduce the two stages in the TSPA-PUMA method in detail.

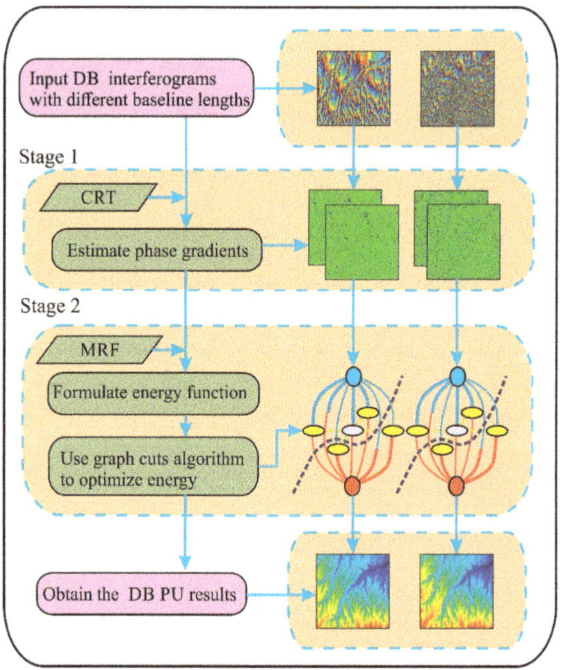

Figure 2. Schematic representation of the proposed TSPA-PUMA method.

3.1. Stage 1: Estimating the Phase Gradient

The DB InSAR measurement of a pixel case can be given by:

$$\varphi_r(s) = \psi_r(s) - 2k_r(s)\cdot\pi, \tag{4}$$

where $\varphi_r(s)$, $\psi_r(s)$ and $k_r(s)$ are the wrapped phase, absolute phase, and ambiguity number of the sth pixel in interferogram r ($r = 1, 2$), respectively. $\varphi_r(s)$ can be measured by the DB InSAR system, but $\psi_r(s)$ and $k_r(s)$ are the unknowns in one equation that need to be solved. If the ambiguity number $k_r(s)$ of each pixel in two interferograms can be solved, $\psi_r(s)$ can be obtained through Equation (4).

The absolute phases of the two interferograms can be calculated by using the baseline lengths such as [19]:

$$B_2 \cdot (\varphi_1(s) + 2\pi \cdot k_1(s)) = B_1 \cdot (\varphi_2(s) + 2\pi \cdot k_2(s)), \tag{5}$$

where B_1 and B_2 represent two different normal baseline (also known as perpendicular baseline) lengths. In this paper, normal baseline length is abbreviated as baseline length. According to Equation (5), the TSPA-PUMA method maintains the stage 1 of the TSPA, which builds the relationship of phase gradient information in different interferograms with different baseline lengths, given by:

$$B_2 \cdot \left(\Delta_{\varphi_1}(s,s-1) + 2\pi \cdot \hat{\Delta}_{k_1}(s,s-1) \right) = B_1 \cdot \left(\Delta_{\varphi_2}(s,s-1) + 2\pi \cdot \hat{\Delta}_{k_2}(s,s-1) \right), \tag{6}$$

where $\Delta_{\varphi_1}(s,s-1)$ and $\Delta_{\varphi_2}(s,s-1)$ are the wrapped phase differences between neighboring pixels of interferogram r (r = 1, 2), $\hat{\Delta}_{k_1}(s,s-1)$ and $\hat{\Delta}_{k_2}(s,s-1)$ are the ambiguity number gradient between neighboring pixels of interferogram r. Note that there are two directions (vertical and horizontal) of neighboring pixels for $\hat{\Delta}_{k_r}(s,s-1)$ and $\Delta_{\varphi_r}(s,s-1)$. Because $\hat{\Delta}_{k_1}(s,s-1)$ and $\hat{\Delta}_{k_2}(s,s-1)$ belong to the integer, we can obtain the solution to Equation (6) under some special combination of the baseline lengths according to CRT [19]. Equation (6) can be solved by the optimization model given by:

$$\begin{aligned} &\arg \quad \min_{\hat{\Delta}_{k_r}(s,s-1)} \quad \left| h(s,s-1) \right| \\ &\text{s.t.} \quad \hat{\Delta}_{k_r}(s,s-1) \in \text{integer}, r = 1,2, \end{aligned} \tag{7}$$

where $\hat{\Delta}_{k_r}(s,s-1)$ are the decision variables of interferogram r. It is noted that $\hat{\Delta}_{k_r}(s,s-1)$ can be larger than 1 or less than -1, which implies that the phase continuity assumption does not need to be satisfied (the phase continuity assumption only allows $\hat{\Delta}_{k_r}(s,s-1)$ to be ± 1 or 0). $h(s,s-1)$ is the auxiliary variables, defined by:

$$h(s,s-1) = B_2 \cdot \left(\Delta_{\varphi_1}(s,s-1) + 2\pi \cdot k_1(s) \right) - B_1 \cdot \left(\Delta_{\varphi_2}(s,s-1) + 2\pi \cdot k_2(s) \right). \tag{8}$$

It can be seen that $h(s,s-1)$ is the CRT bias, so Equation (8) is to find the ambiguity number gradient $\hat{\Delta}_{k_r}(s,s-1)$ with minimum CRT bias [19]. Under this condition, the phase gradient $\hat{\Delta}_{\psi_r}(s,s-1)$ of interferogram r can be estimated by:

$$\hat{\Delta}_{\psi_r}(s,s-1) = \Delta_{\varphi_r}(s,s-1) + 2\pi \cdot \hat{\Delta}_{k_r}(s,s-1). \tag{9}$$

3.2. Stage 2: Unwrapping the Phase Gradient Using Graph Cuts Algorithm

Based on the gradient information obtained by Equation (9), the energy minimization framework based on the MRF model for TSPA-PUMA respectively obtain the final PU solution of each interferogram r, which is obtained by Equation (10),

$$\arg \quad \min_{k_r(s)} \quad \sum_{(s,s-1)} w_r(s,s-1) \cdot V \left(\Delta_{\psi r}(s,s-1) - \hat{\Delta}_{\psi_r}(s,s-1) \right), \tag{10}$$

where $w_r(s,s-1)$ is the weighted coefficient of interferogram r, and $k_r(s)$ is the decision variable of interferogram r. From Equation (10), it can be seen that the aim of TSPA-PUMA is to minimize the difference between the absolute phase gradients $\Delta_{\psi r}(s,s-1)$ and the estimated gradients $\hat{\Delta}_{\psi_r}(s,s-1)$ obtained from stage 1 of TSPA-PUMA. Compared with the traditional PUMA algorithm which obeys the phase continuity assumption, the major improvement of TSPA-PUMA is that it does not need to follow the assumption, because the ambiguity number gradient $\hat{\Delta}_{k_r}(s,s-1)$ obtained by Equation (7) can be larger than 1 or less than -1. If we transform the phase gradients $\Delta_{\psi}(s,s-1)$ obtained by the Equation (3) into DB case, we will obtain:

$$\Delta_{\psi r}(s, s-1) = \Delta_{\varphi r}(s, s-1) + 2\pi \cdot (k_r(s) - k_r(s-1)).\tag{11}$$

Then, if we substitute Equations (9) and (11) into Equation (10), the energy minimization framework for TSPA-PUMA can be rewritten to:

$$\arg\min_{k_r(s)} \sum_{(s,s-1)} w_r(s, s-1) \cdot V\big(k_r(s) - k_r(s-1) - \hat{\Delta}_{k_r}(s, s-1)\big),\tag{12}$$

where $k_r(s)$ $(r = 1, 2)$ are solutions to Equation (12) of the two different interferograms r. Because optimization of $k_1(s)$ and $k_2(s)$ is independent of each other, we can optimize them separately. Similar to the PUMA algorithm, the minimization of the energy function of TSPA-PUMA obtained by Equation (12) can be regarded as a jump-move optimization problem. It is worth mentioning that, with respect to TSPA, the innovative part of TSPA-PUMA lies in stage 2, where the graph cuts algorithm is used to optimize the energy function of (12). Initially, the ambiguity number of the sth pixel in interferogram r is set to zero, i.e., $k_r^{t=0}(s) = 0$. At each iteration, every ambiguity number of the sth pixel in interferogram r either increases by one (i.e., the ambiguity number pluses one) or zero (i.e., the ambiguity number remains unchanged), that is, $k_r^{t+1}(s) = k_r^t(s) + \delta_r^{t+1}(s)$, where $\delta_r^{t+1}(s) \in \{0, 1\}$. For each pair of neighboring pixels $(s, s-1)$ in interferogram r, the clique potential to be minimized is defined as:

$$E\big(\delta_r^{t+1}(s), \delta_r^{t+1}(s-1)\big) = V\big(k_r^t(s) - k_r^t(s-1) - \hat{\Delta}_{k_r}(s, s-1) + \delta_r^{t+1}(s) - \delta_r^{t+1}(s-1)\big).\tag{13}$$

For the convex clique potential $(p \geq 1)$, the clique potential obtained by (13) satisfies the regularity condition, so the standard graph cuts algorithm can be used to optimize them. For the non-convex clique potential $(0 < p < 1)$, the MM concept [27] is employed to make the non-convex clique potential obtained by (14) obey the regularity condition, so they can also be optimized by the standard graph cuts algorithm. According to (13), we have:

$$\begin{aligned}
E(0,0) &= V\big(k_r^t(s) - k_r^t(s-1) - \hat{\Delta}_{k_r}(s, s-1)\big) \\
E(1,1) &= V\big(k_r^t(s) - k_r^t(s-1) - \hat{\Delta}_{k_r}(s, s-1)\big) \\
E(0,1) &= V\big(k_r^t(s) - k_r^t(s-1) - \hat{\Delta}_{k_r}(s, s-1) - 1\big) \\
E(1,0) &= V\big(k_r^t(s) - k_r^t(s-1) - \hat{\Delta}_{k_r}(s, s-1) + 1\big).
\end{aligned}\tag{14}$$

Considering all pairs of neighboring pixels, the energy minimization function of each binary iteration is given by:

$$\arg\min_{\delta_r^{t+1}(s)} \sum_{(s,s-1)} w_r(s, s-1) \cdot E\big(\delta_r^{t+1}(s), \delta_r^{t+1}(s-1)\big).\tag{15}$$

The minimization of (15) can be achieved through a cut on the weighted graph $\sigma = \langle v, \epsilon \rangle$ with two terminals α and β. The set of vertices v represent the pixels in each interferogram, and the set of edges ϵ denote the pairs of neighboring vertices in each interferogram. An $\alpha - \beta$ cut is a set of edges such that the terminals are separated into two disjoint sets $\alpha \in 1$, i.e., the ambiguity number pluses one, and $\beta \in 0$, i.e., the ambiguity number remains unchanged. The cost of the cut equals the sum of its clique potential between α and β. Then, we construct the elementary graph for each clique potential, as shown in Figure 3a,b. From Figure 3a,b, it can be seen that the directed edge $(s, s-1)$ is assigned a weight of $E(0,1) + E(1,0) - E(0,0) - E(1,1)$. Moreover, for vertex s, if $E(1,0) - E(0,0) > 0$, then the edge (α, s) is assigned a weight of $E(1,0) - E(0,0)$; otherwise, the edge (s, β) is assigned a weight of $E(0,0) - E(1,0)$. Similarly, for the neighboring vertex $s-1$, if $E(1,1) - E(1,0) > 0$, then the edge $(\alpha, s-1)$ is assigned a weight of $E(1,1) - E(1,0)$; otherwise, the edge $(s-1, \beta)$ is assigned a weight of $E(1,0) - E(1,1)$. Finally, the two elementary graphs are merged to obtain a main graph, as shown in Figure 3c. At every jump-move iteration, the minimum cut problem attempts to find the cheapest cut among all cuts separating the terminals, which can be obtained using the max-flow algorithm. That is to say, every jump-move iteration is intended to reduce the value of the energy function of (15) as much as possible. When the energy ceases to decrease, the binary jump move is terminated. Finally, we can obtain the DB PU results, i.e., $\psi_r^{t=end}(s)$ $(r = 1, 2)$, which is equal to $\varphi_r(s) + 2\pi \cdot k_r^{t=end}(s)$ $(r = 1, 2)$.

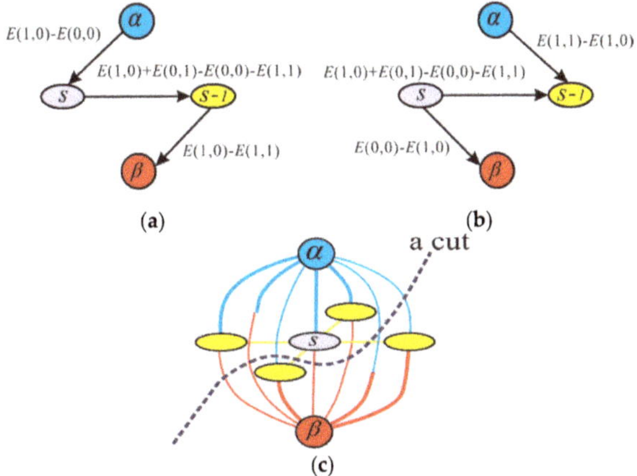

Figure 3. The elementary graph is constructed, where α and β represent two terminals and $(s, s-1)$ represent the two neighboring pixels. (**a**) In the case of $E(1,0) - E(0,0) > 0$ and $E(1,1) - E(1,0) > 0$. (**b**) In the case of $E(1,0) - E(0,0) < 0$ and $E(1,1) - E(1,0) < 0$. (**c**) A main graph is obtained by merging the two elementary graphs, where an $\alpha - \beta$ cut is a set of edges such that the terminals are separated into two disjoint sets $\alpha \in 1$ (the ambiguity number pluses one) and $\beta \in 0$ (the ambiguity number remains unchanged).

3.3. Analysis of the Noise Robustness

It should be noted that stage 1 of TSPA-PUMA is dependent on CRT, which is too sensitive to measurement bias that is potentially caused by some decorrelation factors, e.g., atmospheric effect or co-registration error, etc. Considering the atmospheric artifact, this usually shows a strong spatial correlation [32]. Hence, the effect of atmosphere on the wrapped phases of neighboring pixels should be close to each other. Because Equation (6) uses the information of wrapped phase difference between neighboring pixels, the effect of atmosphere could be counteracted in Equation (6). Therefore, stage 1 of the TSPA-PUMA method does not fear the atmospheric effect. However, it is still sensitive to the noise levels caused by other decorrelation components. Under this condition, the incorrect phase gradient information obtained in stage 1 will reduce the accuracy of final PU result directly. Unlike [24,25] both using filtering-based methods to alleviate the effects of the phase noise on the estimated phase gradients, in this paper, we resist the influence of the noise in stage 1 of TSPA-PUMA through using the MB InSAR dataset with different baseline lengths. To be specific, the more interferograms are involved to estimate the phase gradients based on the CRT formulation, the higher accuracy on ambiguity number gradient estimation will be obtained (it is because that more observed samples of interferometric phases from different interferograms with different baseline lengths are involved, more phase noise can be ignored). Therefore, TSPA-PUMA has good noise robustness if we utilize enough interferograms. In Section 4.2, we will validate the noise robustness of TSPA-PUMA using the MB InSAR system with different baseline lengths.

3.4. Analysis of the Time Complexity

It should be noted that the main running time and memory consumption of TSPA-PUMA lies in stage 2, which is similar to TSPA. In addition, the computational complexity of stage 2 of the TSPA-PUMA method is close to that of the original PUMA method, due to their similar optimization strategy. The time complexity of TSPA-PUMA is $R \cdot K \cdot T(n, m)$, where R is the number of the interferogram (i.e., $R = 2$ in DB case), K is the number 2π of multiples (i.e., the number of iterations) and $T(n, m)$ is the complexity of a max-flow computation in a graph with n nodes and m edges in one interferogram. Regarding memory usage, TSPA-PUMA requires $R \cdot 7n$ bytes. We observe that the computational burden of TSPA-PUMA lies in computing the max-flow algorithm. However, the max-flow solution in the graph cuts algorithm has potential for parallelization, which is suitable for GPU acceleration [33]. Under this condition, the time efficiency of TSPA-PUMA can be increased drastically. Therefore, it can be seen that the total time and space complexities of TSPA-PUMA are practical.

3.5. Analysis of the Parameter Selection

Note that TSPA-PUMA requires only one parameter, i.e., the potential exponent p in stage 2, to be chosen. The potential exponent p in TSPA-PUMA is similar to that in the traditional PUMA method, which defines how the phase of the neighboring pixels in the clique interact [27]. As mentioned earlier, if $p \geq 1$, i.e., using the convex potential, PUMA can find the correct PU result in the flat area. If $0 < p < 1$, i.e., using the non-convex potential, PUMA has phase discontinuity preserving ability in the rugged area. However, in the TSPA-PUMA method, the meaning of potential exponent p seems to be completely different. The reason is that the phase gradients estimated by stage 1 of TSPA-PUMA can violate the phase continuity assumption, so stage 2 of TSPA-PUMA does not need to use non-convex potential to preserve the phase discontinuity. On the contrary, the smaller the potential exponent p is, the lower accuracy on the final PU result will be obtained (it is because that nonconvex potential grows much slower than the convex potential, so it allows strong phase noise not to be penalized too much). Similarly, the larger p the potential exponent is, the accuracy of the final PU result will also be reduced. This is because, when $p > 1$, TSPA-PUMA allows the high-quality regions to share the phase gradient error from the noisy region. According to experimental results, we observe that $p = 1$, i.e., L^1-norm model, is the best parameter for the TSPA-PUMA method not only in the discontinuous area but also in the noisy region. In Section 4.5, some detailed experiments on the effect of the potential exponent p will be presented.

4. Performance Analysis

In this Section, the TSPA-PUMA method is compared with the original PUMA and TSPA methods through five independent experiments from different aspects. The source codes of PUMA and TSPA are both from their algorithm designers [34,35]. The implementation environment of these three methods is MATLAB. Note that the clique potential exponent p of TSPA-PUMA is set to 1 (to be kept in experiments 1–4), and the reason will be given in Section 4.5. The first experiment tests the PU performance of the TSPA-PUMA method using the simulated DB InSAR dataset in the terrain with the abrupt change. The second experiment examines the noise robustness of TSPA-PUMA when applied to a simulated MB InSAR dataset with eight interferograms. The third one verifies TSPA-PUMA through using a real TanDEM-X DB InSAR dataset with two interferograms. The fourth one examines the effectiveness of TSPA-PUMA in a real InSAR MB dataset of ALOS PALSAR with four interferograms. The last one explores the effect of the potential exponent p on the TSPA-PUMA method.

4.1. Experiment 1

The first experiment is also performed on the simulated DB InSAR dataset which is shown in Figure 1c,d. Figure 4a,b illustrate the vertical and horizontal ambiguity number differences between neighboring pixels of Figure 1c, estimated by stage 1 of TSPA-PUMA, respectively. Figure 4c,d are the vertical and horizontal ambiguity number differences between neighboring pixels of Figure 1d, estimated by stage 1 of TSPA-PUMA, respectively. From Figure 4a,b, we can observe that the estimated ambiguity number differences are restricted to ±1 and 0, because the fringe of Figure 1c does not fiercely change, so the phase continuity assumption still holds well. From Figure 4c,d, it can be observed that some ambiguity number differences are larger than 1 or less than −1, meaning that the phase continuity assumption does not hold any more, due to the fringe of Figure 1d with the violent change. Based on these phase gradients, TSPA-PUMA can overcome the limitation of the phase continuity assumption. Figure 4e shows the PU result of Figure 1c obtained by TSPA-PUMA by using the gradient information shown in Figure 4a,b, and Figure 4g shows the errors between Figures 1a and 4e. Figure 4f is the PU result of Figure 1d obtained by TSPA-PUMA by using the gradient information shown in Figure 4c,d, and Figure 4h shows the errors between Figures 1b and 4f. From Figure 4g, we observe that TSPA-PUMA can generate the correct PU result on short baseline as same as the PUMA method. From Figure 4h, it can be noticed that TSPA-PUMA gives us a flawless PU result using the phase gradient information of Figure 4c,d. The statistical information of Figure 1k,j, Figure 4g,h is shown in Table 2, where the root mean-square error (abbreviated as RMSE) of the PU accuracy is given by:

$$\eta = \sqrt{\frac{1}{L} \cdot \| \hat{\Psi} - \Psi \|^2} \tag{16}$$

where Ψ is the vector collecting from the reference unwrapped phase, $\hat{\Psi}$ is the vector collecting from the estimated unwrapped phase, L is the length of the vector Ψ and $\hat{\Psi}$, and $\| \cdot \|^2$ is the quadratic norm. It is worth mentioning that the units of Ψ and $\hat{\Psi}$ are both radian in this paper. From Table 2, it can be seen that RMSEs of Figure 1g,h and Figure 4g are about 0.9, meaning that the three methods all obtain the correct PU solution of an interferogram with short baseline length. In addition, for the interferogram with long baseline length, we observe that TSPA-PUMA generates the lower RMSE of Figure 4h than those of Figure 1k,l obtained by PUMA with potential exponent 0.5 and 1. Therefore, we can conclude that the TSPA-PUMA method is more effective in the terrain with abrupt change than the original PUMA method.

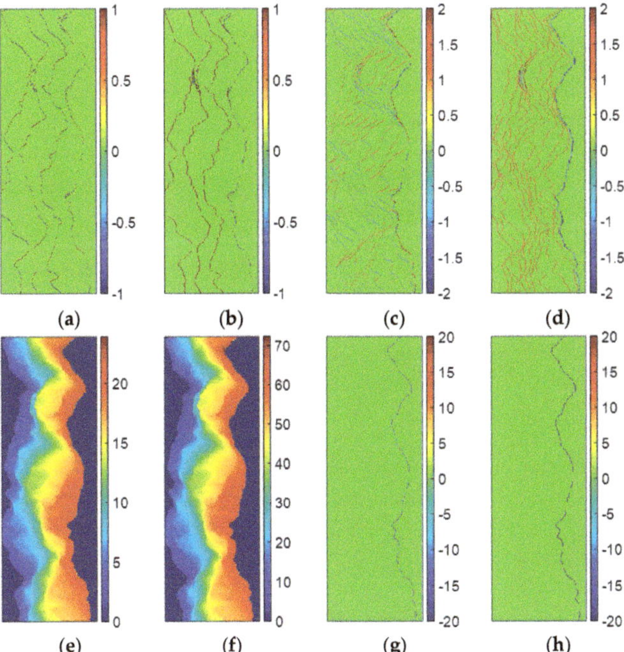

Figure 4. (**a**,**b**) Vertical and horizontal neighboring ambiguity number differences of Figure 1c. (**c**,**d**) Vertical and horizontal neighboring ambiguity number differences of Figure 1d. (**e**) PU results of Figure 1c obtained by TSPA-PUMA. (**f**) PU results of Figure 1d obtained by TSPA-PUMA. (**g**) Errors between Figure 1a and PU results (**e**). (**h**) Errors between Figure 1b and PU results (**f**).

Table 2. Statistical information of PU performance in Figure 1g,h,k,l, and Figure 4g,h.

U Method	Short Baseline		Long Baseline	
	Figure	RMSE	Figure	RMSE
PUMA with potential exponent 1	Figure 1g	0.9505	Figure 1k	3.6438
PUMA with potential exponent 0.5	Figure 1h	0.9505	Figure 1l	3.2356
TSPA-PUMA	Figure 4g	0.9164	Figure 4h	0.9164

4.2. Experiment 2

The second experiment is also conducted on the simulated interferogram (baseline length is 330 m) which is shown in Figure 1d. To examine the noise robustness of TSPA-PUMA, some phase noise is added according to the employed probability density function of the noise wrapped phase in [36]. It is worth mentioning that, in our simulation, we use the general coherence coefficient to jointly express all the decorrelation components, e.g., atmosphere effect or co-registration error, etc. Figure 5a shows the simulated interferogram, and the mean coherence coefficient of Figure 5a is 0.75. The reference unwrapped phase of Figure 5a is Figure 1b. From Figure 5a, it can be found that, due to the low coherence, the phase fringes are very complicated and PU becomes very tough. Figure 5b is the PU result of the original PUMA method with potential exponent 0.5, and the corresponding errors between Figures 5b and 1b are shown in Figure 5c. We can see that several discontinuous variations are seen clearly in Figure 5c, and RMSE of Figure 5c is up to 9.5992. The reason is that low coherence of interferogram of Figure 5a aggravates the fringe blurrier, which destroy the phase continuity assumption, so it is hard for the SB PU methods to perform correctly. Figure 5d is the PU result of the TSPA-PUMA method based on the DB InSAR dataset used in experiment 1, whose parameters are listed in Table 1. Figure 5f is the corresponding errors between Figures 5d and 1b. From Figure 5f, we find that TSPA-PUMA using the DB InSAR dataset also has obvious unwrapping errors in the phase image, and the RMSE of Figure 5f is 7.6592. The reason is that, although TSPA-PUMA does not obey the phase continuity assumption in the rugged area, stage 1 of TSPA-PUMA is sensitive to noise level which produces the incorrect phase gradient information. In this case, TSPA-PUMA cannot obtain the correct PU result where the fringe is polluted by high noise. Unlike [24,25] who apply the filter-based methods to suppress the influence of incorrect phase gradients obtained by stage 1 of TSPA-PUMA, in this paper, we utilize MB InSAR dataset with more interferograms to remove the phase gradient errors. Major parameters of the MB InSAR system are the same as that used in experiment 1 which is listed in Table 1, but this time, eight interferograms with different baseline lengths are used to perform the TSPA-PUMA method (baseline lengths are 70 m, 150 m, 330 m, 471 m, 550 m, 631 m, 753 m and 831 m, respectively). It should be noted that the number of baselines used in TSPA-PUMA could be any value theoretically, if the ratio of baseline lengths of different interferograms satisfies the CRT formulation. However, CRT is sensitive to the baseline length. In other words, different baseline lengths could result in different performances of TSPA-PUMA. A baseline design criterion was proposed by [37] to determine the optimal baseline length for MB PU. In this experiment, the choice of eight baseline lengths satisfies the baseline design criterion proposed in [37]. Figure 5e is the PU result generated by TSPA-PUMA using MB InSAR dataset, and the corresponding errors between Figures 5e and 1b is illustrated in Figure 5g. From Figure 5g, we observe that the TSPA-PUMA method using the MB InSAR dataset alleviates most of unwrapping errors in the PU result, and RMSE of Figure 5g is 3.4297, which is much lower than the former two methods. This is because that more interferograms are involved in stage 1 of TSPA-PUMA, the higher accuracy on ambiguity number gradient estimation will be obtained. Under this condition, the noise robustness of TSPA-PUMA can be improved drastically.

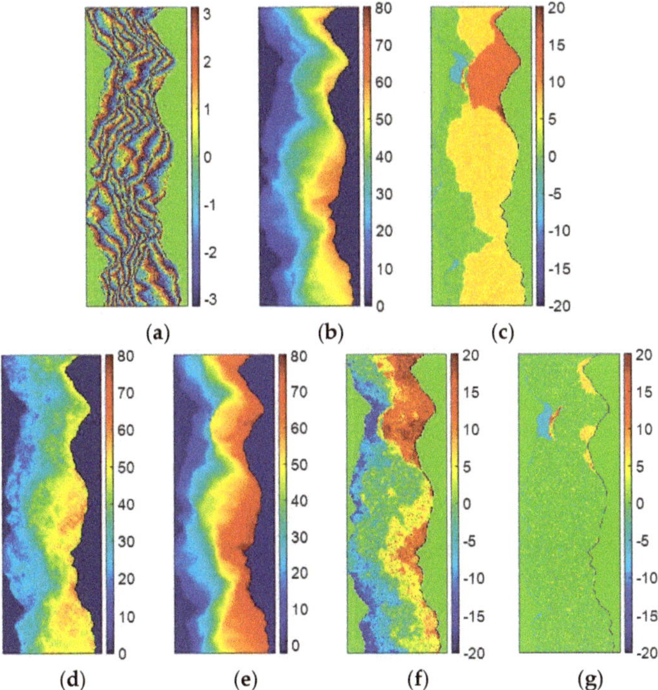

Figure 5. (**a**) Simulated wrapped phases of Figure 1b (coherence coefficient is 0.75). (**b**) PU results of (**a**) obtained by PUMA (clique potential exponent is 0.5). (**c**) Errors between Figure 1b and PU results (**b**). (**d**,**e**) PU results of (**a**) obtained by (**d**) DB TSPA-PUMA and (**e**) MB TSPA-PUMA. (**f**,**g**) Errors between Figure 1b and PU results (**d**,**e**).

To further research into the relationship between the number of interferograms and the noise robustness of TSPA-PUMA, we utilize seven MB InSAR datasets with different number of interferograms between 2 and 8 with an increment of 1 to perform the TSPA-PUMA method. The relationship between the estimation RMSE of TSPA-PUMA and the number of interferograms is tabulated in Table 3. From Table 3, it can be observed clearly that the RMSE of the TSPA-PUMA performance can be decreased with the number of interferograms increasing. That is to say, the noise robustness of TSPA-PUMA can be enhanced through using more interferograms, because when more observed samples of interferometric phases are involved, the phase noise can be reduced. Therefore, we can see that TSPA-PUMA has good noise robustness if we utilize enough interferograms with different baseline lengths.

Table 3. The relationship between the estimation RMSE of TSPA-PUMA and the number of interferograms.

ID	Number of Interferograms	Baseline Length (m)								RMSE
1	2	150	330							7.6592
2	3	70	150	330						6.9732
3	4	70	150	330	471					6.7486
4	5	70	150	330	471	550				6.6240
5	6	70	150	330	471	550	631			4.6023
6	7	70	150	330	471	550	631	753		4.3318
7	8	70	150	330	471	550	631	753	831	3.4297

4.3. Experiment 3

The third experiment is carried out on a real TanDEM-X DB InSAR dataset with two interferograms (single-pass) of Weinan of Shaanxi province, China. Figure 6a is the Google Earth image of the study area (1000×1000 pixels). We can see that Figure 6a is the area whose topography is mountainous and rugged. Under this condition, the phase continuity assumption may not work well, which causes that the SB PU cannot obtain the correct PU solution. Figure 6b,c are the flattened and filtered input interferograms with short and long baseline lengths, respectively. The major interferometric parameters of Figure 6b,c are listed in Table 4. Figure 6c,d are the corresponding reference unwrapped phase of Figure 6b,c, which are generated by the Shuttle Radar Topography Mission (SRTM) digital elevation model (DEM). Figure 6f,g show the PU results of Figure 6b,c obtained by the PUMA method with potential exponent 0.5, and Figure 6h,i are the errors between Figure 6d,e and Figure 6f,g, respectively. Figure 6j,k are the PU results of Figure 6b,c obtained by TSPA, and Figure 6l,m are the errors between Figure 6d,e and Figure 6j,k, respectively. Figure 6n,o are the PU results of Figure 6b,c obtained by TSPA-PUMA, and the corresponding errors between Figure 6d,e and Figure 6n,o is shown in Figure 6p,q, respectively. The statistical information of Figure 6 is listed in Table 5. From Table 5, we can see that when the baseline length is short and the fringe pattern in the interferogram is simple, the PU performance of all three methods is similar to each other. However, for the long baseline interferogram, because the phase variation is rapid, which does not obey the phase continuity assumption, the PU performance of TSPA and TSPA-PUMA are much better than that of PUMA. Although the PU results of TSPA and TSPA-PUMA are mainly the same due to their same L^1-norm model, their performances in terms of running time are not comparable. In this experiment, while TSPA-PUMA only takes 65.81 s for short baseline and 231.72 s for long baseline, the classical TSPA method is ten and eight times slower for the short and long baseline, respectively. Therefore, it can be seen that the running time of TSPA-PUMA is practical.

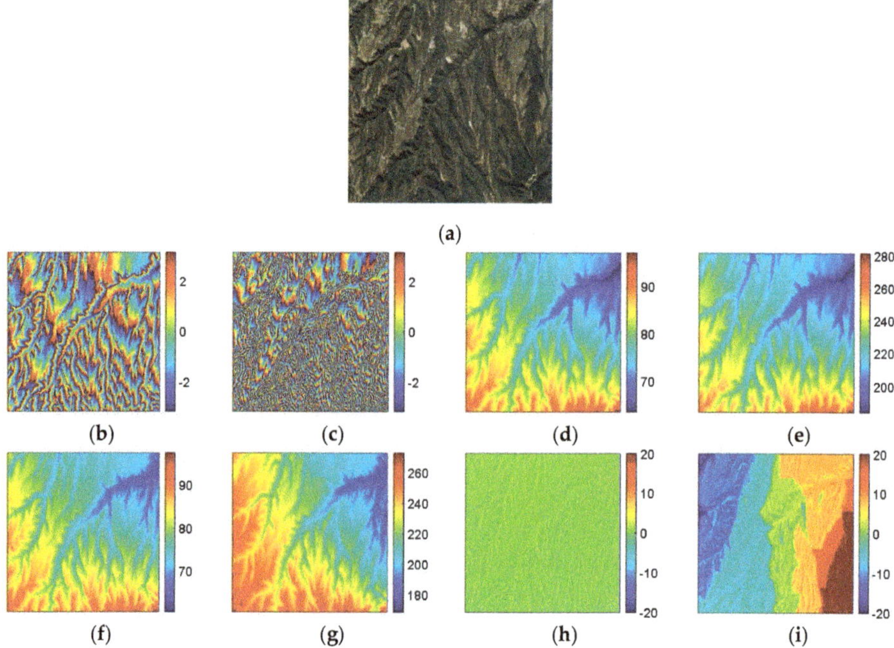

Figure 6. *Cont.*

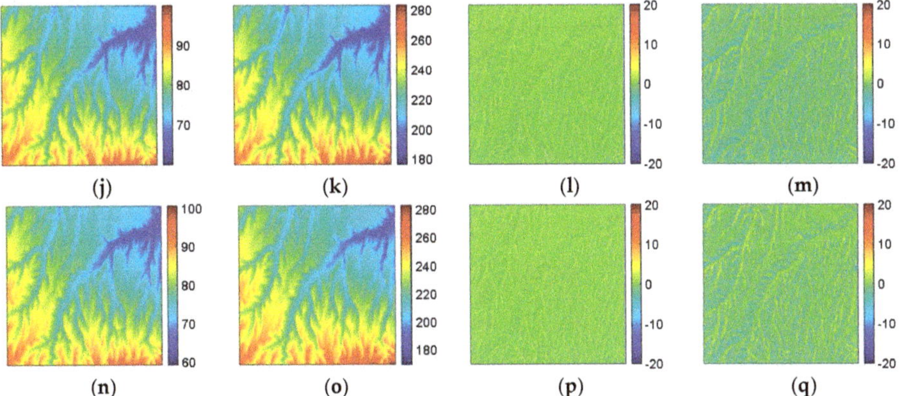

Figure 6. (a) Google Earth image of the study area. (b,c) TanDEM-X interferograms with different baseline lengths ((b) short and (c) long baseline length). (d,e) The reference unwrapped phases of (b,c). (f,g) PU results of (b,c) obtained by PUMA (clique potential exponent is 0.5). (h,i) Errors between (d,e) and PU results (f,g). (j,k) PU results of (b,c) obtained by DB TSPA. (l,m) Errors between (d,e) and PU results (j,k). (n,o) PU results of (b,c) obtained by DB TSPA-PUMA. (p,q) Errors between (d,e) and PU results (n,o).

Table 4. Major interferometric parameters of real DB dataset of TanDEM-X.

Orbit Altitude	Incidence Angle	Wavelength	Latitude	longitude
514.8 km	36.6°	0.032 m	35.82°	109.28°

Interferogram	Figure 6b		Figure 6c	
Date of Master Channel	2 April 2014		21 October 2012	
Date of Slave Channel	2 April 2014		21 October 2012	
Baseline Length	130.62 m		370.45 m	

Resolution	Range (Vertical)	5.46 m	Azimuth (Horizontal)	8.15 m
Image Size	Range	1000 pixels	Azimuth	1000 pixels

Table 5. Statistical information of PU performance in Figure 6h,i,l,m,p,q.

PU Method	Short Baseline			Long Baseline		
	Figure	RMSE	Time (s)	Figure	RMSE	Time (s)
PUMA with potential exponent 0.5	Figure 6h	0.6871	116.18	Figure 6i	10.0892	303.83
TSPA	Figure 6l	0.6114	665.92	Figure 6m	1.91	1941.04
TSPA-PUMA	Figure 6p	0.69	65.81	Figure 6q	1.7616	231.72

4.4. Experiment 4

In the fourth experiment, we will examine the effectiveness of TSPA-PUMA in the real MB dataset of ALOS PALSAR with four interferograms. Figure 7a shows the Google Earth image of the study area in this experiment, which comes from the Himalayan mountain area. Figure 7b–e are four interferograms with different baseline lengths (601×501 pixels). From Figure 7d,e, we can observe that the coherence values of the two interferograms with long baseline are relatively low, because ALOS PALSAR acts as a repeat-pass radar interferometer with the inherent accuracy limitations imposed by temporal decorrelation and atmospheric disturbances. The major interferometric parameters of the ALOS PALSAR dataset are tabulated in Table 6. Figure 7f–i are the corresponding reference unwrapped phase of Figure 7b–e, which are obtained from the PALSAR DEM. Figure 7j–m are the PU results of Figure 7b–e generated by the PUMA method with potential exponent 0.5, and Figure 7n–q are those generated by TSPA-PUMA. Figure 8 show the corresponding errors between the PU results of PUMA, TSPA-PUMA and reference unwrapped phase. The statistical information of Figure 8 is listed in Table 7. From Table 7, it can be seen that, for the short baseline interferogram (Figure 7b,c), the PU results of the two methods are similar to each other. However, for the long baseline interferogram (Figure 7d,e), the PU performance of TSPA-PUMA is much better than that of PUMA, and reason is that TSPA-PUMA can break through the limitation of the phase continuity assumption. Also, TSPA-PUMA can eliminate the effects of low coherence through using the MB InSAR dataset with different baseline lengths.

Table 6. Major interferometric parameters of real MB dataset of ALOS PALSAR.

Orbit Altitude	Incidence Angle	Wavelength	Latitude	longitude
698.51 km	38.75°	0.236m	30.91°	94.23°
Interferogram	Figure 7b	Figure 7c	Figure 7d	Figure 7e
Date of Master Channel	18 August 2007	18 August 2007	18 August 2007	18 August 2007
Date of Slave Channel	3 October 2007	3 July 2007	3 January 2008	8 October 2009
Baseline Length	113.36 m	193.15 m	406.00 m	440.68 m
Resolution	Range (Vertical)	9.37 m	Azimuth (Horizontal)	19.00 m
Image Size	Range	601 pixels	Azimuth	501 pixels

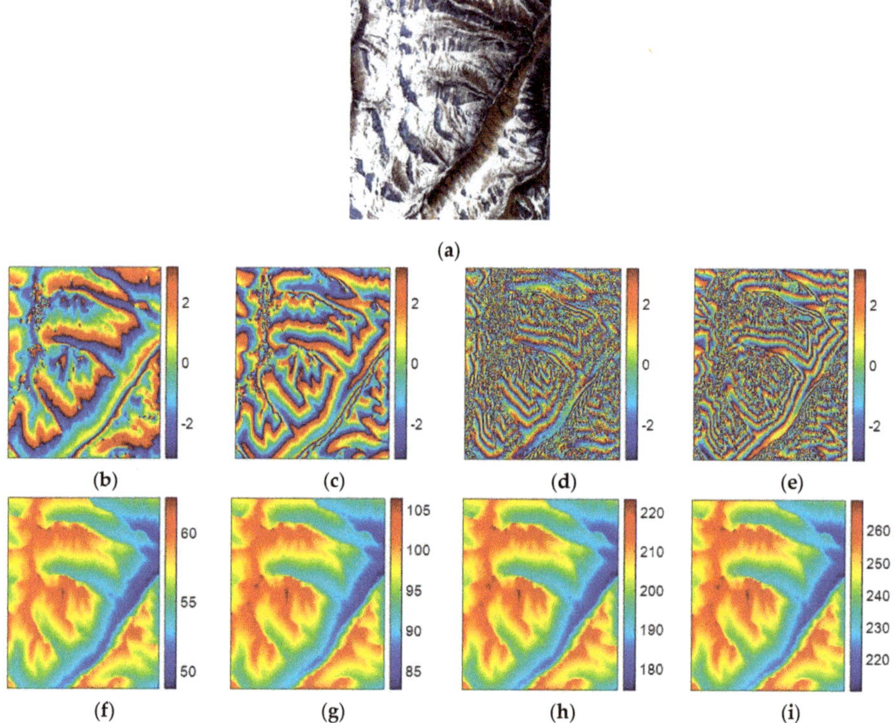

(a)

(b)   (c)   (d)   (e)

(f)   (g)   (h)   (i)

Figure 7. *Cont.*

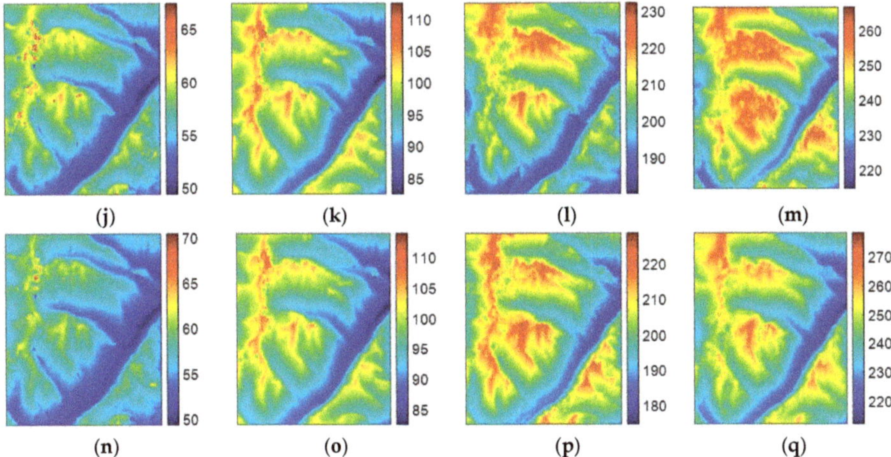

Figure 7. (a) Google Earth image of the study area. (b–e) ALOS PALSAR interferograms with different baseline lengths ((b) interferogram 1, (c) interferogram 2, (d) interferogram 3, and (e) interferogram 4). (f–i) The reference unwrapped phases of (b–e). (j–m) PU results of (b–e) obtained by PUMA (clique potential exponent is 0.5). (n–q) PU results of (b–e) obtained by MB TSPA-PUMA.

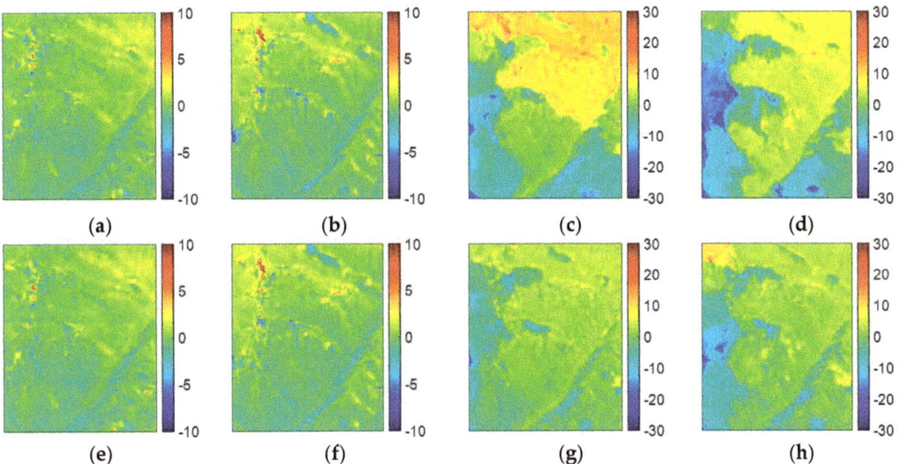

Figure 8. (a–d) PUMA errors of Figure 7j–m. (e–h) MB TSPA-PUMA errors of Figure 7n–q.

Table 7. Statistical information of PU performance in Figure 8a–h.

PU Method	Interferogram 1		Interferogram 2		Interferogram 3		Interferogram 4	
	Figure	RMSE	Figure	RMSE	Figure	RMSE	Figure	RMSE
PUMA with potential exponent 0.5	Figure 8a	0.8719	Figure 8b	1.1576	Figure 8c	7.4004	Figure 8d	7.4951
TSPA-PUMA	Figure 8e	0.8577	Figure 8f	1.0860	Figure 8g	3.4688	Figure 8h	4.6751

4.5. Experiment 5

In the last experiment, we explored the effect of the potential exponent p in stage 2 of TSPA-PUMA method on the simulated MB InSAR dataset. This experiment examined the PU performance of

TSPA-PUMA with different potential exponent p ranging from 0.1 to 3 with an increment of 0.5. A simulated terrain generated by the MATLAB's membrane function was used to test the relationship between the potential exponent p and the terrain change. Figure 9a shows the simulated terrain employed in this experiment (201×201 pixels). According to the simulated terrain, we generated the reference unwrapped phases using $d \times$ membrane, where d is the parameter that determines the height of the terrain. The larger the value of d is, the higher the terrain is, and thus the terrain changes more violently. We considered four MB simulated reference unwrapped phases with different ds (i.e., $d_1 = 17.5$, $d_2 = 35.0$, $d_3 = 52.5$, $d_4 = 70.0$, and unit is radian). Figure 9b–e show four reference unwrapped phases with different ds, respectively. From Figure 9b–e, we observe that, while the value of d is getting larger, and the pattern of the fringe becomes denser, which results in the failure of the phase continuity assumption. We generated two groups of simulated interferograms of Figure 9b–e. In one group, we simulated four noise-free wrapped phase images, as shown in Figure 9f–i, respectively. Under the noise-free condition, the fringe change of Figure 9f–i is only related to the topography changes. In this case, we can test the dependence of the potential exponent p on the steepness of the terrain. In another group, we simulated four noisy wrapped phase images, in which the phase noise was added with using 0.75 mean correlation coefficient [35], as illustrated in Figure 9j–m, respectively. From Figure 9j–m, it can be seen that the pattern of the fringe is destroyed more after the noise is added. Under this condition, we can examine the effect of the potential exponent p in case of high-phase noise. We compare the unwrapped phases obtained by TSPA-PUMA using different potential exponent p with the reference unwrapped phases of Figure 9b–e and obtain the RMSE of each PU result. Figure 10a shows the RMSE curves of Figure 9f–i with different ds, and Figure 10b is the RMSE curves of Figure 9j–m with different ds. From the trends of the curves shown in Figure 10a, we can see that the RMSE curves of the PU results with four different ds are low (below 3×10^{-3}) and identical throughout the whole potential exponent scale, meaning that the potential exponent p is not sensitive to the terrain change. The reason is that, owning to stage 1 of TSPA-PUMA without obeying the phase continuity assumption, no matter what kind of the potential exponent value is chosen in stage 2, it is possible for TSPA-PUMA to perform correctly. From the trends of the curves shown in Figure 10b, we can observe that the PU results with four different ds generate the lowest RMSE when the potential exponent equals to 1 ($p = 1$), while when the potential exponent more than 1 or less than 1 ($p > 1$ or $p < 1$), the PU results both have higher RMSE with the different values ds. This is because that when $p > 1$, TSPA-PUMA introduces the incorrect phase gradients from the low-quality regions into the high-quality regions easily, and when the potential exponent is less than 1 ($p < 1$), the clique potential of TSPA-PUMA grows much more slowly than that when potential exponent equals to 1 ($p = 1$), which allows strong phase noise not be too much penalized. It implies that the potential exponent p which equals to 1 is the optimal parameter for TSPA-PUMA both in the rugged and low-quality regions.

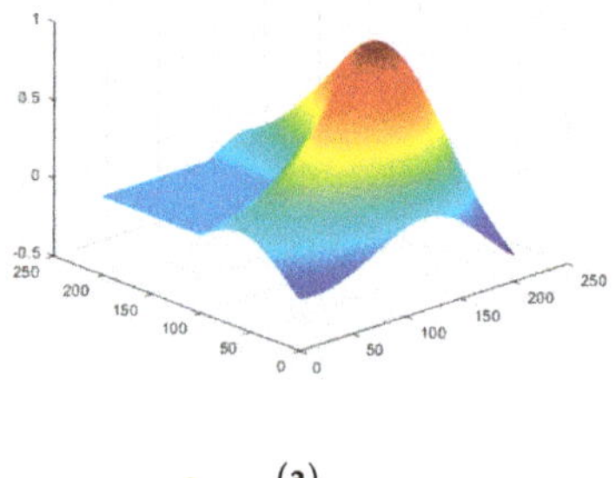

(a)

Figure 9. *Cont.*

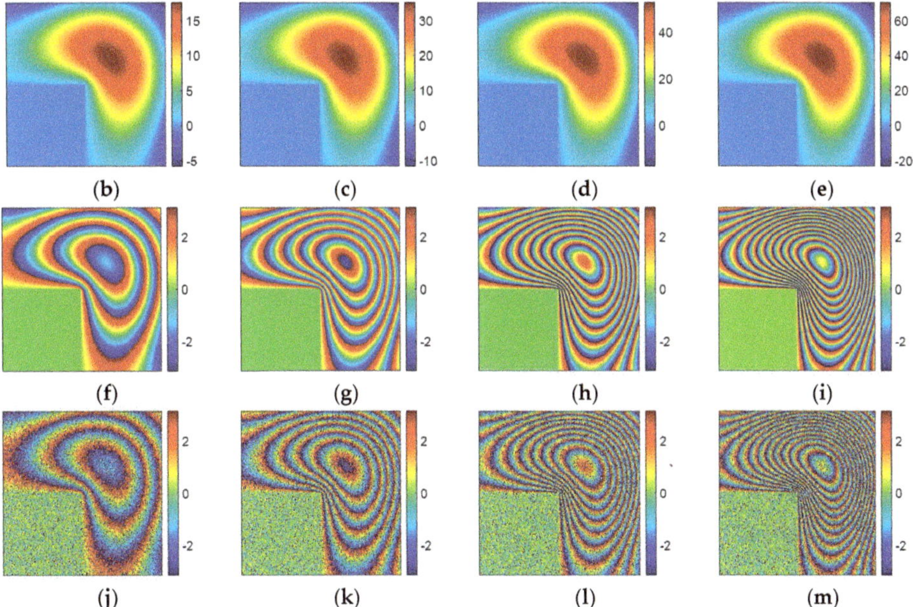

Figure 9. (**a**) The examples of simulated terrain obtained by the function membrane in 3D space. (**b–e**) reference unwrapped phase with four different ds ((**b**) $d_1 = 17.5$, (**c**) $d_2 = 35.0$, (**d**) $d_3 = 52.5$, (**e**) $d_4 = 70.0$, and unit is radian). (**f–i**) The simulated noise-free wrapped phases of (**b–e**). (**j–m**) The simulated wrapped phases of (**b–e**) with mean coherence coefficient 0.75.

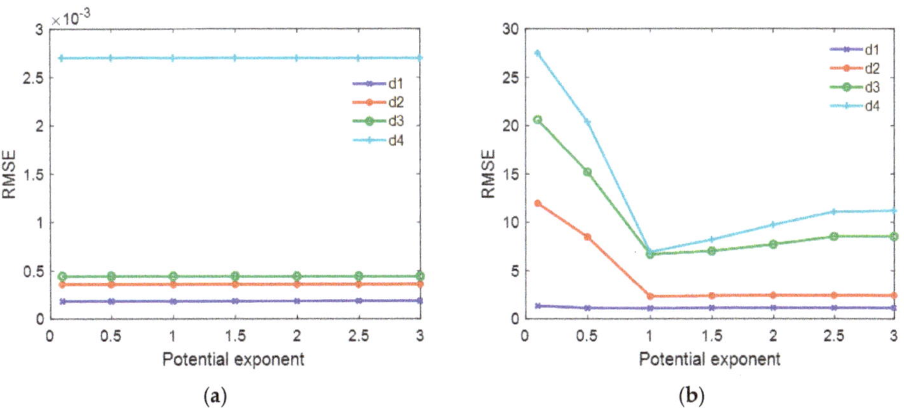

Figure 10. (**a**) RMSE curves of TSPA-PUMA of Figure 9f–i with different potential exponents between 0.1 and 3 with an increment of 0.5. (**b**) RMSE curves of TSPA-PUMA of Figure 9j–m with different potential exponents between 0.1 and 3 with an increment of 0.5.

5. Conclusions

In this paper, we extend the classical PUMA algorithm for MB InSAR using the TSPA approach referred to as TSPA-PUMA, consisting of a two-stage programming procedure. In stage 1 of TSPA-PUMA, the phase gradients are estimated based on CRT, which does not follow the phase continuity assumption. In stage 2, an MRF model of PUMA is designed for modeling local contextual dependence based on the phase gradients obtained by stage 1. Subsequently, the energy of MRF model is minimized by performing a sequence of binary optimizations solved by graph cuts techniques. Results of the simulated and real InSAR data experiments demonstrate that the TSPA-PUMA method can significantly improve the accuracy of the original PUMA method in the area where topography varies drastically due to its ability to overcome the limitation of the phase continuity assumption, and is an efficient MB PU method compared to the original TSPA method. In addition, the noise robustness of TSPA-PUMA can also be improved through adding more interferograms with different baseline lengths.

Author Contributions: L.Z. designed the research and developed the main idea. Y.L. provided crucial guidance and support through the research. L.Z. and Y.X. performed experiments and wrote the manuscript, and S.G. contributed to the revising of the manuscript. All authors have read and agreed to the published version of the manuscript.

Funding: This work was partially supported by National Natural Science Foundation of China (NSFC Grant No.41501461, 61972059), Six talent peaks Project in Jiangsu Province (DZXX-027, XYDXX-057), the sponsorship of Jiangsu Overseas Research and Training Program for University Prominent Young and Middle-aged Teachers and Presidents, Open Foundation of The Suzhou Smart City Research Institute, Suzhou University of Science and Technology (Grant No. SZSCR2019018), "Tiancheng Huizhi" Innovation and Education Fund Project of Ministry of Education in China under grant No. 2018A03008.

Conflicts of Interest: The authors declare no conflict of interest.

References

1.   Moreira, A.; Prats-Iraola, P.; Younis, M.; Krieger, G.; Hajnsek, I.; Papathanassiou, K.P. A tutorial on synthetic aperture radar. *IEEE Geosci. Remote Sens. Mag.* **2013**, *1*, 6–43. [CrossRef]
2.   Itoh, K. Analysis of the phase unwrapping problem. *Appl. Opt.* **1982**, *21*, 2470. [CrossRef] [PubMed]
3.   Yu, H.; Lan, Y.; Yuan, Z.; Xu, J.; Lee, H. Phase Unwrapping in InSAR: A Review. *IEEE Geosci. Remote Sens. Mag.* **2019**, *7*, 40–58. [CrossRef]
4.   Pascazio, V.; Schirinzi, G. Multifrequency InSAR height reconstruction through maximum likelihood estimation of local planes parameters. *IEEE Trans. Image Process.* **2000**, *11*, 1478–1489. [CrossRef]
5.   Fornaro, G.; Guarnieri, A.M. Maximum likelihood multi-baseline SAR interferometry. *IEEE Proc. Radar Sonar Navig.* **2006**, *153*, 279–288. [CrossRef]
6.   Fornaro, G.; Pauciullo, A.; Sansosti, E. Phase difference-based multichannel phase unwrapping. *IEEE Trans. Image Process.* **2005**, *14*, 960–972. [CrossRef]
7.   Ferraiuolo, G.; Pascazio, V.; Schirinzi, G. Maximum a posteriori estimation of height profiles in InSAR imaging. *IEEE Geosci. Remote Sens.* **2004**, *1*, 66–70. [CrossRef]
8.   Ferraioli, G.; Shabou, A.; Tupin, F.; Pascazio, V. Multichannel phase unwrapping with graph cuts. *IEEE Geosci. Remote Sens. Lett.* **2009**, *6*, 562–566. [CrossRef]
9.   Baselice, F.; Ferraioli, G.; Pascazio, V.; Schirinzi, G. Contextual information-based multichannel synthetic aperture radar interferometry: Addressing DEM reconstruction using contextual information. *IEEE Signal Process. Mag.* **2014**, *31*, 59–68. [CrossRef]
10. Ferraiuolo, G.; Meglio, F.; Pascazio, V.; Schirinzi, G. DEM reconstruction accuracy in multichannel SAR interferometry. *IEEE Trans. Geosci. Remote Sens.* **2009**, *47*, 191–201. [CrossRef]
11. Yu, H.; Li, Z.; Bao, Z. A cluster-analysis-based efficient multibaseline phase-unwrapping algorithm. *IEEE Trans. Geosci. Remote Sens.* **2011**, *49*, 478–487. [CrossRef]
12. Liu, H.; Xing, M.; Bao, Z. A cluster-analysis-based noise-robust phase-unwrapping algorithm for multibaseline interferograms. *IEEE Trans. Geosci. Remote Sens.* **2015**, *53*, 494–504.
13. Jiang, Z.; Wang, J.; Song, Q.; Zhou, Z. A refined cluster-analysis-based multibaseline phase-unwrapping algorithm. *IEEE Geosci. Remote Sens. Lett.* **2017**, *14*, 1565–1569. [CrossRef]
14. Liu, H.; Xing, M.; Bao, Z. A novel mixed-norm multibaseline phase-unwrapping algorithm based on linear programming. *IEEE Geosci. Remote Sens. Lett.* **2015**, *12*, 1086–1090. [CrossRef]
15. Yuan, Z.; Deng, Y.; Li, F.; Wang, R.; Liu, G.; Han, X. Multichannel InSAR DEM reconstruction through improved closed-form robust Chinese remainder theorem. *IEEE Geosci. Remote Sens. Lett.* **2013**, *10*, 1314–1318. [CrossRef]
16. Jin, B.; Guo, J.; Wei, P.; Su, B.; He, D. Multi-baseline InSAR phase unwrapping method based on mixed-integer optimisation model. *IET Radar Sonar Navig.* **2018**, *12*, 694–701. [CrossRef]

17. Kim, M.G.; Griffiths, H.D. Phase unwrapping of multibaseline interferometry using Kalman filtering. In Proceedings of the 7th International Conference on Image Processing and its Applications, London, UK, 13–15 July 1999; pp. 251–253.

18. Ferretti, A.; Prati, C.; Rocca, R. Multibaseline InSAR DEM recon-struction: The wavelet approach. *IEEE Trans. Geosci. Remote Sens.* **1999**, *37*, 705–715. [CrossRef]

19. Yu, H.; Lan, Y. Robust two-dimensional phase unwrapping for multibaseline sar interferograms: A two-stage programming approach. *IEEE Trans. Geosci. Remote Sens.* **2016**, *54*, 5217–5225. [CrossRef]

20. Costantini, M. A novel phase unwrapping method based on network programming. *IEEE Trans. Geosci. Remote Sens.* **1998**, *36*, 813–821. [CrossRef]

21. Ahuja, R.; Magnanti, T.; Orlin, J. *Network Flows: Theory, Algorithms and Applications*; Prentice Hall: Upper Saddle River, NJ, USA, 1993; ISBN-13 978-0136175490.

22. Yu, H.; Xing, M.; Bao, Z. A fast phase unwrapping method for largescale interferograms. *IEEE Trans. Geosci. Remote Sens.* **2013**, *51*, 4240–4248.

23. Yu, H.; Lan, Y.; Xu, J.; An, D.; Lee, H. Large-scale L0-norm and L1-norm two-dimensional phase unwrapping. *IEEE Trans. Geosci. Remote Sens.* **2017**, *55*, 4712–4728. [CrossRef]

24. Lan, Y.; Yu, H.; Xing, M. Refined Two-Stage Programming-Based Multi-Baseline Phase Unwrapping Approach Using Local Plane Model. *Remote Sens.* **2019**, *11*, 491. [CrossRef]

25. Gao, Y.; Zhang, S.; Li, T.; Chen, Q.; Zhang, X.; Li, S. Refined Two-Stage Programming Approach of Phase Unwrapping for Multi-Baseline SAR Interferograms Using the Unscented Kalman Filter. *Remote Sens.* **2019**, *11*, 199. [CrossRef]

26. Yu, H.; Zhou, Y.; Ivey, S.; Lan, Y. Large-scale multibaseline phase unwrapping: Interferogram segmentation based on multibaseline envelope-sparsity theorem. *IEEE Trans. Geosci. Remote Sens.* **2019**, *57*, 9308–9322. [CrossRef]

27. Bioucas-Dias, J.; Valadao, G. Phase unwrapping via graph cuts. *IEEE Trans. Image Process* **2007**, *16*, 698–709. [CrossRef] [PubMed]

28. Zhou, L.; Chai, D.; Xia, Y.; Xie, C. An extended PUMA algorithm for multibaseline InSAR DEM reconstruction. *Int. J. Remote Sens.* **2019**, *40*, 7830–7851. [CrossRef]

29. Liu, G.; Wang, R.; Deng, Y.; Chen, R.; Shao, Y.; Yuan, Z. A new quality map for 2-D phase unwrapping based on gray level cooccurrence matrix. *IEEE Geosci. Remote Sens. Lett.* **2014**, *11*, 444–448. [CrossRef]

30. Kolmogorov, V.; Zabih, R. What energy functions can be minimized via graph cuts? *IEEE Trans. Pattern Anal. Mach. Intell.* **2004**, *26*, 147–159. [CrossRef]

31. Ghiglia, D.C.; Pritt, M.D. *Two-Dimensional Phase Unwrapping: Theory, Algorithms, and Software*; Wiley-Interscience: New York, NY, USA, 1998; ISBN 0-471-24935-1.

32. Colesanti, C.; Ferretti, A.; Prati, C.; Rocca, F. Monitoring landslides and tectonic motions with the permanent scatterers technique. *Eng. Geol.* **2003**, *68*, 3–14. [CrossRef]

33. Vineet, V.; Narayanan, P.J. CUDA cuts: Fast graph cuts on the GPU. In Proceedings of the 2008 IEEE Computer Society Conference on Computer Vision and Pattern Recognition Workshops, Anchorage, AK, USA, 23–28 June 2008; pp. 1–8.

34. Bioucas-Dias, J.; Valadão, G. PUMA Algorithm (Phase Unwrapping, Via Graph Cuts) (MATLAB Code). 2007. Available online: http://www.lx.it.pt/~{}bioucas/code.htm. (accessed on 10 November 2019).

35. Lan, Y.; Yu, H. TSPA Multi-Baseline Phase Unwrapping Method (MATLAB Code). 2018. Available online: http://rscl-grss.org/codelibrary/43c468514e6bb1c32994a2f9dbfc5be2.zip (accessed on 10 November 2018).

36. Lee, J.-S.; Hoppel, K.W.; Mango, S.A.; Miller, A.R. Intensity and phase statistics of multilook polarimetric and interferometric SAR imagery. *IEEE Trans. Geosci. Remote Sens.* **1994**, *32*, 1017–1028.

37. Yu, H.; Lee, H.; Cao, N.; Lan, Y. Optimal Baseline Design for Multibaseline InSAR Phase Unwrapping. *IEEE Trans. Geosci. Remote Sens.* **2019**, *57*, 5738–5750. [CrossRef]

Article

Multibaseline Interferometric Phase Denoising Based on Kurtosis in the NSST Domain

Yanfang Liu [1,2,*], **Shiqiang Li** [2] **and Heng Zhang** [2]

[1] School of Electronic, Electrical and Communication Engineering, Chinese Academy of Sciences, Beijing 100190, China

[2] Department of Space Microwave Remote Sensing System, Aerospace Information Research Institute, Chinese Academy of Sciences, Beijing 100190, China; lishq@mail.ie.ac.cn (S.L.); zhangheng@aircas.ac.cn (H.Z.)

[*] Correspondence: liuyanfang171@mails.ucas.ac.cn; Tel.: +86-10-5888-7167

Received: 12 November 2019; Accepted: 16 January 2020; Published: 19 January 2020

Abstract: Interferometric phase filtering is a crucial step in multibaseline interferometric synthetic aperture radar (InSAR). Current multibaseline interferometric phase filtering methods mostly follow methods of single-baseline InSAR and do not bring its data superiority into full play. The joint filtering of multibaseline InSAR based on statistics is proposed in this paper. We study and analyze the fourth-order statistical quantity of interferometric phase: kurtosis. An empirical assumption that the kurtosis of interferograms with different baselines keeps constant is proposed and is named as the baseline-invariant property of kurtosis in this paper. Some numerical experiments and rational analyses confirm its validity and universality. The noise level estimation of nature images is extended to multibaseline InSAR by dint of the baseline-invariant property of kurtosis. A filtering method based on the non-subsampled shearlet transform (NSST) and Wiener filter with estimated noise variance is proposed then. Firstly, multi-scaled and multi-directional coefficients of interferograms are obtained by NSST. Secondly, the noise variance is represented as the solution of a constrained non-convex optimization problem. A pre-thresholded Wiener filtering with estimated noise variance is employed for shrinking or zeroing NSST coefficients. Finally, the inverse NSST is utilized to obtain the filtered interferograms. Experiments on simulated and real data show that the proposed method has excellent comprehensive performance and is superior to conventional single-baseline filtering methods.

Keywords: multibaseline interferometric synthetic aperture radar (InSAR); non-subsampled shearlet transform (NSST); kurtosis; noise level eatimation

---

1. Introduction

Interferometric synthetic aperture radar is an important extension of synthetic aperture radar (SAR), which is extensively adopted to topography surveying [1], surface deformation monitoring [2] and so forth. Multi-baseline interferometry can comprehensively utilize the diversity of interferograms with different baselines in the same scene to effectively extract the height information of difficult topography, particularly under the circumstance in which the interferometric phase does not satisfy phase continuity assumption [3]. The interferometric phase filtering is a critical step in multibaseline interferometric SAR (InSAR). The interferometric phase is contaminated by massively coherent noise brought from thermal noise decoherence, baseline decoherence, time decoherence and many other decoherent factors in practice [4]. The noise directly affects the difficulty of subsequent phase estimation and the accuracy of the final digital elevation model (DEM). The main motivation of the interferometric phase filtering is to eliminate noises as much as possible while preserving most of the detail information.

As matters stand, the filtering methods of multibaseline InSAR are mainly divided into two categories. One applies the filtering method of single-baseline InSAR to denoise multiple interferograms separately. The filtering method of single-baseline InSAR is divided into two parts: the method in spatial domain and the method in transform domain. Some spatial filters, such as boxcar filter [5], Lee filter [6], local frequency estimate algorithm [7], optimal integration-based adaptive direction filter [8], and so forth, denoise along the gradient direction of interferometric phase. Differences between different methods are the process of detecting the direction and the weight of neighborhood pixels. The appearance of the nonlocal InSAR estimator (NL-InSAR) [9] makes the method in spatial domain reach a new stage. It simultaneously estimates the reflectivity, phase, coherence based on maximum likelihood estimation and the non-local similarity of interferograms. In addition, other methods based on non-local similarity have also been proposed successively [10]. What drives the outstanding performance of another part is the different characteristic between signal and noise in transform domain. It comprises of Goldstein method [11,12], wavelet filter [13,14], InSAR-BM3D [15], and so forth Wherein, Xu et al. applied the simultaneously sparse regularized reconstruction of amplitude and interferometric phase to acquire filtered interferograms [16]. InSAR-BM3D, which is the state of art method in transform domain, extends non-local block-matching 3-D (BM3D) to InSAR and reaches a great edge-Preserving performance. This kind of method does not put forward more requirements about the filtering process but focuses on improving the robustness of phase estimation. The filtering performance is not further improved.

Another category is the multibaseline joint filtering method. The strategy of multibaseline InSAR filtering methods is divided into two parts. One works on the SAR data stacks. NL-InSAR can be regarded as a special case of this part, and the number of SAR images is two. The filtered interferogram is extracted from the covariance matrix. And the covariance matrix is estimated with help of the average effect of statistically homogeneous pixels which have a similar statistical distribution with the central pixel [17–20]. The method can obtain despeckled amplitude images, coherence values, filtered interferograms simultaneously. But its performance is affected by the size of the data set. Large data sets are easy to obtain a more accurate estimation. Most methods require at least eight SAR images. Another one works on the InSAR data stacks, that is, a tensor composed of interferograms. In [21], You et al. proposed a tensor-based filter, which perceived the clean multibaseline InSAR data as a tensor with a low-rank matrix and drawn support from the Kronecker Basis Representation (KBR) to transform the filtering process into an estimation of a low-rank matrix. What demonstrates the potentiality of multibaseline joint filtering is that the method is superior to some state of the art single-baseline filter, for example, NL-InSAR, InSAR-BM3D, and so forth. But it still needs many interferograms to ensure the accuracy of the estimation [17,18].

This paper is an exploration of multibaseline interferometric phase filter based on the statistical characteristic. We propose a new filter on the basis of the NSST filter which is a part of the wavelet filter. The interferogram contains a large number of edges, fringes and other high-dimensional anisotropies. The NSST produces a multi-scaled and multi-directional sparse representation to images optimally and drives a more meticulous depiction of the high-dimensional anisotropies. The interferogram is decomposed into coefficient components with various scales and directions. the coefficient component involves little significant information with large amplitude and noise spreading in whole frequency domain. Coefficients, which are considered as noise are removed immediately, while the significant information is retained or shrinked. A pre-thresholded Wiener filter [22] is applied to eliminate noise. Then the inverse NSST is applied to obtain the reconstructed image.

The noise variance, which decides whether the coefficient is zeroed or retained, is a critical parameter of the Wiener filter. The accuracy of noise variance determines whether the performance of the wiener filter is optimal. A noise level estimation framework which is conceptually similar to the method in Reference [23] is proposed based on the kurtosis model in NSST domain and baseline-invariant property of kurtosis that is proposed and confirmed in this paper. The noise variance estimation is converted into a modified non-convex optimization problem. Moreover, the proposed estimator has higher operation efficiency due to skipping the clustering operation in Reference [23]. Considering the noise variance is space-variant, block estimation is applied. With the help of estimated noise variance, the wiener filter eliminates noise more accurately. Last but not least, the result of experiments on simulated data and real data confirms the efficiency and excellent performance of the proposed method.

2. Method

2.1. Signal Model

In the case of single-look, the probability density function of interferometric phase can be represented as (1). The interferometric phase satisfies additive noise model in spatial domain, which is deduced in Reference [13]. It can be expressed as (2).

$$pdf(\phi; \gamma, \phi_0) = \frac{1}{2\pi} \frac{1 - |\gamma^2|}{1 - |\gamma^2|cos^2(\phi - \phi_0)}$$
$$\cdot \left\{ 1 + \frac{|\gamma|cos(\phi - \phi_0)cos^{-1}[-|\gamma|cos(\phi - \phi_0)])}{[1 - |\gamma^2|cos^2(\phi - \phi_0)]^{1/2}} \right\}, -\pi < (\phi - \phi_0) \leq \pi \tag{1}$$

$$y = x + n, \tag{2}$$

where x is the ideal phase deduced by the natural topography. y is the observed phase disturbed by the zero mean noise n, which is assumed to be independent of x. The phase jump ranged from $-\pi$ to π, which is induced by the interferometric phase wrapping, derived a high frequency similarity to noise in frequency domain. Therefore, not surprisingly, it is apt to greatly be confused with noises. It is desirable that we convert the image to the complex domain to get the continuous complex phase and filter the real part and the imaginary part respectively. The signal model in the complex domain can be induced as

$$exp(jy) = cos(y) + jsin(y). \tag{3}$$

The real part and the imaginary part can be expressed as

$$cos(y) = N_c cos(x) + n_c \tag{4}$$

$$sin(y) = N_c sin(x) + n_s, \tag{5}$$

where $N_c = \frac{\pi}{4}|\gamma|F(\frac{1}{2}, \frac{1}{2}; 2; |\gamma|^2)$ and $F(\frac{1}{2}, \frac{1}{2}; 2; |\gamma|^2)$ is the Gaussian hypergeometric distribution function. n_c and n_s are zero-mean random variables, which are generally assumed to additive Gaussian white noises in the filtering process.

2.2. Denoising Based on NSST

2.2.1. The Nonsubsampled Shearlet Transform

Wavelet is prone to deal with 1-D signals existing pointwise singularities. Nevertheless, it is weak to handle multidimensional data dominated by distributed discontinuities, such as edges and fringes. In an effort to solve this problem, the wavelet basis with much higher directional sensitivity and more flexible shapes is encouraged for effectively capturing the singularity features of multidimensional data, involving composite wavelets [24], contourlets [25], and so forth. The shearlet transform is an important part of composite wavelet theory, which merges classical geometry and multiscale analysis [26–29]. The shearlet provides nearly optimal nonlinear approximation performance and produces an optimal sparse representation of images with distributed discontinuities. Thanks to its time-frequency local feature and directional sensitivity, the shearlet transform can be applied in image processing, for example, image denoising, image fusion, texture feature extraction, and so forth. In the context of composite wavelet, the discrete shearlet is defined as

$$SH(\psi) = \left\{ \psi_{j,l,k} = 2^{\frac{3j}{2}} \psi(G^l S^j x - k) : j \geq 0, -2^j \leq l \leq 2^j, k \in \mathbb{Z}^2 \right\} \tag{6}$$

$$S = \begin{pmatrix} 4 & 0 \\ 0 & 2 \end{pmatrix}, G = \begin{pmatrix} 1 & 1 \\ 0 & 1' \end{pmatrix} \tag{7}$$

where $\psi \in L^2(\mathbb{R}^2)$. S is the anisotropic dilation matrix related to scale transformation. j denotes the scale parameter in particular, which dominates the refinement of frequency and the redundancy of basis elements. G is the shear matrix related to geometrical transformation. l denotes the shear parameter which restricts the orientation of each shearlet element. Moreover, k indicates the shift parameter to locate distributed discontinuities in spatial domain. Calculating the Fourier transform to elements $\psi_{j,l,k}(x)$, we get

$$\hat{\psi}_{j,l,k}(w) = 2^{-\frac{3j}{2}} \psi(wS^{-j}G^{-l})e^{2\pi i w S^{-j}G^{-l}k}. \tag{8}$$

It has frequency support as (9). The frequency division produced by the shearlet transform is illustrated in Figure 1.

$$supp\hat{\psi}_{j,l,k} \subset \left\{ (w_1, w_2) : w_1 \in [-2^{2j-1}, -2^{2j-4}] \cup [2^{2j-4}, 2^{2j-1}], \left| \frac{w_2}{w_1} - l2_{-j} \right| \leq 2^{-j} \right\} \tag{9}$$

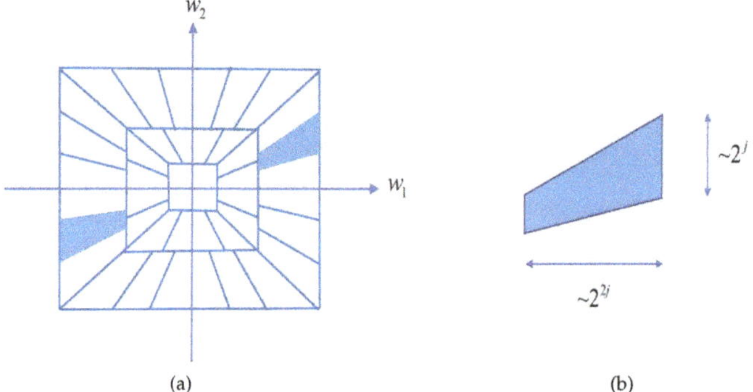

Figure 1. (**a**) The partition of frequency domain; (**b**) Frequency structure of the shearlet $\hat{\psi}_{j,l,k}(w_1, w_2)$, for $w_1 > 0, w_2 > 0$.

The asymptotic approximation error of the shearlet transform is $N^{-2}(logN)^3$ when $N \to \infty$ [29]. So it precisely depicts the interferometric fringe. Besides, the shearlet forms Parseval frames in frequency domain. Its elements are trapezoidal pairs whose area is $2^j \times 2^{2j}$ and oriention is along the zero-crossing line with slope of $-l2^{-j}$ [29]. The corresponding orientation in spatial domain is along the line with slope of $l2^{-j}$. the shearlet elements can be discriminated by scales, locations and orientations. In addition, it apace decays in spatial domain. The aforementioned content indicates the highly directional sensitivity of shearlet, which makes a huge difference in the interferogram filtering.

In practice, the shearlet is shift-variant. The shearlet transform adopts the shift operation of the window function to realize the directional filtering. It involves a subsampling operation, which causes spectral aliasing in frequency domain. Thereby the Gibbs distortion occurs in the reconstructed image. To solve this problem, Easley and Labate proposed the nonsubsampled shearlet transform (NSST) which is enlightened by the great performance of the nonsubsampled contourlet transform. The NSST replaces the subsampled operation with convolution in the directional filtering. It is shift-invariant and efficiently eliminates the pseudo-Gibbs phenomenon in reconstructed images. Hereby the reconstructed image is more effective and intuitive. The decomposition procedure primarily contains two steps as shown in Figure 2.

Step 1: Multiscale Decomposition

The image is decomposed into a high-frequency component and a low-frequency component by means of non-subsampled pyramid (NSP). Then iteratively execute this step till image is decomposed into the j scales.

Step 2: direction localization

The core of direction localization is non-subsampled shearing filter banks (NSSFB), which impose the 2-D convolution of the shearing filter and the high-frequency component on the cartesian domain. The convolution averts subsampled operation, thereby the NSST is shift-invariant.

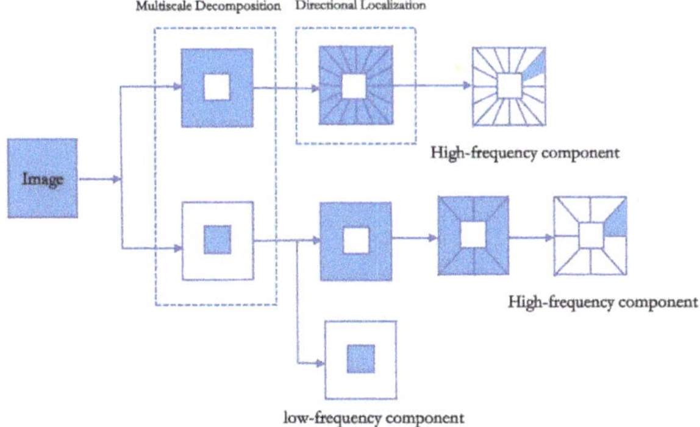

Figure 2. Decomposition process of non-subsampled shearlet transform (NSST).

2.2.2. Pre-Thresholded Wiener Filter

On account of the shift-invariant property, the NSST displays great performance in image denoising, particularly for the texture image. It is also desirable that the NSST filter exploits the coefficient shrinkage method which is consistent with the wavelet filter. Shearlet gives a sparse expression to images. That is to say, the intrinsical information of image is concentrated on few coefficients spreading over each scale with a considerable large amplitude. By contrast, shearlet coefficients generated by noise widely distribute in shearlet domain and its amplitude is small. Owing to this feature, a more accurate pre-thresholded Wiener filtering method with known noise variance is employed to remove the noise component. It consists of two steps:the pre-thresholded operation and Wiener filter. The pre-thresholded operation ensures the smaller local expected square error (LESE) of linear approximation, which is represented as

$$\tilde{c}(i,j) = \begin{cases} c(i,j) & \sigma^2_{c_{i,j}} > k\sigma^2_n; \\ 0 & otherwise. \end{cases} \tag{10}$$

$$k = 1 + \sqrt{\frac{2}{(2N+1)^2}} \tag{11}$$

$$\sigma^2_{c_{i,j}} = \frac{1}{(2N+1)^2} \sum_{m,n=-N}^{N} c^2_{i-m,j-k}. \tag{12}$$

Then the Wiener filter obtains the best linear estimation of clean images. It is represented as

$$\hat{c}(i,j) = a(i,j)\tilde{c}(i,j) \tag{13}$$

$$a(i,j) = \frac{max(\sigma_{\tilde{c}_{i,j}}^2 - \sigma_n^2, 0)}{\sigma_{\tilde{c}_{i,j}}^2}. \tag{14}$$

2.3. Noise Level Estimation Based on Kurtosis

The noise variance is the crux of the Wiener filter. The robust noise level estimator [30] designed by Donoho et al. regards scales median of absolute coefficients as noise variance and is commonly used in many papers. It is straightforward and expedient but tends to over-filter in interferograms with high signal-noise ratio (SNR). A more precise noise variance estimator is extremely urgent to improve the filtering performance of the NSST filter. The most primary innovation of this paper is to introduce the kurtosis-based noise level estimator in Reference [23] to multibaseline InSAR. The kurtosis and the noise variance have a certain relationship in the additive Gaussian white noise model. There exist two unknown variables in the kurtosis model, the number of unknown variables is larger than the number of equations. The result of the minimization method, such as l1-minimization [31], l2-minimization [32] and so forth, exists great errors. To solve the problem, Dong et al combine the kurtosis model with a constraint, in which the kurtosis of images with different structures or statistical behaviors should be unequal, to improve the estimation accuracy of the noise variance. The K-means clustering process is applied to classify the whole image into non-overlapping image patches with different structures.

In this section, the kurtosis of the interferometric phase is introduced. And a special property of the kurtosis is proposed in multibaseline InSAR and is named as the baseline-invariant property. Along with the idea in Reference [23], the baseline-invariant property is regarded as a constraint to ensure the accurate estimation of the noise level. The modified method omits the clustering process and eliminates errors introduced by the fault of the cluster. Efficiency and performance get promoted. Next, the noise level estimator is introduced in two parts. The first one introduces the kurtosis of the interferometric phase and two important properties of the kurtosis. In the other part, the noise level estimation is introduced in detail.

2.3.1. Kurtosis

The image is decomposed into various coefficient components at different scales and directions by NSST. The research on the distribution of NSST coefficient components is conducive to the further analysis of images and is a significant topic in image processing. Among them, the research on its statistics is of great potential. The low-order statistic is weak, even invalid in interferograms which involve a large amount of textures and detail informations. Consequently, the scholar begins with the study of its higher-order statistic, such as kurtosis and skewness. In this section, we study and analyse the kurtosis of interferograms and NSST coefficient components. The kurtosis of a random variable Y is defined as

$$\kappa(Y) = \frac{C_4(Y)}{C_2^2(Y)} - 3, \tag{15}$$

where $C_k(\bullet)$ is the kth cumulant function. The kurtosis reveals the concentration level of the probability density function. Intuitively, the kurtosis reflects the sharpness of the probability density distribution, wherein the kurtosis of the Gaussian distribution is 0. Based on the single-look probability density function of the interferometric phase, the kurtosis is calculated numerically as a function of the coherence, as shown in Figure 3. It indicates that the kurtosis is proportional to the coherence. Obviously, the clean interferogram emerges higher kurtosis when compared with interferograms disturbed by coherent noise with variable degrees. When the noise is strong enough to destroy the fringe structure of interferograms, the kurtosis is smaller due to the influence of the noise. In addition, the kurtosis tends to a negative number when the coherence is close to zero owing to the impact of non-Gaussian noises. In contrast, when the coherence is high, the fringe structure of interferograms plays a primary role. So the kurtosis increases with the improvement of coherence. It should be noted that NSST coefficient components of interferograms are sparse and its kurtosis is greater than zero.

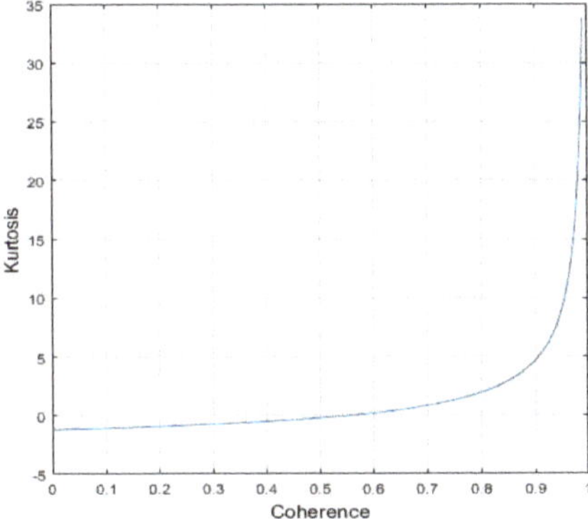

Figure 3. The kurtosis of interferometric phase.

The proposed method takes advantage of two vital properties of the kurtosis of interferograms. One is the scale-invariant property, which works well on all natural images, that is, the kurtosis of coefficient components in the Linear Transform Domain should be held constant at different scales. It is verified and revised by some work in References [31–33]. A modified description for the scale-invariant kurtosis assumption is that the stability is effective in clean images throughout all scales and the variation is the specific impact of noise [32]. The scale-invariant property in the nonsubsampled shearlet transform can be formalized as (16), where Y_i is the ith NSST coefficient component of the clean image Y.

$$\kappa(Y_i) = \kappa(Y_j), i, j = 1, 2, \ldots, N. \tag{16}$$

Another property, which is particular to interferograms and can be yielded from an empirical summary, is proposed in this paper. The kurtosis of images with similar structure or statistical behavior is assumed to be a constant [23]. Along this line, we suggest that the kurtosis of interferograms with different baseline keeps constant and denote it as the baseline-invariant property. Then an interpretation from two perspectives should be introduced. First, It will be explicated further in terms of the probability density function (pdf). The pdf of the interferometric phase is independent of baselines, so does the kurtosis. It can be revealed in (1). In other words, the kurtosis of the interferometric phase is baseline-invariant. However, in virtue of impacts of discretization operation such as sampling, numerical calculation and so forth, the kurtosis of interferograms with different baseline fluctuates around a constant in reality. Fortunately, the fluctuation variance is small enough. So the negative effect of the fluctuation variance can be ignored. On the other hand, as far as the image is concerned, interferograms of the same topography with different baseline intuitively have similar texture trends which show similar statistical behaviors. Correspondingly, the kurtosis maintains invariant in images with similar statistical behaviors, which gives strong support for the baseline-invariant property of kurtosis in multibaseline InSAR.

Some simulated analyses prove the baseline-invariant property. In order to verify the validity of that property for various types of interferograms (generated by various topographies), we select the DEM of five common topography, including Cone, Building, the Northeast plain, China (the elevation below 500 m, the relief is not more than 200 m), the Sichuan Basin, China, Mangkang Mountain, Tibet, China (Plateau, the elevation above 500 m, the relief is more than 200 m). All primordial elevation data are derived from simulated data and Shuttle Radar Topography Mission DEM (SRTM-DEM) elevation data which is provided by Geospatial Data Cloud site, Computer Network Information Center, Chinese Academy of Sciences (http://www.gscloud.cn). The elevation data and typical interferograms of them are shown in Figure 4.

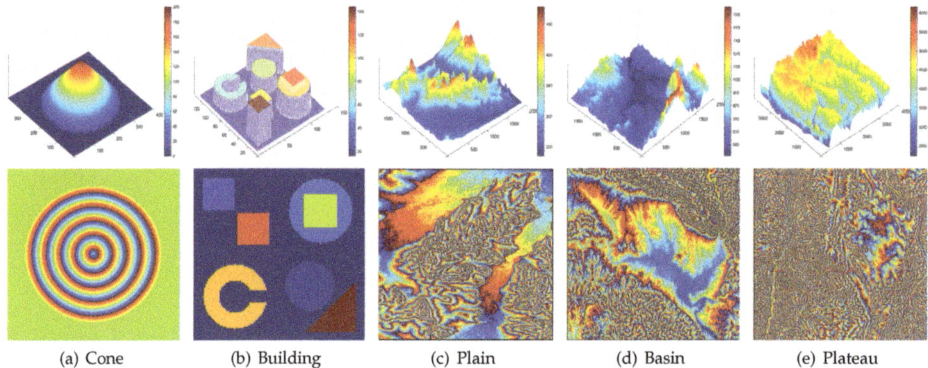

| (a) Cone | (b) Building | (c) Plain | (d) Basin | (e) Plateau |

Figure 4. Five different topography (top) and their typical interferogram (bottom): (**a**)–(**e**) represent cone, building, plain, basin and plateau respectively.

In the light of the DEM of ground scenes and parameters of multibaseline InSAR simulation systems as shown in Table 1, we start with projecting the elevation data into the slant coordinate system. Then we calculate 81 ideal interferograms for each topography when the baseline varies from 50 m to 500 m based on the interferometry principle.

Table 1. Parameters of multibaseline interferometric synthetic aperture radar (InSAR) simulation system.

Parameters	Value
Height	642 km
Cental Frequency	3~9.6 GHz
Bandwidth	100 MHz
Baseline	50~500 m
Look Angle	34.5°
Baseline Orientation Angle	5°

The kurtosis of interferograms with different baseline corresponding to each topography is calculated and its boxplot is shown in Figure 5. In order to approach to the actual situation, the fringe density of simulated interferograms should not be too sparse or too dense. Therefore, the system central frequency for different topography is different. However, in this section, the baseline-invariant property of the kurtosis of interferograms with different baselineswe is confirmed. In other words, the fact that the kurtosis of interferograms with different fringe density remains constant is verified. So the change of the central frequency is not taken into account. We observe the kurtosis standard deviation of each topography numerically in Table 2. The result implies that the maximum standard deviation of kurtosis is 0.1056. Considered the influence of numerical calculation and sampling, it is interpreted that the kurtosis of interferograms with different baseline keeps constant.

Table 2. The standard deviation of kurtosis corresponding to various topography (including Cone, Building, Plain, Basin, Plateau).

Topograpgy	Cone	Building	Plain	Basin	Plateau
Standard Deviation	0.006	0.0298	0.1056	0.0442	0.0064

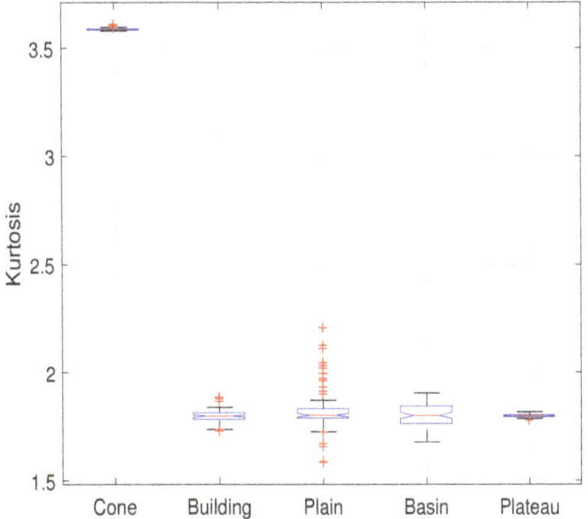

Figure 5. Boxplot of kurtosis corresponding to various topography.

2.3.2. Noise Level Estimation

In this section, the principle and process of the noise level estimator proposed in this paper is introduced in detail. The real part and the imaginary part of interferograms are handled respectively. Taking the real part as an example, it is decomposed into M components in NSST domain. In addition, the additive noise model applies to all shearlet components.

$$y_i = x_i + n_i \tag{17}$$

where y_i, x_i, n_i represents the ith NSST coefficient of the observed phase, ideal phase and noise, respectively. The variance of y_i is represented as

$$\sigma_{y_i}^2 = \sigma_{x_i}^2 + \sigma_{n_i}^2 \tag{18}$$

$$\sigma_{n_i}^2 = \sigma_n^2 \cdot \sigma_{\lambda_i}^2 \tag{19}$$

$$C_4(y_i) = C_4(x_i) + C_4(n_i), \tag{20}$$

where $\sigma_{y_i}^2$, $\sigma_{x_i}^2$, $\sigma_{n_i}^2$ is the variance of y_i, x_i, n_i, respectively. $\sigma_{\lambda_i}^2$ indicates the estimated noise level of the ith NSST coefficient for a white Gaussian noise of standard deviation 1. It is calculated by the Monte Carlo Estimation Method. Then we deduce (21) from (18), (19) and (20):

$$\sigma_{y_i}^4 \kappa(y_i) = \sigma_{x_i}^4 \kappa(x_i) + \sigma_{n_i}^4 \kappa(n_i). \tag{21}$$

Since the assumption that n_i obeys the Gaussian distribution, $\kappa(n_i) = 0$. Besides, the coefficient distribution of subband components is generally more centralized than the Gauss distribution, that is, $\kappa(x_i), \kappa(y_i) \geq 0$, because the interferogram is subdivided into subband components with different scales and directions. The deterministic relationship between noise variance and kurtosis is deduced from (18) and (21), as shown in (22).

$$\sqrt{\kappa(y_i)} = \sqrt{\kappa(x_i)} - \frac{\sigma_n^2 \cdot \sigma_{\lambda_i}^2}{\sigma_{y_i}^2} \sqrt{\kappa(x_i)}. \tag{22}$$

Equation (22) is the kurtosis model which describes the deterministic relationship between the kurtosis and the noise variance. The kurtosis and variance of the observed phase y_i can be calculated directly but the kurtosis of the ideal phase and the noise variance are unknown. The number of unknown variables is larger than the number of equations, so the noise variance cannot be determined directly by Equation (22). A large number of texture structures appearing in interferograms represent similar characteristics with noise in the frequency domain. The existence of texture structure leads to great errors of the noise variance estimation achieved by the minimization method of (22), that is, l1-minimization [31], l2-minimization [32]. To solve this problem, Equation (22) and the baseline-invariant property which acts as another equation are used to jointly estimate the noise variance. With the help of the new information from the baseline-invariant property, the estimation with higher accuracy is realized. the baseline-invariant property is represented as

$$\sqrt{\kappa(x^k)} = \sqrt{\kappa(x^l)}, k, l = 1, 2, \ldots, N. \tag{23}$$

The form of sqrt is adopted for the convenience of the subsequent solution of optimization model. The following optimization model is proposed from (22) and (23).

$$\left\{\hat{\sigma}_n^2, \left\{\hat{\kappa}(x^j)\right\}_{j=1}^N\right\} = \underset{\sigma_n^2, \left\{\hat{\kappa}(x^i)\right\}_{i=1}^N}{\arg\min} \left\{ \sum_{k=1}^N \sum_{l=1}^N \left(\sqrt{\kappa(x^k)} - \sqrt{\kappa(x^l)}\right)^2 \right.$$
$$\left. + \sum_{j=1}^N \sum_{i=1}^M \left(\sqrt{\kappa(y_i^j)} - \sqrt{\kappa(x^j)} + \frac{\sigma_n^2 \cdot \sigma_{\lambda_i}^2}{\sigma_{y_i}^2}\sqrt{\kappa(x^j)}\right)^2 \right\} \tag{24}$$
$$subject\ to : \kappa(x^j) \leq \frac{1}{M}\sum_{i=1}^M \kappa(y_i^j), for\ j = 1, 2, 3, \ldots, N,$$

where the superscript j denotes the jth baseline and the subscript i denotes the ith coefficient component. The first term of optimization function is deduced by the baseline-invariant property of kurtosis and another one is added for fitting the kurtosis model in (22). Then the constraint is derived from the fact that the kurtosis of coefficients decreases owing to the noise disturbance as shown in Figure 3.

The aforementioned optimization function is constrained and non-convex optimization problem with two variables: σ_n^2 and $\kappa(x^j)_{j=1}^N$, which should be considered and optimized simultaneously. This constrained and non-convex optimization problem can be decomposed into two continuous and convex optimization sub-problems by fixing one variable to optimize another variable. Firstly, fix the noise variance σ_n^2 and then update $\kappa(x^j)_{j=1}^N$. The optimization model 1 to be solved is

$$\left\{\kappa(x^j)\right\}_{j=1}^N = \arg\min\left\{ \sum_{k=1}^N \sum_{l=1}^N \left(\sqrt{\kappa(x^k)} - \sqrt{\kappa(x^l)}\right)^2 \right.$$
$$\left. + \sum_{j=1}^N \sum_{i=1}^M \left(\sqrt{\kappa(y_i^j)} - \sqrt{\kappa(x^j)} + \frac{\left(\sigma_n^2 \cdot \sigma_{\lambda_i}^2\right)_t}{\sigma_{y_i}^2}\sqrt{\kappa(x^i)}\right)^2 \right\}$$
$$= \arg\min\left\{ \sum_{k=1}^N \sum_{l=1}^N \left(\sqrt{\kappa(x^k)} - \sqrt{\kappa(x^l)}\right)^2 + \sum_{j=1}^N \sum_{i=1}^M \kappa(y_i^j) \right. \tag{25}$$
$$\left. + 2\sum_{j=1}^N \sum_{i=1}^M \left[\sqrt{\kappa(y_i^j)}\left(\frac{\left(\sigma_n^2 \cdot \sigma_{\lambda_i}^2\right)_t}{\sigma_{y_i}^2} - 1\right)\sqrt{\kappa(x^j)}\right] + \sum_{j=1}^N \sum_{i=1}^M \left(\frac{\left(\sigma_n^2 \cdot \sigma_{\lambda_i}^2\right)_t}{\sigma_{y_i}^2} - 1\right)^2 \kappa(x^j) \right\}.$$

Ignoring the second item which is independent with $\kappa(x^j)_{j=1}^N$. Let

- the vector $k \in \mathbb{R}^N$

$$k = \left[\sqrt{\kappa(x^1)}, \sqrt{\kappa(x^2)}, \ldots, \sqrt{\kappa(x^N)}\right]^T$$

- A is a diagonal matrix of $N \times N$ and the diagonal element is

$$A_{ii} = \sum_{i=0}^{M} \left(\frac{\left(\hat{\sigma}_n^2 \cdot \sigma_{\lambda_i}^2 \right)_t}{\sigma_{y_i}^2} - 1 \right)^2$$

- B is a symmetric matrix

$$B_{ij} = \begin{cases} N-1 & i=j; \\ -1 & otherwise. \end{cases}$$

- the vector $C \in \mathbb{R}^N$

$$c_i = \sum_{i=0}^{M} 2\sqrt{\kappa(y_i^j)} \left(\frac{\left(\hat{\sigma}_n^2 \cdot \sigma_{\lambda_i}^2 \right)_t}{\sigma_{y_i}^2} - 1 \right).$$

Then the optimization function can be simplified as

$$\arg\min \; k^T(A+B)k + c^T k. \tag{26}$$

Because $A+B$ is a positive definite matrix, it is a standard convex optimization for quadratic programming with constraints and can be solved directly.

Similarly, fix $\kappa(x^j)_{j=1}^N$ and update σ_n^2, we deduce optimization model 2 as shown in (27).

$$(\hat{\sigma}_n^2)_{t+1} = \underset{\{\kappa(x^i)\}_{j=1}^N}{\arg\min} \sum_{j=1}^{N} \sum_{i=1}^{M} \left(\sqrt{\kappa(y_i^j)} - \sqrt{\hat{\kappa}_t(x^j)} + \frac{\sigma_n^2 \cdot \sigma_{\lambda_i}^2}{\sigma_{y_i}^2} \sqrt{\hat{\kappa}_t(x^j)} \right)^2. \tag{27}$$

Let the partial derivative of function (27) equals to 0 and then we get the noise variance.

$$2\sum_{j=1}^{N}\sum_{i=1}^{M} \left(\sqrt{\kappa(y_i^j)} - \sqrt{\hat{\kappa}(x^j)} + \frac{\sigma_n^2 \cdot \sigma_{\lambda_i}^2}{\sigma_{y_i}^2} \sqrt{\hat{\kappa}_t(x^j)} \right) \frac{\sqrt{\hat{\kappa}_t(x^j)}}{\sigma_{y_i}^2} = 0$$

$$\hat{\sigma}_n^2 = \frac{\sum_{ij} \left(\sqrt{\hat{\kappa}_t(x^j)} - \sqrt{\kappa(y_i^j)} \right)}{\sigma_{\lambda_i}^2 \cdot \sum_{ij} \frac{\sqrt{\hat{\kappa}_t(x^j)}}{\sigma_{y_i}^2}}. \tag{28}$$

Iteratively update $\kappa(x^j)_{j=1}^N$ and σ_n^2, until convergence.

A pivotal assumption of the aforementioned method is that the noise variance remains constant spatially. Namely, the noise is homogeneous throughout the image space. Yet the interferogram suffers the coherent noise with spatially variable characteristic. The consistent noise variance induces unbalanced filtering results. So we further advance the global noise variance to the local noise variance. Specifically, we divide the image into a certain amount of non-overlapping patches with the same size and assume the stability of noise variance in each patch and estimate its local noise variance simultaneously. The noise level estimation procedure can be summarized in Algorithm 1.

Algorithm 1 Estimating the local noise variance $\left\{(\hat{\sigma}_n^2)_l\right\}_{l=1}^{L}$

Input: $N \times M$ NSST coefficients $\left\{Y_j^i\right\}_{i=1,j=1}^{M,N}$ of the observed interferograms with N different baselines, the size of patch $m \times n$ and the maximum iteration number N_{iter}.

Initialization: $\left\{(\hat{\sigma}_n^2)_l\right\}_{l=1}^{L}=0$.

1: Divide all coefficients into L patches whose size is $m \times n$ and calculate the kurtosis $\left\{\kappa_l(y_i^j)\right\}_{i=1,j=1,l=1}^{N,M,L}$ and variance $\left\{(\sigma_{y_i^j}^2)_l\right\}_{i=1,j=1,l=1}^{N,M,L}$ of eath patch.

2: Repeat.

3: Let $\left\{(\sigma_n^2)_l\right\}_{l=1}^{L}$ equals the solution of the last optimization, update $\left\{\hat{\kappa}_l(x)\right\}_{l=1}^{L}$ by optimization function 1.

4: Let $\left\{\kappa_l(x)\right\}_{l=1}^{L}$ equals the solution of the step 3, update $\left\{(\hat{\sigma}_n^2)_l\right\}_{l=1}^{L}$ by optimization function 2.

5: Until $\left\{(\hat{\sigma}_n^2)_l\right\}_{l=1}^{L}$ and $\left\{\hat{\kappa}_l(x)\right\}_{l=1}^{L}$ converges or N_{iter} is reached.

6: Return $\left\{(\hat{\sigma}_n^2)_l\right\}_{l=1}^{L}$.

---

3. Results

In this section, we validate the efficiency and validity of the proposed method via extensive experiments on simulated and real interferograms. Experiments consist of three parts. First of all, we demonstrate the estimation accuracy of noise level on simulated noisy interferograms. It is the crux of the proposed method. Then, the simulated experiments are conducted. As a promotion of NSST filter, the developed method is compared with five state of the art single-baseline filters, including: Goldstein method, local frequency estimate (LFE) algorithm, optimal integration-based adaptive direction filter (OADF), iterative NL-InSAR and InSAR-BM3D. Finally, the proposed method on real interferograms will be tested. For simplicity, the proposed method is termed as NSST in the following sections. The parameters of various filters are set as

- Goldstein: the filtering window size is 32×32, α equals 0.9;
- OADF: the filtering window size is 7×7;
- LFE: the local frequency estimation window and filtering window are set to 9×9;
- NL-InSAR: the iterative number is 10;
- InSAR-BM3D: the parameters are consistent with [15];
- NSST: the decomposition scale equals 5. Each scale contains 16 different directions.

3.1. Noise Estimation Experiments

The key of the proposed method is noise variance estimation. In this section, we verify the accuracy of the estimated noise variance on simulated data. The original elevation model is a cone, as shown in Figure 4a. As noted before, the interferometric phase accords with the additive noise model in complex domain, the real part and the imaginary part are denoised respectively. We generate three clean interferograms with different baseline. Noisy interferograms are disturbed by the circular complex standard Gaussian noise. The coherence of noisy interferograms is set to 0.1, 0.3, 0.5, 0.7, 0.9, respectively. The true noise variance of the real part, for example, is calculated numerically. Compared with the true value, the estimated noise variance is generated by the proposed method. In order to further test the robustness of the proposed method, 100 Mont-Calo simulations are conducted. Statistics of its results are shown in Figure 6, where the black dotted line is the mean of true value in 100 experiments.

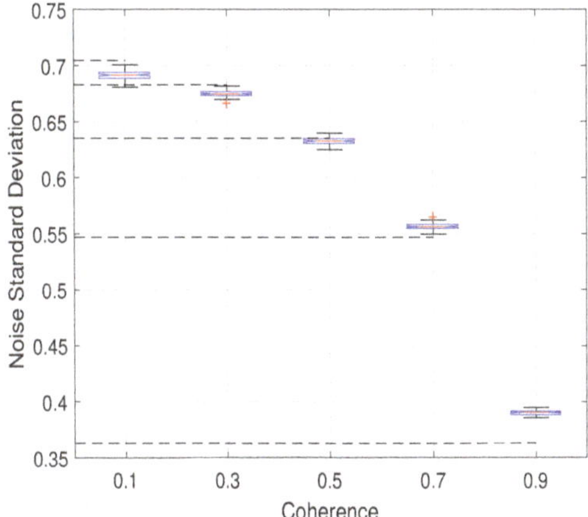

Figure 6. Boxplot of 100 noise level estimation experiments corresponding to each coherence (the black dotted line is the mean of true value).

The comparison between the estimation and the mean implies the accuracy and stability of the proposed method. The maximum error rate is calculated to evaluate the estimation accuracy and is defined as

$$R_M = \frac{max(|\hat{\sigma} - \bar{\sigma}|)}{\bar{\sigma}} \times 100 \tag{29}$$

Where $\hat{\sigma}$ denotes the estimated value in 100 experiments, $\bar{\sigma}$ indicates the mean of true value. The result is shown in Table 3. It is obvious that some errors exist in the estimation. The higher the coherence is, the larger the SNR is. In the case of low coherence, the significant noise level engenders the confusion of the high-frequency information and the noise, which results in a slight overestimation. On the contrary, the noise near fringe in interferograms is mistaken for significant pixels owing to its weak effect to fringes in the case of high coherence. So the estimation is lower than the true value. We must emphasize that the maximum error rate is controlled within 8.76%. The underestimation is compensated by the excellent performance of Wiener filter.

Table 3. Maximum error rate of noise level estimation.

Coherence	0.1	0.3	0.5	0.7	0.9
Actual Value	0.7044	0.6830	0.6351	0.5469	0.3632
Maximum Error Rate(%)	3.35	2.37	1.62	3.31	8.76

3.2. Simulated Experiments

In this section, we simulate three interferograms of cone and mountain to assess the performance of the proposed method. The noise environment comprises two situations: constant and variable noise variance in spatial domain. It is necessary for comparative experiments within each section. The experiments in interferograms with unitary noise variance are conducted to inspect the reconstructed performance for phase jump and phase gradient mutation. We select interferograms with 400 × 400 pixels, which are generated by cone and contain both phase jump and phase gradient mutation. Its clean interferograms and noisy interferograms can be shown in Figure 7. Coherence is set to 0.5. The block operation is omitted because of the constant noise variance. The comparable results of the interferogram with the shortest baseline are shown in the first row of Figure 8.

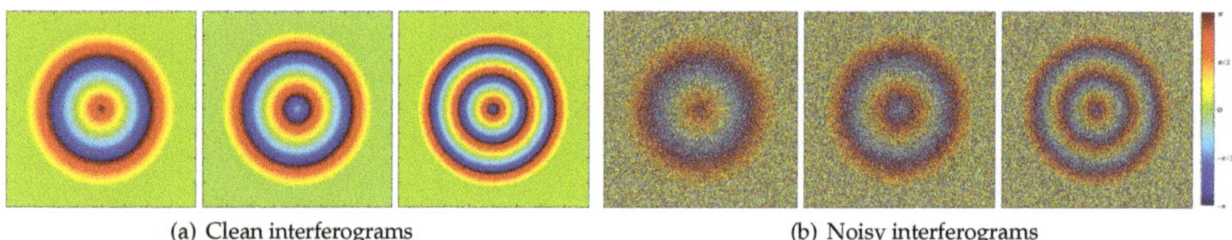

(a) Clean interferograms   (b) Noisy interferograms

Figure 7. Clean interferograms and noisy interferograms generated by a cone with coherence of 0.5.

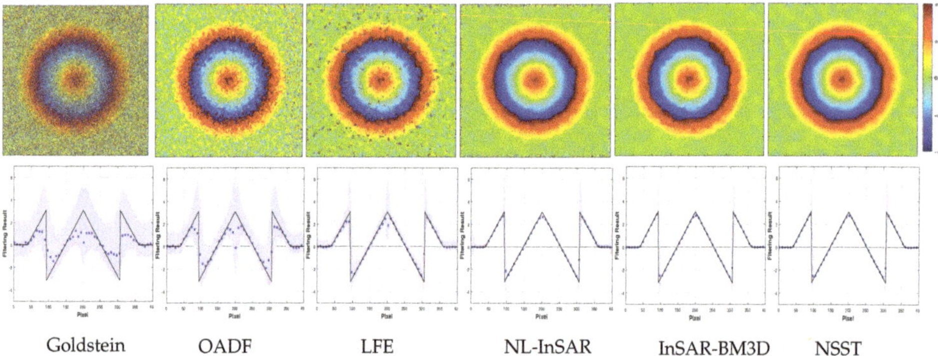

Goldstein   OADF   LFE   NL-InSAR   InSAR-BM3D   NSST

Figure 8. The filter results of the interferogram generated by a cone with coherence of 0.5 (top) and the statistical result of pixels at the center row (bottom, the black solid line is the true value; the blue dotted line denotes the mean of 100 experiments; the pale blue shadow is the range of three times standard deviation near the mean.).

Intuitively, the result of the Goldstein method is incorrect. There are obvious errors in OADF and LFE. NL-InSAR, InSAR-BM3D and the proposed method all obtain appreciable results. The mean square error (MSE) between the clean interferogram and the filtered interferogram confirms above statements. What is more, Table 4 lists the number of residues in the filtered interferogram and the computation time. Note that the bold font indicates the best performance in the table. Table 4 exhibits that the proposed method outperforms to others with a running time that is second only to Goldstein method. The similar MSE are found in InSAR-BM3D but its computation time is about twice as long as our method. The results in NL-InSAR is superior to Goldstein method, OADF, LFE but its operation efficiency is the worst due to iterative operation. By and large, a combination of minimum MSE, minimum number of residues and high efficiency has taken in our method.

Table 4. Performance of various methods.

	MSE	Residues	Times (s)
Noisy Image	1.7897	34492	–
Goldstein	1.853	21041	**0.32**
LFE	0.7699	1454	113.39
OADF	0.8951	389	59.45
NL-InSAR	0.6577	290	459.33
InSAR-BM3D	0.6014	**0**	38.02
NSST	**0.4954**	**0**	12.68

The second row of Figure 8 displays the mean and standard deviation of 100 Monte Carlo experiments at the central row of results. Thereinto, the black solid line is the true value. The blue dotted line denotes the mean of 100 experiments. The pale blue shadow is the range of three times standard deviation near the mean. The poorest result in Goldstein method is interrelated with fixed

α and its boundedness to lower SNR. The result of NL-InSAR, InSAR-BM3D and our method for the stationary phase is close to unbiased estimation, while other methods emerge distinct deviation. The basic idea of OADF and LFE is the estimation to local direction and frequency of interferometric phase. So the invalid estimation has contributed to a heavy fluctuation near phase jump and phase gradient mutation. NL-InSAR produces excellent performance in phase jump but its non-local mean operation induces the outlier which can be observed on both sides. Nevertheless, it produces excellent performance in phase jump. Generally, InSAR-BM3D and our method outperform other methods but our method has higher operation efficiency.

Considering a more complex noise level model in the second experiment, in which the coherence ranges from 0.1 to 0.9 and increases from left to right at regular intervals. Figure 9 shows clean interferograms and noisy interferograms with the size of 240×240. The image is divided into 9 non-overlapping patches in noise variance estimation procedure. The size of each patch is 80×80.

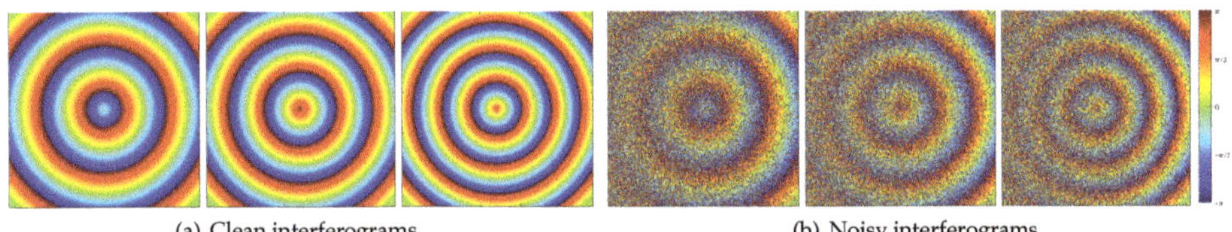

(a) Clean interferograms   (b) Noisy interferograms

Figure 9. Clean interferograms and noisy interferograms generated by a cone with coherence ranging from 0.1 to 0.9.

In this part, a new evaluation index, which is expressed as the pixel-wise Gradient Magnitude Similarity (GMS) [34,35] between the reference and filtered images, is adhibited to evaluate the filtering results of various methods. Gradient magnitude is an apparent indication of the difference between adjacent pixels. The gradient of interferometric phase consists of two parts: the gradient of the local stationary phase reflects the local slope of topography and the similarity of gradient casts light upon the similarity of local topography. In addition, the outlier implies phase discontinuity within a phase period. Similar to the well-known Structure SIMilarity (SSIM) index, the gradient similarity of phase jump can also reflect the edge-preserving ability of methods. Therefore, it is worth to use GMS as a new evaluation index. GMS is defined as

$$GMS(i) = \frac{2G_o(i)G_f(i) + \lambda}{G_o^2(i) + G_f^2(i) + \lambda}, \tag{30}$$

where λ is set to 0.0026 to ensure numerical stability. G_o and G_f indicate the gradient magnitudes of o and f. The gradient magnitudes is derived from (31) and (32).

$$G_o(i) = \sqrt{(o \otimes h_x)^2 + (o \otimes h_y)^2} \tag{31}$$

$$G_f(i) = \sqrt{(f \otimes h_x)^2 + (f \otimes h_y)^2}, \tag{32}$$

where o and f indicate the original images and filtered images respectively. h_x and h_y indicate the Prewitt filter along the horizontal and vertical direction. They are derived from (33).

$$h_x = \begin{bmatrix} \frac{1}{3} & 0 & -\frac{1}{3} \\ \frac{1}{3} & 0 & -\frac{1}{3} \\ \frac{1}{3} & 0 & -\frac{1}{3} \end{bmatrix}, h_y = \begin{bmatrix} \frac{1}{3} & \frac{1}{3} & \frac{1}{3} \\ 0 & 0 & 0 \\ -\frac{1}{3} & -\frac{1}{3} & -\frac{1}{3} \end{bmatrix} \tag{33}$$

It should be noted that the larger the GMS value is, the higher the quality of the restored image is. When $GMS = 1$, the reference image is fully recovered. The mean of GMS map (GMSM) refers to the overall performance of GMS map.

Figure 10 shows the filtering results, residual graph and GMS map corresponding to six different filters. Results show that all methods can correctly restore the original phase in the high-coherence region. However, only InSAR-BM3D and the proposed method get considerable results in the low coherence region. Besides, as far as the GMS map is concerned, our method has better ability to maintain the phase gradient, especially in the low-coherence region. The estimated phase of the proposed method tends to be more stationary and closes to the original phase. Table 5 shows the MSE, GMSM and computation time. Note that the bold font indicates the best performance in the table. The performance of various methods can be expressed as (where > denotes better performance):

- MSE: NSST>InSAR-BM3D>NL-InSAR>OADF>LFE>Goldstein
- GMSM: NSST>InSAR-BM3D>NL-InSAR$\geq$LFE>OADF>Goldstein
- Computation efficiency: Goldstein>NSST>InSAR-BM3D>OADF>LFE>NL-InSAR

As a whole, our method is superior to other methods.

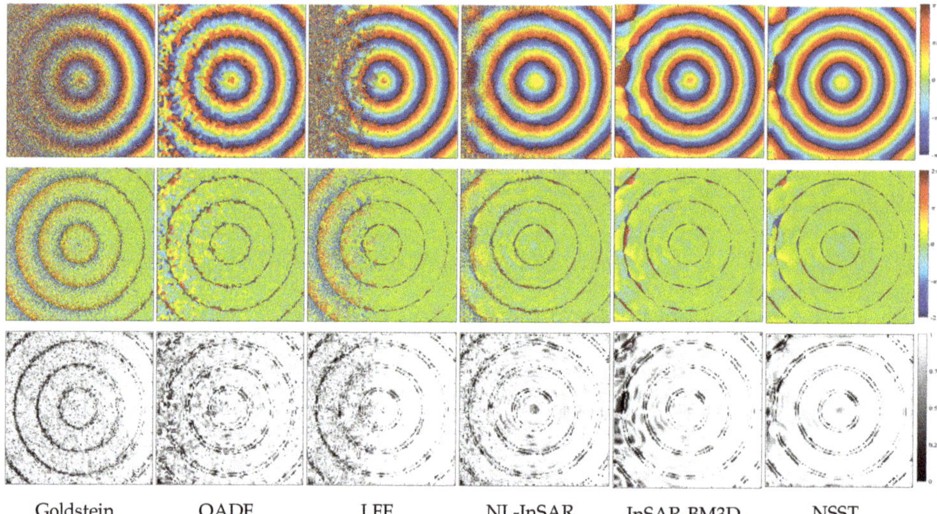

Figure 10. The filter results of the interferogram generated by a cone with variable coherence (**top**), the residuals graph (**middle**) and the Gradient Magnitude Similarity (GMS) map (**bottom**).

Table 5. Performance of various methods.

	MSE	Residues	GMSM	Times (s)
Noisy Image	2.1571	11679	0.8297	–
Goldstein	1.9478	8490	0.8562	**0.15**
LFE	1.5504	4660	0.8975	39.00
OADF	1.4209	317	0.8684	20.95
NL-InSAR	1.3339	1211	0.8994	96.66
InSAR-BM3D	1.0631	14	0.9103	10.59
NSST	**0.9841**	**9**	**0.9343**	5.07

We consider a more complex topography to test the performance of various methods. The elevation data of a steep mountain in Shaanxi Province, China is selected to generate three interferograms. Coherence is consistent with last experiment. The size of interferograms is 1600×1600. The image is divided into 25 non-overlapping patches in noise variance estimation procedure. The size of each patch is 320×320. Interferograms involve dense and sparse fringes. Dense fringes are mostly located in the region with low coherence, which can better verify the effectiveness of the proposed method. Figure 11 shows clean interferograms and noisy interferograms.

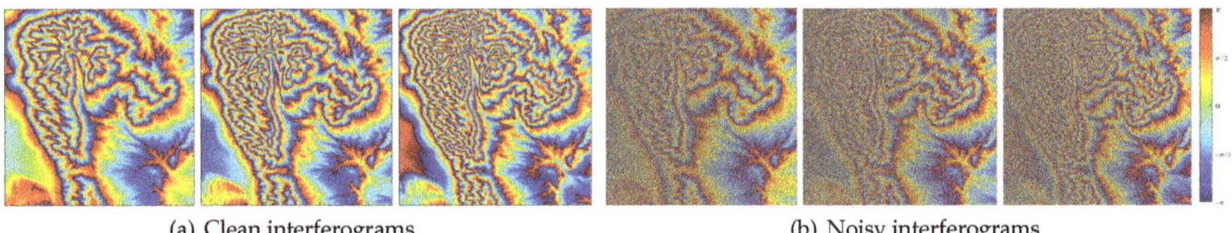

(a) Clean interferograms   (b) Noisy interferograms

Figure 11. Clean interferograms and noisy interferograms generated by a mountain with coherence ranging from 0.1 to 0.9.

Figure 12 shows the filtered results. Table 6 shows the results are similar to the results of the previous experiment. Note that the bold font indicates the best performance in the table. The proposed method produces minimum MSE and maximum GMSM , which prove the prominent filtering performance of it. The minimum MSE of the proposed method proves that the result of the proposed method is closer to the true interferometric phase. And the maximum GMSM implies that the result of the proposed method has fewer outliers and better local stability. The number of residues of the proposed method is second only to InSAR-BM3D and is very close to InSAR-BM3D. The reduction of residues is up to 99.97% compared with the residues of noisy image. Moreover, the computation time of the proposed method is half of the time of InSAR-BM3D. In general, the proposed method not only has outstanding filtering performance but also has high operation efficiency.

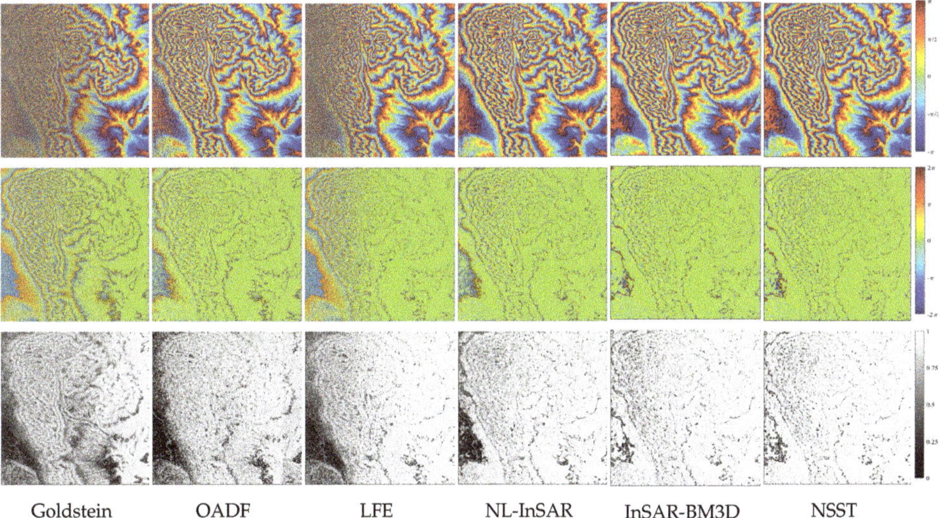

Goldstein   OADF   LFE   NL-InSAR   InSAR-BM3D   NSST

Figure 12. The filter results of the interferogram generated by a complex topography with variable coherence (**top**), the residuals graph (**middle**) and the GMS map (**bottom**).

Table 6. Performance of various methods.

	MSE	Residues	GMSM	Times (s)
Noisy Image	2.1870	518970	0.6876	–
Goldstein	1.9655	371334	0.7534	**4.30**
LFE	1.5598	199348	0.8284	2784.72
OADF	1.4402	9138	0.7828	901.24
NL-InSAR	1.2885	18963	0.8648	5491.40
InSAR-BM3D	1.0564	**126**	0.9026	543.13
NSST	**1.0386**	148	**0.9213**	272.79

3.3. Experiments on Real Interferograms

The original data set is three repeat-orbit InSAR data at Colorado Grand Canyon, USA, which is obtained by Alos-1 satellite. Figure 13 shows its interferograms. Baselines are 738.182, 1241.066 and 1827.02 m, respectively. The size of interferograms is 6000×5910. In noise variance estimation procedure, each interferogram is divided into 225 non-overlapping patches whose size is 400×394.

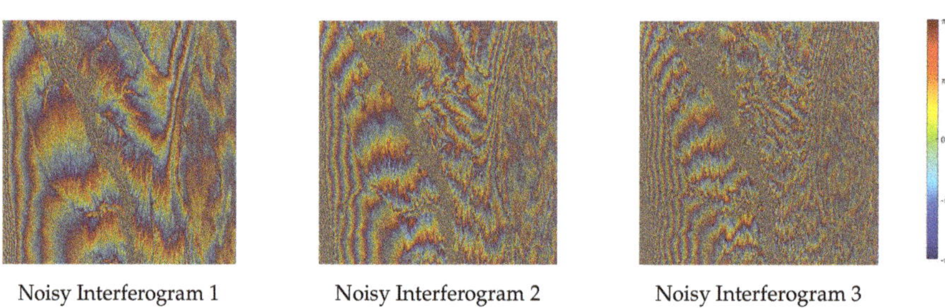

Noisy Interferogram 1   Noisy Interferogram 2   Noisy Interferogram 3

Figure 13. The real interferograms with different baseline (the length of baseline increase form left to right).

Results are shown in Figure 14. Intuitively, the Goldstein method has completely failed. And the apparent noise remains in the result of OADF and LFE. The excellent results of NL-InSAR, InSAR-BM3D and the proposed method are similar.

In order to further compare various methods, the low-coherence region in the upper right corner (row: 1:1000, column: 4910:5910) is cropped to further analysis. Figure 15 presents denoising results of different methods. Table 7 lists the number of residues, the reduction rate of resides and computation time. Note that the bold font indicates the best performance in the table. The excellent performance of the proposed method can be confirmed directly by visual observation. In the proposed method, the reduction rate of residues (up to 99%) is remarkable and the result is more conducive for the subsequent phase unwrapping.

Table 7. Performance of various methods.

	Residues	Residues Reduction Rate	Times(s)
Noisy Image	174198	–	–
Goldstein	124397	28.59	**0.9**
LFE	60714	65.15	720.00
OADF	10899	93.74	433.47
NL-InSAR	15866	90.89	2542.23
InSAR-BM3D	16672	90.42	171.69
NSST	**1374**	**99.21**	91.57

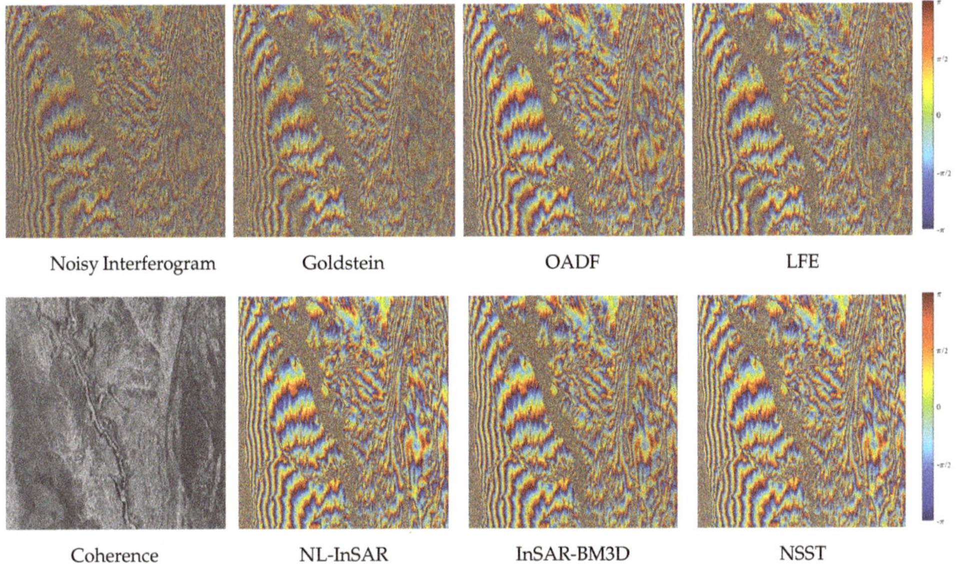

Noisy Interferogram   Goldstein   OADF   LFE

Coherence   NL-InSAR   InSAR-BM3D   NSST

Figure 14. The filtered results of the real interferogram with the longest baseline.

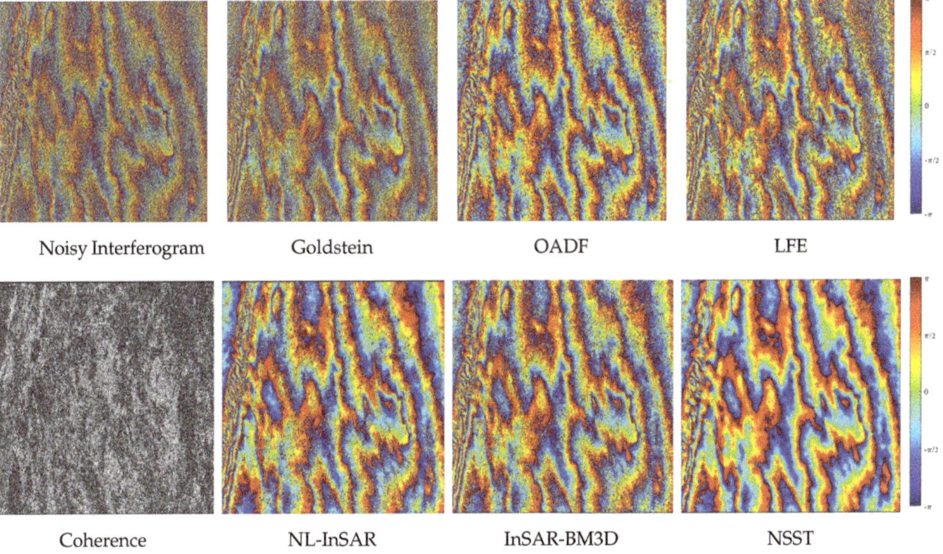

Noisy Interferogram   Goldstein   OADF   LFE

Coherence   NL-InSAR   InSAR-BM3D   NSST

Figure 15. The filtered results of the low-coherence reagion (the upper right corner of real interferogram with the longest baseline (row: 1:1000, column: 4910:5910)).

Eight phase profiles along the phase gradient direction, which involve intact phase period and satisfy local stationarity, are extracted for contrast. White lines in Figure 16 represent the phase profile at low-coherence region (line 2 and line 8), high-coherence region (line 1 and line 6), complex topography region (line 3), the region corresponding to steep topography (line 5), and so forth. As shown in Figure 17, results of phase profiles are arranged in the order of its position (increase from left to right, from top to bottom). For simplicity, Figure 17 only exhibits results of NL-InSAR, InSAR-BM3D and the proposed method, which are superior to other methods intuitively. As shown in line 3, none of three methods can recover the real phase correctly at complex topography region.The difficulty is inherent defect of interferogram with too long baseline. The phase profile at flat region with high-coherence, which corresponds to line 1 and line 6, is estimated appropriately by NL-InSAR and the proposed method. However, a few abnormal values arise in the result of InSAR-BM3D. The comparison result of the number of abnormal values at high-coherence region corresponding to steep topography (line 5) can be expressed as: the proposed method$\geq$NL-InSAR$>$InSAR-BM3D. For the low-coherence region (line 2 and line 8), the proposed method outperforms NL-InSAR and InSAR-BM3D. The proposed method produces a more stationarity and authentic result. It is consistent with the result in Figure 15. On balance, the proposed method has the best comprehensive performance.

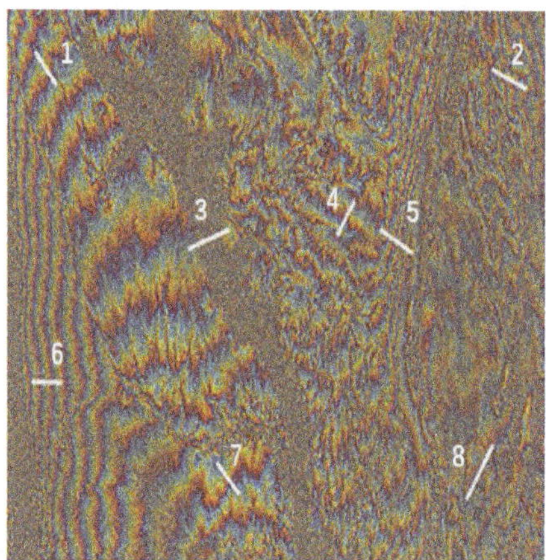

Figure 16. The real interferogram with the longest baseline(the order of white lines increases from left to right, from top to bottom).

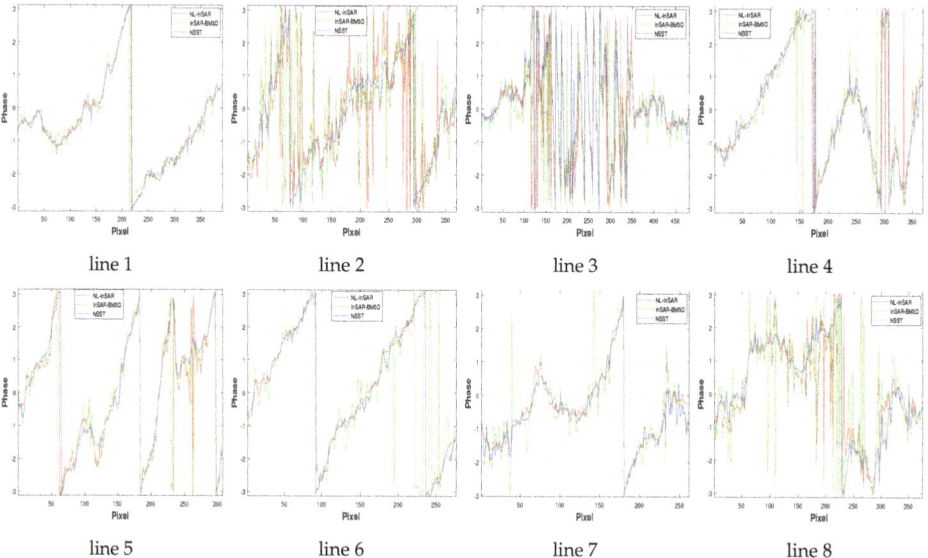

Figure 17. The phase profile of white lines in Figure 16 (the red, green and blue solid line represent the result of NL-InSAR, InSAR-BM3D and NSST, respectively.).

4. Conclusions

An attempt to the joint filtering method in multibaseline InSAR based on the statistical property of interferometric phase is proposed in this paper. This paper analyses the high-order statistical property of interferograms with different baseline and proposes an empirical assumption: the kurtosis of interferograms with different baseline keeps invariant. Simulated experiments give numerical support to it. The filtering process of the proposed method involves four parts: the NSST decomposition, the noise level estimation, pre-thresholded Wiener filter and inverse NSST. NSST gives an optimal sparse representation of distributed discontinuities, such as fringes of interferograms. We obtain a series of NSST coefficients at different scales and directions after NSST decomposition. Based on the kurtosis model in NSST domain and baseline-invariant property of interferograms, the noise variance of interferograms is represented as the solution of a constrained non-convex optimization problem. The clean NSST coefficient is estimated by the Wiener filter with the local noise variance derived by block estimation. The noise estimation experiments prove the validity of the noise level estimator. Experiments on simulated data and real data prove the edge-preservation performance and excellent filtering performance of the proposed method. Many coefficient components with the same kurtosis are obtained by NSST. Sufficient data means that the filtering performance of the proposed method is not affected by the number of interferograms. The great performance can be acquired when the number of interferograms is small. However, a large amount of memory is occupied by a large number of coefficient components. The algorithm has some requirements for memory performance. But this problem can be alleviated by adjusting the scale of NSST decomposition according to the actual computer performance.

Author Contributions: Conceptualization, Y.L. and S.L.; methodology, Y.L. and S.L.; software, Y.L.; validation, Y.L., S.L. and H.Z.; formal analysis, Y.L.; investigation, Y.L.; resources, H.Z.; data curation, H.Z.; writing–original draft preparation, Y.L.; writing–review and editing, Y.L., S.L. and H.Z.; visualization, Y.L.; supervision, S.L.; project administration, S.L.; funding acquisition, S.L. All authors have read and agreed to the published version of the manuscript.

Funding: This research was funded by the National Key Research and Development Program under Grant 2017YFB0502700.

Acknowledgments: The author would like to thank a anonymous reviewers and the editors for their significant and constructive comments.

Conflicts of Interest: The authors declare no conflict of interest.

References

1. Zebker, H.A.; Goldstein, R.M. Topographic Mapping From Interferometric Synthetic Aperture Radar Observations. *J. Geophys. Res. Solid Earth* **1986**, *91*, 4993–4999. [CrossRef]
2. Teng, W.; Perissin, D.; Liao, M.; Rocca, F. Deformation Monitoring by Long Term D-InSAR Analysis in Three Gorges Area, China. In Proceedings of the IEEE International Geoscience & Remote Sensing Symposium (IGARSS 2008), Boston, MA, USA, 8–11 July 2008.
3. Gini, F.; Lombardini, F. Multibaseline cross-track SAR interferometry: A signal processing perspective. *IEEE Aerosp. Electron. Syst. Mag.* **2005**, *20*, 71–93. [CrossRef]
4. Bamler, R.; Hartl, P. Synthetic aperture radar interferometry. *Inverse Probl.* **1999**, *14*, 12–13. [CrossRef]
5. Roscoe, A.J.; Blair, S.M. Choice and properties of adaptive and tunable digital boxcar (moving average) filters for power systems and other signal processing applications. In Proceedings of the 2016 IEEE International Workshop on Applied Measurements for Power Systems (AMPS), Aachen, Germany, 28–30 September 2016; pp. 1–6. doi:10.1109/AMPS.2016.7602853. [CrossRef]
6. Lee, J.S.; Papathanassiou, K.P.; Ainsworth, T.L.; Grunes, M.R.; Reigber, A. A new technique for noise filtering of SAR interferometric phase images. *IEEE Trans. Geosci. Remote Sens.* **1998**, *36*, 1456–1465.
7. Trouve, E.; Nicolas, J.M.; Maitre, H. Improving phase unwrapping techniques by the use of local frequency estimates. *IEEE Trans. Geosci. Remote Sens.* **1998**, *36*, 1963–1972. [CrossRef]

8. Yin, H.J.; Li, Z.W.; Ding, X.L.; Jiang, M.; Sun, Q.; Wang, P. Optimal integration-based adaptive direction filter for InSAR interferogram. *J. Remote Sens.* **2009**, *13*, 1092–1098.

9. Deledalle, C.A.; Denis, L.; Tupin, F. NL-InSAR: Nonlocal interferogram estimation. *IEEE Trans. Geosci. Remote Sens.* **2010**, *49*, 1441–1452. [CrossRef]

10. Gao, Y.; Zhang, S.; Li, T.; Guo, L.; Chen, Q.; Li, S. A novel two-step noise reduction approach for interferometric phase images. *Opt. Lasers Eng.* **2019** *121*, 1–10.

11. Goldstein, R.M.; Werner, C.L. Radar interferogram filtering for geophysical applications. *Geophys. Res. Lett.* **1998**, *25*, 4035–4038. [CrossRef]

12. Mestre-Quereda, A.; Lopez-Sanchez, J.M.; Selva, J.; Gonzalez, P.J. An Improved Phase Filter for Differential SAR Interferometry Based on an Iterative Method. *IEEE Trans. Geosci. Remote Sens.* **2018**, *56*, 4477–4491. [CrossRef]

13. Lopez-Martinez, C.; Fabregas, X. Modeling and reduction of SAR interferometric phase noise in the wavelet domain. *IEEE Trans. Geosci. Remote Sens.* **2002**, *40*, 2553–2566. [CrossRef]

14. Chang, S.G.; Yu, B.; Vetterli, M. Adaptive wavelet thresholding for image denoising and compression. *IEEE Trans. Image Process.* **2000**, *9*, 1532–1546. [CrossRef] [PubMed]

15. Sica, F.; Cozzolino, D.; Zhu, X.X.; Verdoliva, L.; Poggi, G. INSAR-BM3D: A nonlocal filter for SAR interferometric phase restoration. *IEEE Trans. Geosci. Remote Sens.* **2018**, *56*, 3456–3467. [CrossRef]

16. Xu, G.; Xing, M.D.; Xia, X.G.; Zhang, L.; Liu, Y.Y.; Bao, Z. Sparse Regularization of Interferometric Phase and Amplitude for InSAR Image Formation Based on Bayesian Representation. *IEEE Trans. Geosci. Remote Sens.* **2015**, *53*, 2123–2136. [CrossRef]

17. Parizzi, A.; Brcic, R. Adaptive InSAR Stack Multilooking Exploiting Amplitude Statistics: A Comparison Between Different Techniques and Practical Results. *IEEE Geosci. Remote Sens. Lett.* **2011**, *8*, 441–445. [CrossRef]

18. Schmitt, M.; Stilla, U. Adaptive Multilooking of Airborne Single-Pass Multi-Baseline InSAR Stacks. *IEEE Trans. Geosci. Remote Sens.* **2014**, *52*, 305–312. doi:10.1109/TGRS.2013.2238947. [CrossRef]

19. Ferretti, A.; Fumagalli, A.; Novali, F.; Prati, C.; Rocca, F.; Rucci, A. A New Algorithm for Processing Interferometric Data-Stacks: SqueeSAR. *IEEE Trans. Geosci. Remote Sens.* **2011**, *49*, 3460–3470. doi:10.1109/TGRS.2011.2124465. [CrossRef]

20. Schmitt, M.; Schönberger, J.L.; Stilla, U. Adaptive Covariance Matrix Estimation for Multi-Baseline InSAR Data Stacks. *IEEE Trans. Geosci. Remote Sens.* **2014**, *52*, 6807–6817. doi:10.1109/TGRS.2014.2303516. [CrossRef]

21. You, Y.; Wang, R.; Zhou, W. A Phase Filter for Multi-Pass InSAR Stack Data by Hybrid Tensor Rank Representation. *IEEE Access* **2019**, *7*, 135176–135191. doi:10.1109/ACCESS.2019.2942008. [CrossRef]

22. Kazubek, M. Wavelet domain image denoising by thresholding and Wiener filtering. *Signal Process. Lett. IEEE* **2003**, *10*, 324–326. [CrossRef]

23. Dong, L.; Zhou, J.; Tang, Y.Y. Noise level estimation for natural images based on scale-invariant kurtosis and piecewise stationarity. *IEEE Trans. Image Process.* **2016**, *26*, 1017–1030. [CrossRef] [PubMed]

24. Guo, K.; Labate, D.; Lim, W.Q.; Weiss, G.; Wilson, E. Wavelets with composite dilations and their MRA properties. *Appl. Comput. Harmon. Anal.* **2006**, *20*, 202–236. [CrossRef]

25. Starck, J.L.; Candès, E.J.; Donoho, D.L. The curvelet transform for image denoising. *IEEE Trans. Image Process.* **2002**, *11*, 670–684. [CrossRef] [PubMed]

26. Häuser, S.; Steidl, G. Fast finite shearlet transform. *arXiv* **2012**, arXiv:1202.1773.

27. Lim, W.Q. The discrete shearlet transform: A new directional transform and compactly supported shearlet frames. *IEEE Trans. Image Process.* **2010**, *19*, 1166–1180. [PubMed]

28. Easley, G.; Labate, D.; Lim, W.Q. Sparse directional image representations using the discrete shearlet transform. *Appl. Comput. Harmon. Anal.* **2008**, *25*, 25–46. [CrossRef]

29. Labate, D.; Lim, W.Q.; Kutyniok, G.; Weiss, G. Sparse multidimensional representation using shearlets. *Proc. SPIE* **2005**, *5914*, 59140U.

30. Donoho, D.L.; Johnstone, J.M. Ideal spatial adaptation by wavelet shrinkage. *Biometrika* **1994**, *81*, 425–455. [CrossRef]

31. Tang, C.; Yang, X.; Zhai, G. Noise estimation of natural images via statistical analysis and noise injection. *IEEE Trans. Circuits Syst. Video Technol.* **2014**, *25*, 1283–1294. [CrossRef]

32. Zoran, D.; Weiss, Y. Scale invariance and noise in natural images. In Proceedings of the 2009 IEEE 12th International Conference on Computer Vision, Kyoto, Japan, 29 September–2 October 2009; pp. 2209–2216.

33. Lam, E.Y.; Goodman, J.W. A mathematical analysis of the DCT coefficient distributions for images. *IEEE Trans. Image Process.* **2000**, *9*, 1661–1666. [CrossRef]

34. Kim, D.; Han, H.; Park, R. Gradient information-based image quality metric. *IEEE Trans. Consum. Electron.* **2010**, *56*, 930–936. doi:10.1109/TCE.2010.5506022. [CrossRef]

35. Xue, W.; Zhang, L.; Mou, X.; Bovik, A.C. Gradient magnitude similarity deviation: A highly efficient perceptual image quality index. *IEEE Trans. Image Process.* **2013**, *23*, 684–695. [CrossRef] [PubMed]

Letter

Safe Helicopter Landing on Unprepared Terrain Using Onboard Interferometric Radar

Pavel E. Shimkin [1], Alexander I. Baskakov [1], Aleksey A. Komarov [1] and Min-Ho Ka [2,*]

[1] Department of Radio Engineering Devices and Antenna Systems, Moscow Power Engineering Institute, National University, 111250 Moscow, Russia; shimkinpy@mpei.ru (P.E.S.); baskakovai@mpei.ru (A.I.B.); komarovaa@mpei.ru (A.A.K.)
[2] School of Integrated Technology, Yonsei University, Seoul 21983, Korea
* Correspondence: kaminho@yonsei.ac.kr; Tel.: +82-32-749-5840

Received: 4 March 2020; Accepted: 22 April 2020; Published: 24 April 2020

Abstract: This letter proposes a radar interferometric survey system for the ground surface of helicopter landing sites. This system generates high-quality three-dimensional terrain surface topography data and estimates the slope of the site with the required accuracy. This study presents the processing algorithms of the radar system for safe helicopter landing using an interferometric method and also demonstrates the efficiency of the proposed approach based on computer simulation results. The results of the calculated potential accuracy characteristics of such a system are presented, as well as one of the variants of the algorithmic implementation of a simulation computer model implemented on MATLAB. Visual results of modeling using an example of a helicopter landing on a non-uniform surface relief similar to a real case are shown. The study focuses on the simulation of a unique on-board radar system, which allows helicopters to land on an unprepared site with a high degree of safety, having previously determined the presence of dangerous irregularities, inclines, foreign objects, and mechanisms on the site.

Keywords: interferometric radar; helicopter landing; simulation model

---

1. Introduction

One of the main causes of helicopter accidents [1,2] is the unreliability of means to ensure their landing on unprepared landing sites (LSs) in adverse weather conditions during the day and at night with poor visual visibility. Even in good weather conditions, owing to the dusty surface of the earth, the pilot and crew are at risk during landing. Massive dust clouds formed by air swirls owing to the helicopter's screws substantially mask the LS. At the same time, irregularities with a height of 0.5 m and more and LSs with slopes more than 15° [2] already represent a danger to the landing of the helicopter, especially in strong winds. Existing on-board systems (satellite navigation systems, on-board radio altimeters) that most helicopters are equipped with cannot provide necessary information about the state of the terrain, slopes of the LS, and presence of foreign objects.

Until now, studies have been focused on two main areas of research in this field [3–9]. The first is the use of laser locators in the safe landing systems of a helicopter (SLSHs). High relief detailing is achieved and information about the LS relief is displayed on the screen in the cockpit. The main disadvantages of laser SLSHs are their strong dependence on weather conditions, i.e., it is impossible to survey the surface of the LS in the conditions of rain, fog, and snow, as well as their high cost compared to radar systems. The second is the use of radar systems in combination with special processing of signals reflected from the landing pad. Both continuous and pulsed systems with complex signals are used. There are several methods that allow information about the elevations of the surface relief to be isolated from radar data: stereoscopic, interferometric, clinometric, and polarimetric. Stereoscopic and interferometric methods require two images of the same surface area from different positions, the clinometric method works with only one image, and the polarimetric method requires a set of images taken with different signal polarizations.

Owing to a number of features of these methods, as well as flight regulation requirements [10], which discuss the need for mandatory flight of the proposed landing zone from several perspectives, a combination of the stereoscopic and interferometric methods is considered to be suitable for practical use when evaluating the surface topography.

The purpose of this work is to show the main stages of one of the options for the algorithmic implementation of a simulation model of the radar SLSH (RSLSH) interferometric method and also to demonstrate the performance of the proposed solution for safe landing of the helicopter based on the results of computer simulation.

2. Description of the RSLSH

To ensure a safe landing of the helicopter, a flight test is carried out when approaching it at a speed not exceeding 15 ms^{-1}, according to flight regulations [10] from a height of approximately 50 to 100 m. During the flyby, a radar survey of the LS is carried out in the form of manual system. As the carrier moves, line-by-line scanning of the viewing area is performed using a narrow beam of a receiving-transmitting waveguide slot antenna in the azimuth plane without aperture synthesis; a wide beam in the slope plane is used to highlight the required area of view of the LS (Figure 1). It is important to note that during radar observation of the flight station, the helicopter must fly at a constant height with a constant speed.

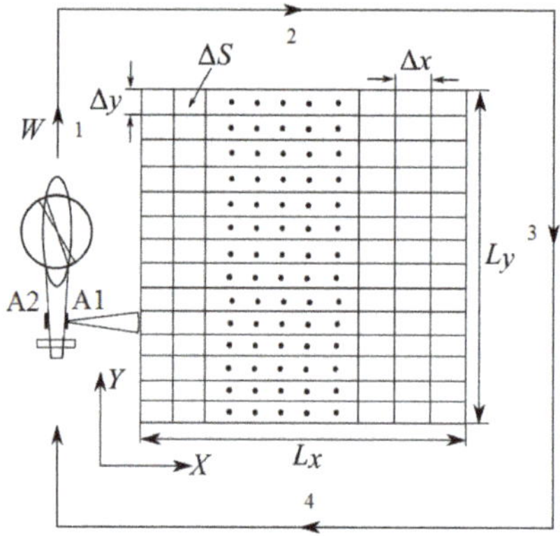

Figure 1. Imaging geometry of radar safe landing system of a helicopter (RSLSH).

An interferometer with a fixed base [11–13] in the form of a pair of antennae spatially separated by an interferometric base of a waveguide slot antenna mounted on a tail beam is used as a tool for measuring the relief of the LS and estimating the presence of foreign objects on the LS. One of the antennae works for reception and transmission, and the other only for reception.

The operating frequency of the system is selected in the Ka-band, which is caused by minimizing the size of the antennae, ensuring high resolution of the on-board radar, as well as reducing the effect of losses on radio wave propagation.

High horizontal resolution $\Delta x = c/(2\Delta f \sin \theta_1)$ is provided by the use of a signal with a nanosecond duration, where Δf is the bandwidth of sensing signal; c is the speed of light; $\theta_1 = \arccos(h/r_1)$ is the look angle; h is the flight altitude; and r_1 is the slant range.

The resolution in the azimuthal direction is determined by the size of the antenna, since at selected altitudes of the helicopter and the size of the LS, the radar operates in the azimuth plane in the near zone of the antennae.

A one-to-one relationship between the observation parameters and the interferometric phase difference (IPD) ϕ at the input of spatially separated receivers, which carries information about the resolution element, is determined by the relation [12,13]:

$$z_i = h - r_1 \cos \omega \sqrt{1 - \left(\left(r_1^2 + b^2 - \left(r_1 - \tfrac{\lambda}{4\pi}\phi\right)^2\right)/(2r_1 b)\right)^2} - r_1 \sin \alpha \cdot \left(\left(r_1^2 + b^2 - \left(r_1 - \tfrac{\lambda}{4\pi}\phi\right)^2\right)/(2r_1 b)\right). \tag{1}$$

where α is the inclination of the baseline from horizontal; λ is the wavelength; and b is the baseline.

In accordance with Equation (1), the resolution of the resolution element is a function of many variables and theoretically, provided that the individual components are uncorrelated, the resulting error in estimating the relief of the LS is determined by the sum of the errors of each of the parameters included in Equation (1), e.g., Equation (2):

$$\sigma_z^2 = \sigma_{z\phi}^2 + \sigma_{zh}^2 + \sigma_{zr_1}^2 + \sigma_{zb}^2 \tag{2}$$

where $\sigma_{z\phi}^2$, σ_{zh}^2, $\sigma_{zr_1}^2$, and σ_{zb}^2 are the variance in the resolution element height due to the estimation error of the phase difference σ_ϕ, the measurement error of the altitude σ_h, the measurement error of the slant range σ_{r_1}, and the measurement error of the baseline length σ_b, respectively.

In order to determine the potential accuracy characteristic of the measurement of the relief of the LS with the help of the RSLSH, it is necessary to obtain a ratio only for the fluctuation error $\sigma_{z\phi}$, since the remaining errors are inherently systematic and can be compensated for. The determining error of measuring the relief of the LS, as is known [12–15], is associated with the evaluation of the IPD $\hat{\phi}$ as seen in Equation (3):

$$\sigma_{z\phi} = \frac{\lambda h \tan \theta_1}{4\pi b \cos(\theta_1 - \alpha)} \sigma_{\hat{\phi}}; \; \sigma_{\hat{\phi}} = \frac{1}{\sqrt{2N}} \frac{\sqrt{1 - \gamma^2}}{\gamma} \tag{3}$$

where $\sigma_{\hat{\phi}}$ is the root mean square (RMS) error of IPD estimate; N is the number of incoherent integration; and γ is the correlation coefficient for two received signals in the interferometer.

The used interferometer with a fixed baseline is characterized by the decorrelation of paired echoes coming to the spatially separated antennae of two receivers $\gamma_{spatial}$ and due to thermal noise in system γ_{noise}.

For each of the factors, analytical expressions are derived and the resulting correlation coefficient is determined by using Equation (4), under the assumption that the real surface is a distributed radar target consisting of a set of independent partial reflectors inside the resolution element whose applets are distributed according to the normal law [12]:

$$\gamma = \gamma_{spatial} \cdot \gamma_{noise}; \; \gamma_{noise} = \frac{1}{1 + snr^{-1}};$$
$$\gamma_{spatial} = \left(1 - \frac{2b \cos(\theta_1 - \alpha)}{\lambda r_1 \tan \theta_1} \Delta r\right) \cdot \exp\left[-2\pi^2 \left(\frac{\sigma_h b \cos(\theta_1 - \alpha)}{\lambda r_1 \sin \theta_1}\right)^2\right] \tag{4}$$

where Δr is the slant range resolution; σ_h is the RMS of small irregularities on the surface of a large relief; and snr is the single-to-noise ratio.

The final expressions for the standard deviation of the estimate of the applicability of the relief through the standard deviation of the estimates of the IPD are obtained by substituting Equation (4) into Equation (3).

As a result, with the parameters of the RSLSH: $f_c = 35$ GHz, $h = 75$ m, $\theta_1 = 30° \sim 60°$, $N = 4$, $\Delta r = 0.5$ m, $\Delta y = 0.8$ m, $\sigma_h = 7,77 \cdot 10^{-3}$ m, and $snr = 13$ dB, we have the following dependence of the standard deviation of the relief estimate on the size of the interferometer base at different look angles (Figure 2).

According to Figure 2, it is preferable to choose the size of the fixed baseline of the interferometer to be from 0.48 to 0.57 m, at which the potential values of the accuracy of the measurements of the LS surface will be in the range of approximately from 6 to 10 cm.

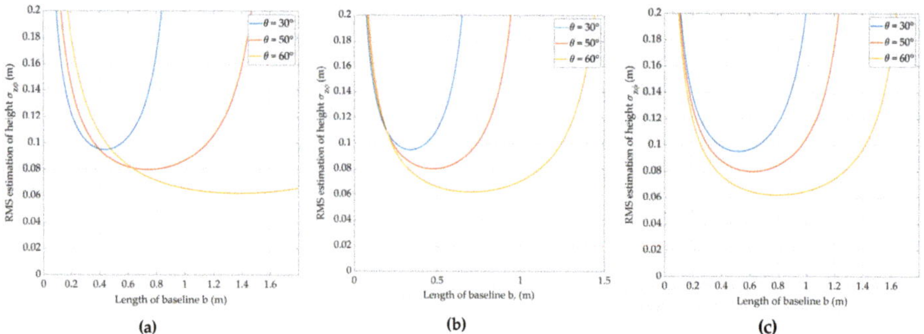

Figure 2. Height estimation error due phase estimation error on the baseline length at different look angles: (**a**) $\alpha = 0°$; (**b**) $\alpha = 45°$; (**c**) $\alpha = 90°$.

3. Numerical Simulation

3.1. Structure of the Simulation Model

The software package MATLAB is used as the simulation platform; the primary toolset for radar simulation with this software is the Phased Array System Toolbox, as in [16]. The simulation process can be divided into the following stages: (1) setting the Digital Elevation Model (DEM) and its parameters; (2) setting the parameters of the interferometric system;(3) simulation of the trajectory signal, its processing, and synthesizing the radar images; (4) calculation of the IPD; (5) interferometric processing to obtain an elevation map as the final output.

3.2. Digital Elevation Model

At this stage, the terrain features are generated according to the phenomenological surface model [11,12]. Each resolution element on the Earth's surface is represented by a set of normally distributed partial scatterers, on which scattering conditions known from the experimental results are imposed. Illustrative simulation results are shown on the example of a user-defined DEM shown in Figure 3a. By type, the surface consists of water, sand, soil, grass, and snow, the optical image of which is shown in Figure 3b.

As a model of the radar cross section (RCS) for surfaces such as grass, trees and snow, an experimentally obtained full-scale model RCS for various types of surfaces is used, which is valid for the microwave frequency range from 3 to 95 GHz. It takes into account the standard deviation of fine surface roughness σ_h, the look angle on the surface θ, and the wavelength λ, and has the following form [17] (Equation (8)):

$$\sigma^0(\theta, \sigma_h, \lambda) = A\left(\frac{\pi}{2} - \theta + C\right)^B \exp\left[-D / \left(1 + \frac{0.1\sigma_h}{\lambda}\right)\right] \tag{5}$$

where A, B, C, and D are empirical model coefficients. In [14], the values for these constants are given in the frequency range from 3 to 95 GHz for the indicated types of surfaces.

The RCS model for surfaces such as soil, sand, and stone is a semi-empirical model of backscattering of the earth's surface [18–20] for three types of polarization. For them, backscatters from the four surfaces are simulated using the semi-empirical model for the backscattering coefficient σ^0 in three polarizations: horizontal (HH), vertical (VV), and cross-polarization (HV) [18]:

$$\sigma^0_{VV} = g\frac{\cos^x \theta}{\sqrt{p}}[\Gamma_{VV}(\theta) + \Gamma_{HH}(\theta)], \quad \sigma^0_{HH} = p\sigma^0_{VV}, \quad \sigma^0_{HV} = q\sigma^0_{VV} \tag{6}$$

where $p = \left[1 - \left(\frac{2\theta}{\pi}\right)^{\frac{1}{3\Gamma_0}} \exp(-0,4k\sigma_h)\right]^2$; $g = 2,2[1 - \exp(-0,2k\sigma_h)]$; $\Gamma_0 = \left|\frac{1-\sqrt{\varepsilon_r}}{1+\sqrt{\varepsilon_r}}\right|^2$ is the reflection coefficient for normal incidence; $\Gamma_{VV}(\theta)$ and $\Gamma_{HH}(\theta)$ are Fresnel's reflection coefficients for oblique incidence at angle θ; ε_r is the relative permittivity $q = 0,23\Gamma_0^{0,5}[1 - \exp(-0,5k\sigma_h \sin\theta)]$; and $x = 3,5 + \frac{1}{\pi}\tan^{-1}[10(1,64 - k\sigma_h)]$.

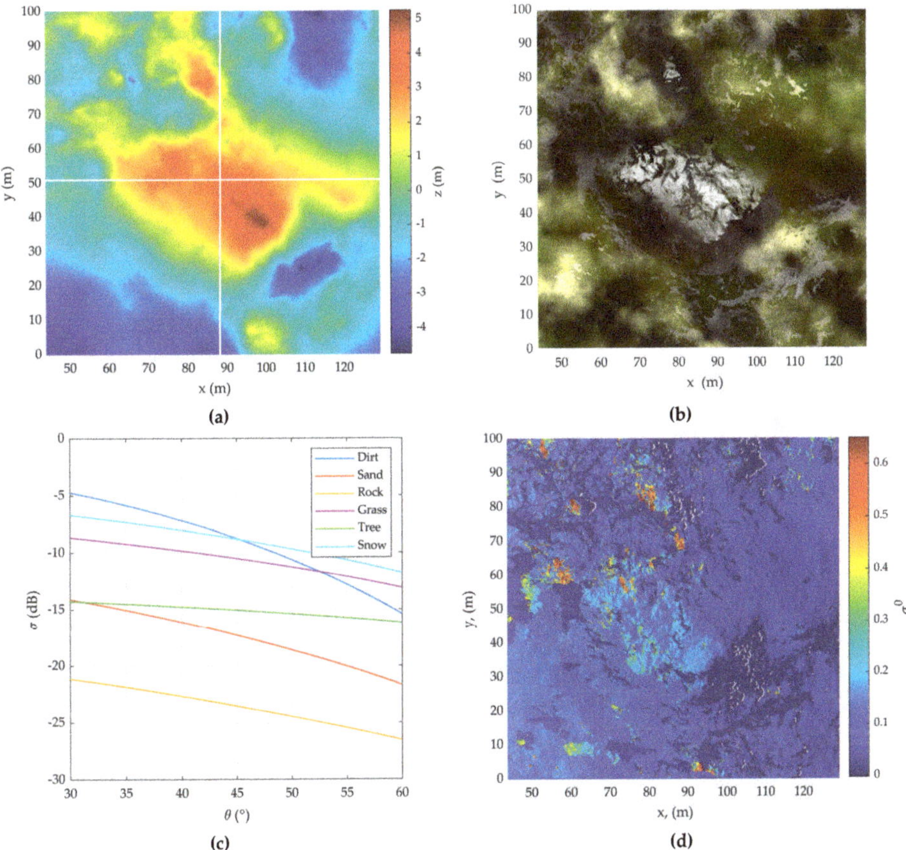

Figure 3. (a) Digital elevation model (DEM) for acquisition 1; (b) optical image of DEM for acquisition 1; (c) radar cross section (RCS) for some types of surface; (d) RCS of DEM for acquisition 1.

3.3. Synthesis of Radar Images and IPD Processing

If we denote the radar images obtained during two intervals or sub-intervals of observations as $\dot{P}_1$ and $\dot{P}_2$, we can then obtain an interferogram from their pixel-by-pixel complex conjugate multiplication using Equation (7):

$$I_{P_1 P_2}(x,y) = \dot{P}_1(x,y)P_2^*(x,y) = \left|P_1(x,y)\right| \cdot \left|P_2(x,y)\right| \exp\left\{j\left[\phi_{P_1}(x,y) - \phi_{P_2}(x,y)\right]\right\} \tag{7}$$

and the interferometric phase difference can be defined as the argument of the multiplication result as seen in Equation (8):

$$\phi_{P_1 P_2}(x,y) = \arg\left\{\sum_{n=1}^{N} I_{P_1 P_2}(x,y)\right\} = \sum_{n=1}^{N}\left[\phi_{P_1}(x,y) - \phi_{P_2}(x,y)\right] \tag{8}$$

Figure 4 illustrates the interferometric phase difference (IPD) for DEMs, which are rotations by 90, 180, and 270° anticlockwise from the acquisition 1 simulation model, respectively.

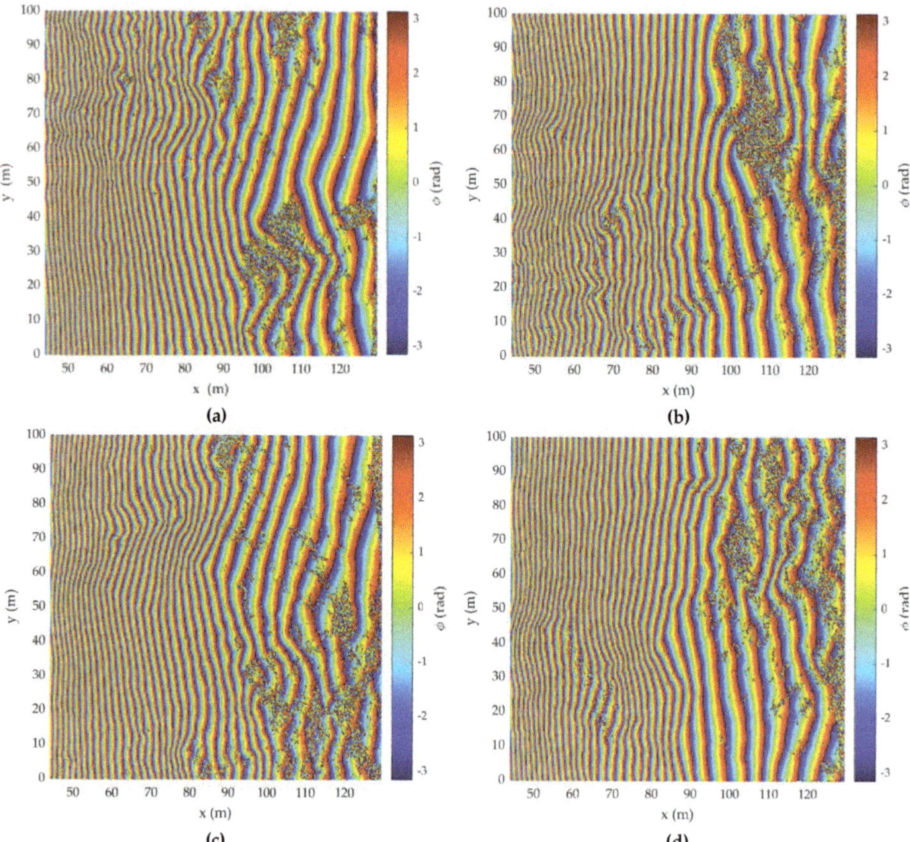

Figure 4. Interferometric phase difference (IPD) of DEM of (**a**) acquisition 1; (**b**) acquisition 2; (**c**) acquisition 3; (**d**) acquisition 4.

The standard interferometric processing followed that described in References [16,17,21–23] and included: elimination of the linear phase component along the range by subtracting the phase of the flat Earth from the IPD of the DEM; removing to the effects of the flat surface of the Earth (Figure 5); and elimination of the phase ambiguity.

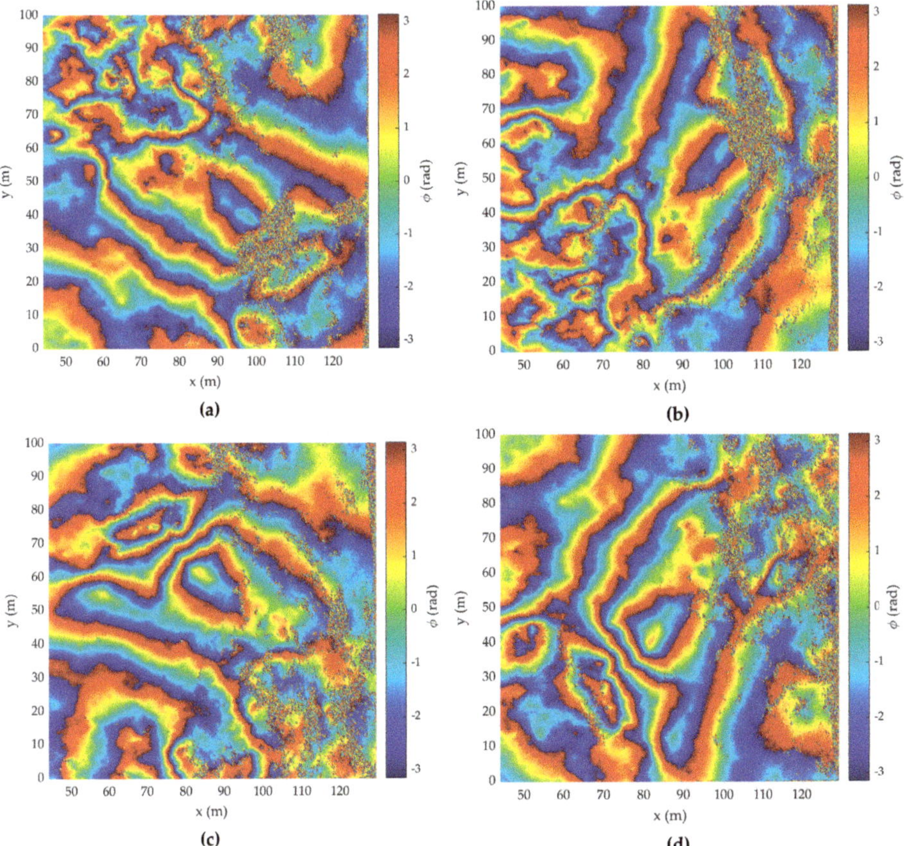

Figure 5. IPD after removing the flat Earth IPD of (**a**) acquisition 1; (**b**) acquisition 2; (**c**) acquisition 3; (**d**) acquisition 4.

As the IPD may significantly exceed two during elevation changes, the recovery of the true phase difference from the IPD reduces to the interval $(-\pi, \pi]$ and must be processed in an approach known as phase unwrapping (Goldshtein et al. (1988)). The scaling of the unwrapped IPD and generation of the DEM according to the unambiguous relationship between terrain elevation and change of IPD is shown in Figure 6.

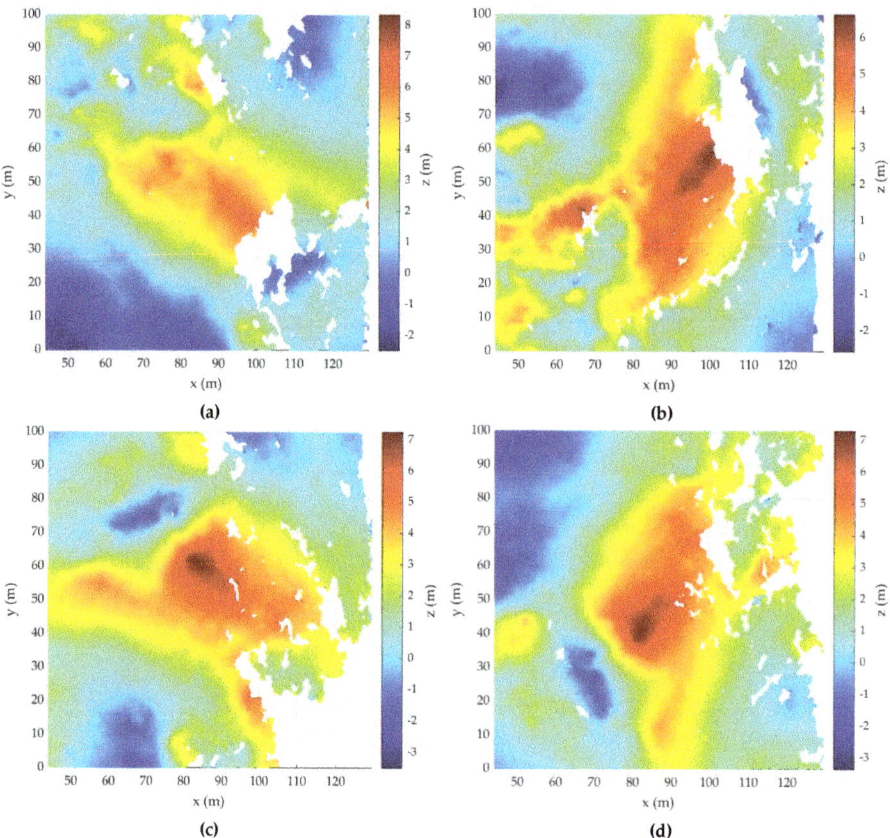

Figure 6. Estimation of DEM of (**a**) acquisition 1; (**b**) acquisition 2; (**c**) acquisition 3; (**d**) acquisition 4.

Taking into account only the phase component, the error in estimating the topography of the surface and its histogram in the selected sections (white vertical and horizontal lines in Figure 3a) for four observations are added in Figure 7a. Here, in order to prevent the graphs from merging into one, the value errors added a constant component multiple of 0.75 m depending on the observation number. The standard deviations of the estimation errors are in the range from 0.082 to 0.086 m, which is consistent with the theoretically calculated value.

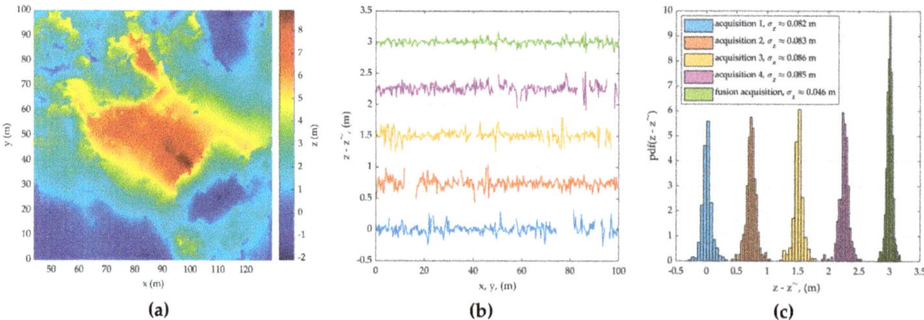

Figure 7. (**a**) combined DEM; (**b**) standard deviation of the errors; and (**c**) its histograms.

4. Conclusions

In this article, we analyzed the algorithm of a proposed radar interferometric remote sensing system for a helicopter LS surface using an airborne radar. This algorithm helps in obtaining a high-quality radar image of the LS, which shows surface variation characteristics with sufficient accuracy to confidently determine the type of LS and the presence of unknown objects on it, by using a linear 3D model of the surface.

First, after illuminating the LS surface using electromagnetic waves, we obtained radio contrast patterns according to the backscattering from the resolution elements. Then, the resulting contrast pattern was superimposed with the phase difference information, covering the resolution elements of LS. This was the starting point of reconstructing the LS terrain topography. In this approach, the visualization of man-made objects on the LS is significantly improved. Our work shows that the measurement accuracy of the variations in the z coordinate was most significantly affected by the error in the measured phase differences of the interferometer signals. Therefore, the detection probability increased with an increase in the number of measurements.

Therefore, the proposed method demonstrates the ability to significantly improve the visualization of man-made objects at a helicopter LS using the phase difference information of the reflected signals reaching both antennae. However, the detection of sharp variations in LS terrain, such as hills and ravines, must be performed by considering the background-to-noise ratio. Phase-difference information helps to highlight large surface roughness in radar images and determine their relative heights.

According to the results of this research, the proposed algorithm can be applied for the safe landing of a helicopter under conditions of insufficient a priori information on the LS. According to flight regulations, helicopters fly around an expected landing site to determine the topography, slopes, and presence of unknown objects; then, the pilot makes a decision about landing.

The results can be a theoretical and implementation basis for the safe landing of a helicopter for building perspective onboard radar systems, choosing the geometry of LS illumination, and calculating the optimal performance of the system. This can detect the roughness and disturbing objects on the LS and increase the reliability of a safe landing in a dusty environment under day and night conditions, as well as under harsh weather conditions.

In this work, the algorithm of the radar interferometric recording of the surface for the on-board radar was simulated, which made it possible to obtain a high-quality 3D image of the relief with the definition of the nature of the relief with the required accuracy.

The results of the simulation of interferometric signal processing RSLSH confirmed the possibility of its use as a promising tool in determining hazardous irregularities and foreign objects at the landing site from the resulting differential-phase interferometric images from the helicopter.

The main advantage of using RSLSH compared with other methods of safe landing of a helicopter on an unprepared site is that it is independent of the weather conditions and time of day.

Author Contributions: P.E.S., M.-H.K. and A.I.B. performed the mathematical derivations, edited and revised the manuscript. P.E.S. and A.A.K. performed the simulation, analysis of the experiments. All authors have read and agreed to the published version of the manuscript.

Funding: The work was supported by the ADD, DAPA Korea under the "New Generation SAR Research" program and was funded by CONTRACT No. RF-MPEI-08096/19 and by RFBR according to the research project No. 18-37-00184.

Conflicts of Interest: The authors declare no conflict of interest.

References

1. *Rotary–Wing Brownout Mitigation: Technologies and Training*; Technical Report; NATO Science and Technology Organization: Paris, France, 2012.

2. Analysis of the State of Flight Safety in the Civil Aviation of the Russian Federation in the First Half of 2016. In *Flight Safety Inspectorate*; Federal Air Transport Agency of the Russian Federation: Moscow, Russia, 2016.

3. Cheung, B.W. Nonvisual Spatial Orientation Mechanisms. Spatial Disorientation in Aviation. *Prog. Astronaut. Aeronaut.* **2004**, *203*, 37–94. [CrossRef]

4. Cross, J.; Schneider, J.; Cariani, P. MMW radar enhanced vision systems: The Helicopter Autonomous Landing System (HALS) and Radar-Enhanced Vision System (REVS) are rotary and fixed wing enhanced flight vision systems that enable safe flight operations in degraded visual environments. In Proceedings of the Degraded Visual Environments: Enhanced, Synthetic, and External Vision Solutions 2014, Baltimore, MD, USA, 7–8 May 2014. [CrossRef]

5. Sykora, B. BAE systems brownout landing aid system technology (BLAST) system overview and flight test results. In Proceedings of the Airborne Intelligence, Surveillance, Reconnaissance (ISR) Systems and Applications IX, 83600M, Baltimore, MD, USA, 24–26 April 2012. [CrossRef]

6. Savage, J.; Harrington, W.; McKinley, A.R.; Burns, H.M.; Braddom, S.; Szoboszlay, Z. 3D–LZ helicopter lidar imaging system. In Proceedings of the Laser Radar Technology and Applications XV, Orlando, FL, USA, 5–9 April 2010. [CrossRef]

7. Murray, J.T.; Seely, J.; Plath, J.; Gotfreson, E.; Engel, J.; Ryder, B.; Van Lieu, N.; Goodwin, R.; Wagner, T.; Fetzer, G.; et al. Dust-Penetrating (DUSPEN) "see-through" lidar for helicopter situational awareness in DVE. In Proceedings of the Degraded Visual Environments: Enhanced, Synthetic, and External Vision Solutions 2013, Baltimore, MA, USA, 16 May 2013. [CrossRef]

8. Rangwala, J.M.; Srabandi, L.K. Design of FMCW Millimeter-Wave Radar for Helicopter Assisted Landing. In Proceedings of the 2007 IEEE International Geoscience and Remote Sensing Symposium, Barcelona, Spain, 23–28 July 2007; pp. 4183–4186. [CrossRef]

9. Braun, H.M.; Baessler, H.; Jackson, B.; Jonas, C.; Lentz, H.; Rhein, R.V.; Essen, H. Helicopter Flight and Landing RADAR—A New Technology Developed in the European EUROSTARS Program. In Proceedings of the 2013 14th International Radar Symposium (IRS), Dresden, Germany, 19 June 2013; Volume 1, pp. 1214–1217.

10. Manual on the flight operation of the Mi-8 helicopter. M.: Air Transport Department of the Ministry of Transport of the Russian Federation.

11. Baskakov, A.I.; Jutyaeva, T.S.; Lukashenko, Y.I. *Locational Methods for Studying Objects and Environments. A Textbook for University Students*; Baskakov, A.I.M., Ed.; Academy: Katy, TX, USA, 2011.

12. Baskakov, A.I.; Ka, M.-H. Analysis of the Effect of Phase Noise on the Accuracy Characteristics of Interferometric Fixed-Baseline SARs. In *Earth Observation and Remote Sensing*; Russian Academy of Science: Moscow, Russia, 2000; pp. 247–256. ISBN 1024-5251.

13. Ka, M.H.; Kononov, A.A. Effect of Look Angle on the Accuracy Performance of Fixed-Baseline Interferometric SAR. *IEEE Geosci. Remote Sens. Lett.* **2007**, *4*, 65–69. [CrossRef]

14. Rodriguez, E.; Martin, J.M. Theory and design of interferometric synthetic aperture radars. *IEE Proc. Radar Signal Process.* **1992**, *139*, 147–159. [CrossRef]

15. Bamler, R.; Hartl, P. Synthetic aperture radar interferometry. *Inverse Probl.* **1998**, *14*, R1. [CrossRef]

16. Ka, M.-H.; Shimkin, P.E.; Baskakov, A.I.; Babokin, M.I. A New Single-Pass SAR Interferometry Technique with a Single-Antenna for Terrain Height Measurements. *Remote Sens.* **2019**, *11*, 1070. [CrossRef]

17. Richards, M.A.; Scheer, J.A.; Holm, W.A. *Principles of Modern Radar: Basic Principles*; SciTech Publishing: Raleigh, NC, USA, 2010; p. 934.

18. Adib, N.; Fawwz, T.; Kamal, S. Measurement and modeling of the millimeter-wave backscatter response of soil surfaces. *IEEE Trans. Geosci. Remote Sens.* **1996**, *34*, 561–572.

19. Gatesman, A.J.; Goyette, T.M.; Dickinson, J.C.; Waldman, J.; Neilson, J.; Nixon, W.E. Physical Scale Modeling the Millimeter-Wave Backscattering Behavior of Ground Clutter. In *Targets and Backgrounds VII: Characterization and Representation*; International Society for Optics and Photonics: Bellingham, WA, USA, 2001.

20. Li, E.S. Millimeter Wave Polarimetric Radar System as an Advanced Vehicle Control and Warning Sensor. Ph. D. Thesis, University of Michigan, Ann Arbor, Michigan, 1998.

21. Richards, M.A. A beginner's guide to interferometric SAR concepts and signal processing. *IEEE Aerosp. Electron. Syst. Mag.* **2007**, *22*, 5–29. [CrossRef]

22. Melvin, W.L.; Sheer, J.A. Interferometric SAR and coherent exploitation. In *Principles of Modern Radar: Advanced Techniques*; SciTech Publisihng: Edison, NJ, USA, 2013; pp. 337–398.
23. Goldshtein, R.M.; Zebker, H.A.; Werner, C.L. Satellite radar interferometry: Two-dimensional phase unwrapping. *Radio Sci.* **1988**, *23*, 713–720. [CrossRef]

 *sensors*

Article

Monitoring the Land Subsidence Area in a Coastal Urban Area with InSAR and GNSS

Bo Hu *, Junyu Chen and Xingfu Zhang

Surveying Engineering, Guangdong University of Technology, Guangzhou 510006, China
* Correspondence: hubo@asch.whigg.ac.cn; Tel.: +86-20-3932-2530

Received: 21 May 2019; Accepted: 14 July 2019; Published: 19 July 2019

Abstract: In recent years, the enormous losses caused by urban surface deformation have received more and more attention. Traditional geodetic techniques are point-based measurements, which have limitations in using traditional geodetic techniques to detect and monitor in areas where geological disasters occur. Therefore, we chose Interferometric Synthetic Aperture Radar (InSAR) technology to study the surface deformation in urban areas. In this research, we discovered the land subsidence phenomenon using InSAR and Global Navigation Satellite System (GNSS) technology. Two different kinds of time-series InSAR (TS-InSAR) methods: Small BAseline Subset (SBAS) and the Permanent Scatterer InSAR (PSI) process were executed on a dataset with 31 Sentinel-1A Synthetic Aperture Radar (SAR) images. We generated the surface deformation field of Shenzhen, China and Hong Kong Special Administrative Region (HKSAR). The time series of the 3d variation of the reference station network located in the HKSAR was generated at the same time. We compare the characteristics and advantages of PSI, SBAS, and GNSS in the study area. We mainly focus on the variety along the coastline area. From the results generated by SBAS and PSI techniques, we discovered the occurrence of significant subsidence phenomenon in the land reclamation area, especially in the metro construction area and the buildings with a shallow foundation located in the land reclamation area.

Keywords: time-series InSAR; subsidence; GNSS; coastal urban area

1. Introduction

Traditional geodetic methods represented by levelling or total station geometry are point-based measurements. In the task of deformation monitoring, the usual operation process is: Firstly, select representative sites in the monitoring area and set up monitoring stations (observation station or observation markers), and then set the monitoring period for a single or continuous observation. From the original geodetic workflow, we can find that the traditional geodetic survey method has two significant characteristics: The first is the necessity of setting up the observation stations and observation markers. The other one is the long working time-consuming of each period of observations. The former characteristic leads to an insufficient spatial sampling rate of monitoring results which is unable to discover new deformation areas from the monitoring results effectively, especially when the monitoring scope is relatively extensive, and the latter one limits the temporal sampling rate of monitoring results of the station or the observation marker. Therefore, in the task of early identification and monitoring of a deformation area, we need a method that can conduct extensive observation to the monitoring region in a short period, and provide a reliable result for early identification and continuous observation of the region where geological disasters occurred. Interferometric Synthetic Aperture Radar (InSAR) has the advantages of an extensive monitoring area, high temporal resolution, and all-weather monitoring capability [1–6]. The time-series InSAR (TS-InSAR) technology effectively weakens the key factors that affect the reliability of InSAR monitoring results: Atmospheric effects and time-space decoherence [1,7]. InSAR has been widely used by researchers in different fields of research over the past decade. InSAR shows significant advantages in various fields of geoscience, including seismicity research [8–11], volcanic research [12], artificial building deformation monitoring [13], and surface deformation caused by underground resource exploitation [14,15]. In the case of urban areas research, Yang Zhang et al. used RADARSAT-2 satellite data to monitor the land subsidence of Wuhan, China, with Small BAseline Subset (SBAS) [16]. Xiaoqiong Qin et al. used the TerraSAR-X image to monitor land subsidence along the subway in the urban area of Shanghai, China [17]. Matthew North et al. studied the response of seasonal soil movement along the British railway and road with the PSI method [18]. Bing Xu et al. studied the land reclamation area with ENVISAT data [19]. In these examples, InSAR technology provides researchers with a reliable way of generating land deformation velocity fields.

The SBAS method is a kind of InSAR time series surface deformation inversion technology proposed by Berarndino [1] in 2002. To the InSAR differential interferogram, the stability of the interferometric phase is affected by the temporal baseline of the interferometric pair and the spatial baseline. Different from the traditional Differential InSAR (D-InSAR) method, SBAS is characterized by reducing spatial decoherence and screening out the interferometric pairs with higher coherence by limiting the time baseline and the spatial baseline. One applies the small baseline differential interferogram to the surface deformation inversion model, which can obtain the deformation time series of the coherent target, while ensuring that the factors affecting the interferometric quality of the interferogram are estimated and removed. The PSI method mainly focuses on the target that maintaining a high coherence in the interferometric data set, which is called "permanent scatterers (PS)" [9,20,21]. PS usually targets on the artificial buildings and bare stones. PS provides stable interferometric phase coherence. The PSI method directly executes the inversion model on the unwrapped differential interferometric phase, so that it can effectively avoid the unwrapping error caused by the traditional InSAR technology in the phase unwrapping step.

We use SBAS and PSI to study ground deformation in Shenzhen and HKSAR. In Shenzhen, we mainly study the Qianhai area with the risk of subsidence [19] using both the SBAS and PSI methods. In HKSAR, we use the SBAS method and Global Navigation Satellite System (GNSS) static network which consists of a series station that belongs to Hong Kong Continuously Operating Reference Stations, CORS (SatRef), run by Lands Department of HKSAR. The observation data observed by the GNSS station are transformed into Receiver INdependent EXchange (RINEX) format. Hi-Target Geomatics Office (HGO) software was used to generate the geodetic coordinates and the geodetic height under the WGS-84 coordinate system using the GNSS static network observation mode. The baseline solution solving step uses the SP3 post-accuracy ephemeris data released from International GNSS Service (IGS) to provide satellite coordinate information to maximize the accuracy of the baseline solution.

2. Overview of Study Area

The study area includes Shenzhen, China and the HKSAR bordering the south (Figure 1). Shenzhen is located in the southern part of Guangdong Province, China. As one of the special economic zones of China, Shenzhen has carried out much urban construction since 1980. In recent years, the construction area of the city has been transferred from Luohu district in the central part of Shenzhen city to Nanshan district in Qianhai district in the west. While HKSAR started urbanization earlier, it also carried out a large number of land reclamation projects to acquire land used for construction. A famous example is the Hong Kong International Airport, HKIA (Rose Garden Project). The construction project of the HKIA started in 1992 and was commissioned in 1998. The third runway and associated facilities will be constructed on the north side of the existing airport. The land reclamation project for the runway expansion began in 2016 [22]. In the rapid urbanization construction of these two cities, the land reclamation area is widely distributed in the study area. Since the special economic zone was built in Shenzhen in the 1980s, the land reclamation area reached 69 km^2, while the land reclamation area in HKSAR is 70 km^2. Qianhai district is the hot-spot area for infrastructure construction in

Shenzhen [23–25]. Since 2006, the Qianhai district has begun reclamation. The land reclamation area of Qianhai district reached 9.9 km^2 in 2009.

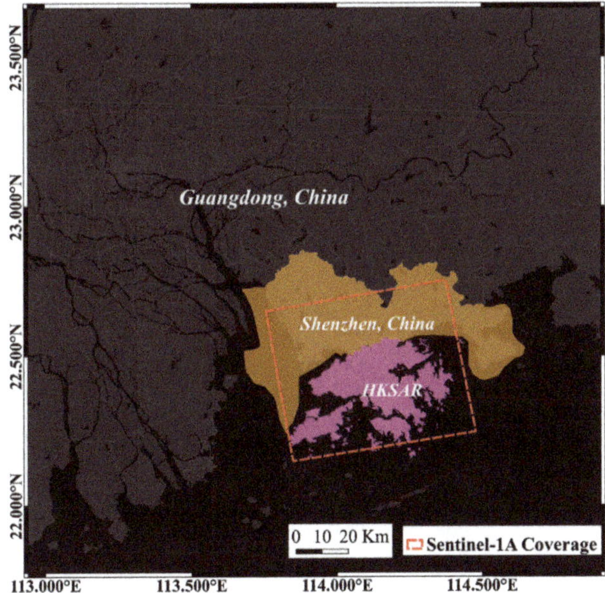

Figure 1. Location of the study area and the coverage of cropped Sentinel-1A SAR image.

Shenzhen runs eight subway line operators, with a total mileage of 285 km. It is estimated that by 2022, the Shenzhen Metro will reach 16 operating lines with a mileage of 596 km. The HKSAR metro system has a mileage of 264 km with 11 operating lines and average daily passenger flow of 5.59 million (2016) [26].

3. Data and Methods

Both GNSS technology and InSAR technology are space geodetic techniques which are capable of all-weather observation. InSAR technology uses space-based SAR sensor to obtain SAR images, then generates the displacement phase through performing a differential operation by acquiring the interferometric phase of the coherence target in the SAR dataset [12,14]. Therefore, InSAR technology has the advantage of a high spatial resolution, which is suitable for locating the area that assumes land deformation. However, the satellite has a revisit cycle which is usually about ten days, so InSAR technology cannot achieve high time resolution monitoring. In principle, the PSI method and the SBAS method are used for different kinds of targets in the inversion model, the spatial distribution characteristics between SBAS and PSI inversion results would be different [7]. In areas with extensive urbanisation, more PS points will be selected in this type of study area.

The GNSS technology can reach a very high temporal resolution observation by setting the sampling interval up to the second level due to the observed relationship using the ground-mounted GNSS antenna [27]. CORS stations which are deployed in urban areas can be observed in unattended conditions by deploying permanent GNSS antennas and receivers. The GNSS technology can form a certain degree of complementarity with the InSAR technology, which has a satellite revisit cycle.

The GNSS observation data we used in this paper was published by the satellite positioning reference station network in Hong Kong name "SatRef" (Figure 2) operated by the Survey and Mapping Office (SMO), Hong Kong Land Department. The reference station network consists of 18 continuous operating reference station (CORS) distributed in the HKSAR, including 16 reference station and two integrity monitoring station. The data services of The Hong Kong Satellite Positioning Reference Station was launched on February 4, 2010.

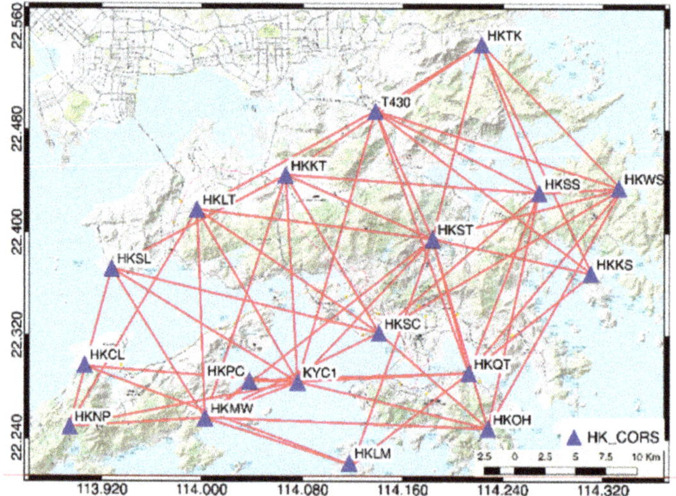

Figure 2. HK Continuous Operating Reference Station (CORS) network (SatRef).

The SAR image dataset used in this research is a total of 34 frames of SLC (Single Look Complex) SAR images (Table 1) which generate in the Interferometric Wide Swath (IW) mode from the Sentinel-1A satellite operated by European Space Agency (ESA). To minimize the orbital error in the interferogram, the SLC image data orbital parameters are provided with the post-accurate orbit data (POE product) provided by the ESA. Moreover, the SAR image data were resampled from the original 5 m × 20 m (range resolution × azimuth resolution) to a ground resolution of 20 m × 20 m (range resolution × azimuth resolution) using multi-looks processing.

Table 1. Acquisition date and the relative orbital position of each image relative to the master image (2017/4/17) of the Sentinel-1A SAR image we used in this paper.

No.	Acquisition Date	Relative Position (m)	No.	Acquisition Date	Relative Position (m)
1	2016/1/29	164.056	18	2016/12/6	35.492
2	2016/2/10	150.344	19	2017/1/11	123.824
3	2016/3/5	46.199	20	2017/2/4	86.124
4	2016/3/29	27.642	21	2017/2/28	135.102
5	2016/4/22	91.172	22	2017/3/12	89.354
6	2016/5/4	101.625	23	2017/3/24	114.535
7	2016/5/16	59.264	24	2017/4/5	58.529
8	2016/5/28	61.442	25	2017/4/17	0.000
9	2016/6/9	77.456	26	2017/5/11	37.645
10	2016/7/3	53.314	27	2017/5/23	100.012
11	2016/8/20	67.135	28	2017/6/4	38.256
12	2016/9/13	71.373	29	2017/6/28	71.516
13	2016/9/25	15.231	30	2017/7/10	125.536
14	2016/10/7	47.799	31	2017/7/22	118.556
15	2016/10/19	123.272	32	2017/8/3	56.789
16	2016/10/31	106.400	33	2017/9/8	121.120
17	2016/11/12	106.080	34	2017/10/2	64.370

We intercept the sub-region SAR image data set for PSI data processing (Figure 3). The velocity distribution map, displacement time series distribution of the SBAS inversion results in the sub-region part are export as a vector data for spatial analysis.

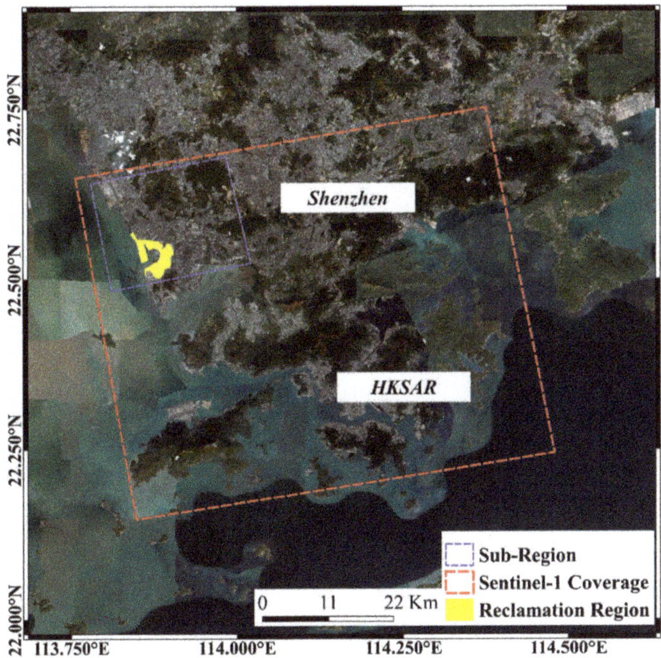

Figure 3. SAR data coverage of the study area.

In this research, the AW3D digital elevation model was used to remove the topographic phase, which is one of the critical steps in the interferogram generation workflow. The AW3D digital elevation product is produced using panchromatic images acquired from the Panchromatic Remote-sensing Instrument for Stereo Mapping (PRISM) sensor of ALOS satellites. If one uses AW3D data to provide elevation data for inversion to generate the topographical phase of the study area; then a differential operation would be executed on the inverted topographic phase and the interferometric phase to achieve the purpose of removing the topographic phase in the interferometric phase.

Because there is a large amount of vegetation coverage in the study area, the SAR sensor mounted on the Sentinel-1 satellite works in the C-band with medium ground penetrability. Using the branch cutting or least square phase unwrapping method, it is possible to generate significant unwrapping errors, hence we use the minimum cost flow method for phase unwrapping in the SBAS model (Figure 4). In this research, we use the Delauny Minimum Cost Flow (Delauny-MCF) method to phase unwrap the coherent targets in each interferogram [28–32]. This method constructs a Delaunay triangulation as the unwrapped network. Defining the costs on arcs between two adjacent coherent targets, according to the phase gradient of this pair of adjacent coherent targets, the optimal solution of the cost function of this network can be solved. Moreover, the optimal phase unwrapped path can be generated at the same time [28].

$$\min\left\{\sum_{i,j}\omega_{i,j}^{(x)}\left|\Delta\varphi_{i,j}^{x}-\Delta\phi_{i,j}^{x}\right|^{p}+\omega_{i,j}^{(y)}\left|\Delta\varphi_{i,j}^{y}-\Delta\phi_{i,j}^{y}\right|^{p}\right\} \tag{1}$$

In (1), Δ denotes a gradient along the azimuth and line of sight (LOS) respectively, $\omega_{i,j}^{(y)}$ and $\sum_{i,j}\omega_{i,j}^{(x)}$ defined as the weight and the sum of all suitable rows i and column j. The Delaunay-MCF method is used to enable the unwrapping path to avoid the more low-coherence regions that may contain phase jumps, which could lead to unwrapping error.

$$\delta\varphi_m^{(LP)}(x,r) \approx \frac{4\pi}{\lambda}\Big[d^{(LP)}\big(t_{IE_m},x,r\big) - d^{(LP)}\big(t_{IS_m},x,r\big)\Big] + \frac{4\pi}{\lambda}\frac{b_m\Delta z^{(LP)}(x,r)}{r\sin\vartheta}$$
$$+\big[\varphi_{atm}\big(t_{IE_m},x,r\big) - \varphi_{atm}\big(t_{IS_m},x,r\big)\big] + \Delta n_m^{(LP)}(x,r) \tag{2}$$

We can express the low-frequency part of a coherent pixel in interferogram no.m as Equation (2). λ stands for wavelength, b_m stands for the space baseline of the interferometric pair and ϑ stands for the incidence angle of the radar signal of the coherent pixel. $d^{(LP)}$ represents the deformation component of the interferometric phase $\delta\varphi_m^{(LP)}$ of the coherent pixel which image coordinate is (x,r). $\varphi_{atm}\big(t_{IE_m},x,r\big) - \varphi_{atm}\big(t_{IS_m},x,r\big)$ represents the atmospheric component between the master image and the slave image and $\Delta n_m^{(LP)}$ stand for the noise term of the interferometric phase. In the process of radar signal propagation in the air, due to the influence of ionosphere and resonance caused by water vapor in the air, a signal propagation delay or path bending in the range is caused. The regional phase changes in each SAR image due to this reason are called atmospheric phase screens (APS). To the interferogram, both the influence of the atmospheric phase and the error in satellite orbit error are low-frequency signals with a typical size of 1km [1]. Therefore, a high-pass filter with the ground range of 1km will be executed in the SBAS deformation inversion model to weaken the errors from the atmosphere and orbit. In this research, we use a linear model to invert the deformation.

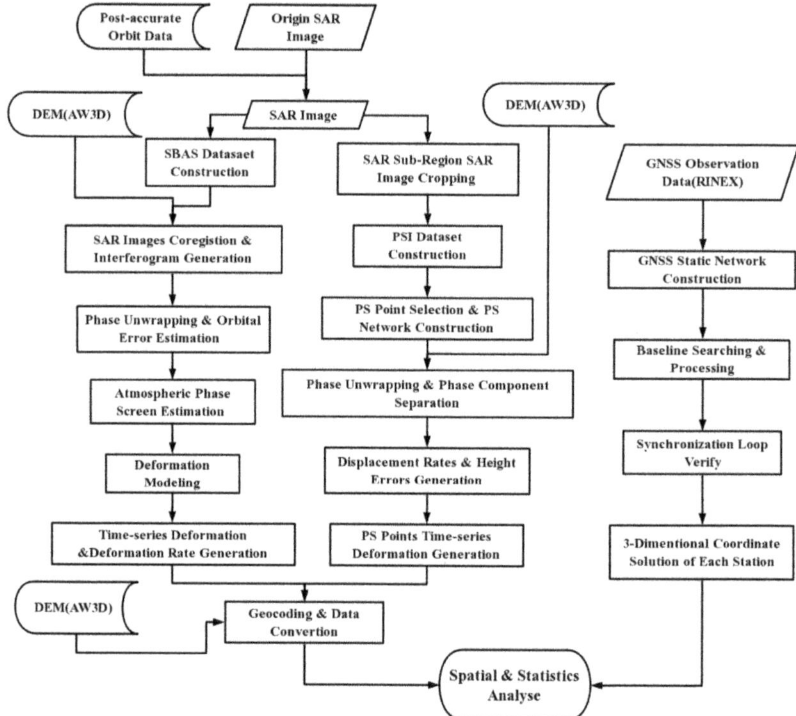

Figure 4. Data processing workflow with SBAS, PSI, and GNSS.

4. Data Processing and Result

In this research, RINEX observation data was collected from 18 stations with 26 periods in the CORS network (Figure 2). The sampling interval is 30 s, and the observation time is 24 h. In this research, we used the observation mode of GNSS static network for data processing. The average length of the searched baseline is set to 20 km; all the baseline solutions are none ionosphere combination fixed solutions. The tolerance of the 3-dimension components of the synchronization loop is:

$$W_x = W_y = W_z \leq \frac{\sqrt{3}}{5}\delta \tag{3}$$

Finally, we calculate the general change trend of the reference station during the monitoring period through the static network (Figure 5).

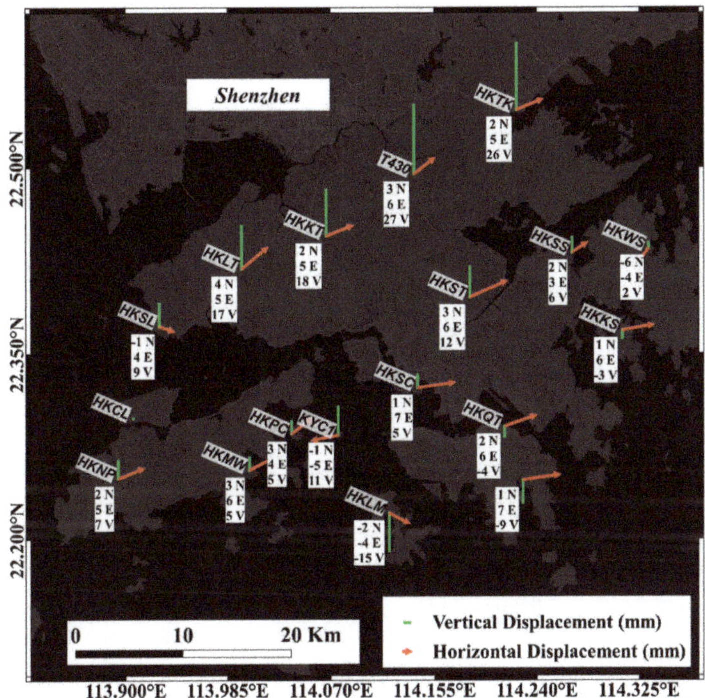

Figure 5. Horizontal and vertical displacement of the Hong Kong CORS network. The arrows and the vertical lines represent the direction and the magnitude of displacement of the station.

The Sentinel-1A satellite SAR image we used in this research had a revisit period of 15 days. Therefore, the temporal baseline of the interferometric data set used for the inversion of the SBAS method was set to 40–360 days. The threshold of the spatial baseline is 55% of the critical baseline, which is 5889.640 m. In the data processing step, we used the Delauny-MCF method with the phase unwrapping threshold set to 0.47. After removing the interferogram which has significant unwrapping error and an orbit error from the initial interferometric data set, 62 pairs of interferometric pairs were reserved for SBAS data process (Figure 6a). The average elevation accuracy of the SBAS inversion results (Figure 7) is 3.6 mm.

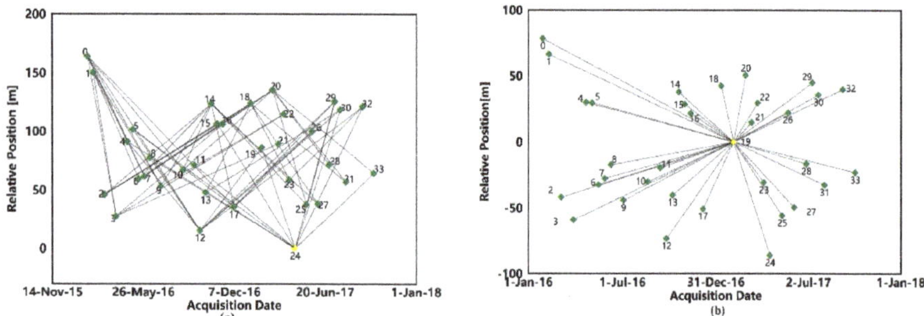

Figure 6. (**a**) Is SBAS interferometric dataset and (**b**) is a PSI interferometric dataset.

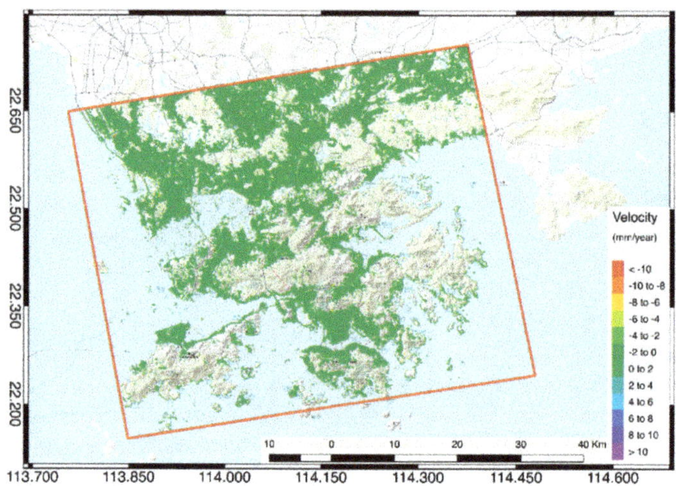

Figure 7. Average vertical velocity distribution of SBAS inversion.

In this research, we use Sentinel-1A SAR data with the PSI method to investigate the coastal area of Qianhai District of Shenzhen and the eastern part of Shenzhen as the research area at the same time. In the PSI method, as long as the baseline length of the interferometric pair is not greater than the critical baseline the baseline of the interferometric pair is not limited, the method of composing the interferometric data set (Figure 6b) using the same main image is adopted in this paper. Alternative PS points are filtered using a coherence threshold method, setting 0.75 as the coherence threshold to extract the PS point in the interferogram of the interferometric data set, a total of 44,464 PS points (Figure 8) were screened out in the study area with an average inversion elevation accuracy of 6.6 mm.

In the inversion model of SBAS, the inversion of the deformation variables in the model was performed while inverting the atmospheric influences in the inversion model (Figure 9).

The three-dimensional free network adjustment was performed with the HKCL station as a fixed point after all the simultaneous observation closure differences met the tolerance requirements, and the three-dimensional coordinates of the remaining stations were obtained. We projected the InSAR results from Line of Sight (LOS) to the normal direction of WGS-84 ellipsoid on T430 station. From the time series of variations on this station (Figure 10), we found that because of the lack of accuracy, the GNSS result shows more jumps, although it has advantages of a higher temporal resolution. However, the InSAR result which was generated by two different methods shows consistency both in magnitude and the trend of variations.

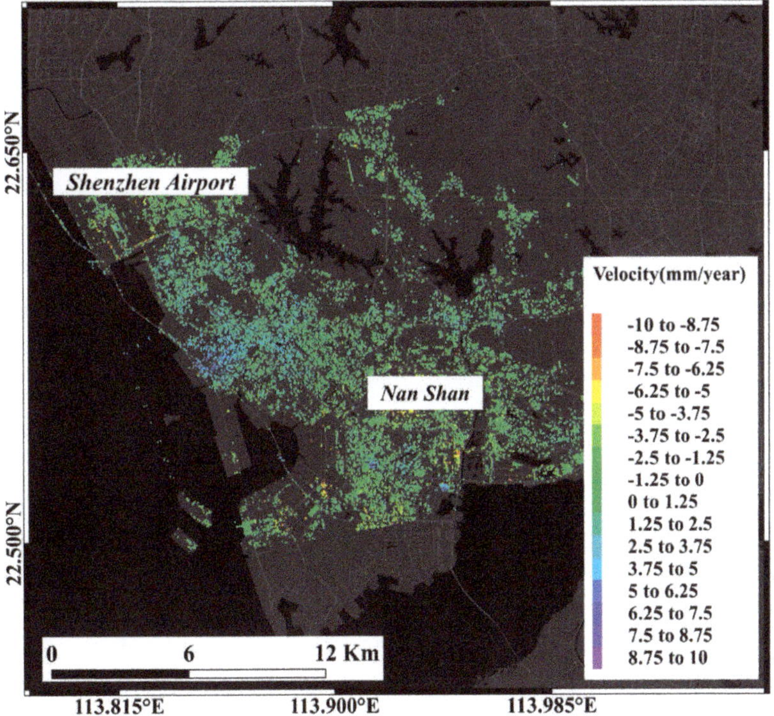

Figure 8. Permanent Scatterer (PS) point in Qianhai district, Shenzhen.

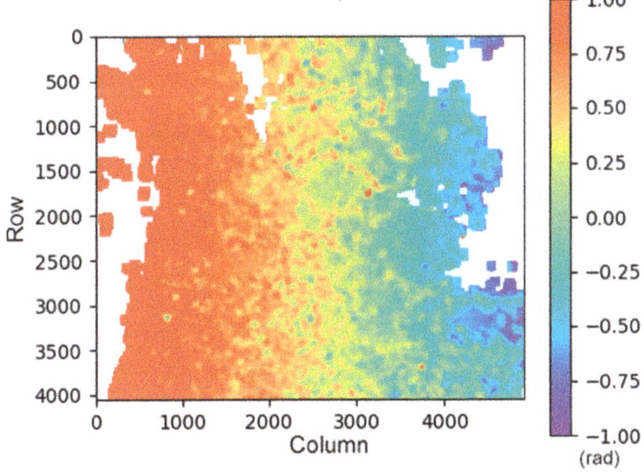

Figure 9. Atmospheric Phase Screens (APS) and orbit error component (low-frequency component) of the master image (2017/4/17) of SBAS dataset.

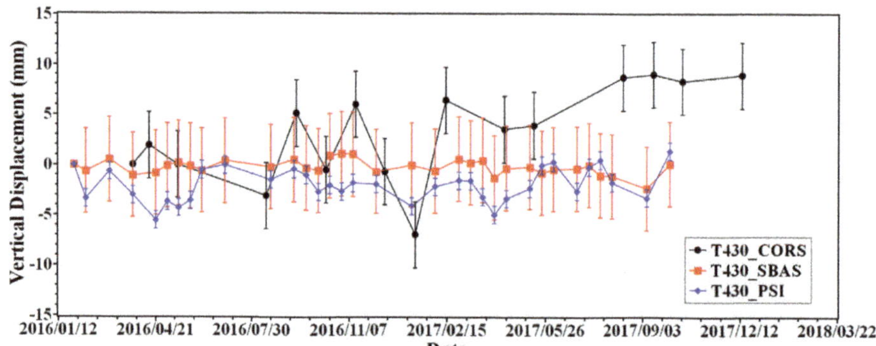

Figure 10. Time series generated by three different solutions on T430 station.

5. InSAR in Coastal Area

We used the historical archival image data from optical satellite imagery to determined the reclamation area between the coastline of 1999 which before the land reclamation project and the coastline of 2016 which the land reclamation project was completed in the sub-study area. The sub-region covers 455 km^2; the land reclamation region covers 22.3 km^2, which covers about 5% of the sub-region area [19]. There are 255,433 targets in the SBAS inversion results, of which 702 targets have a displacement velocity greater than 10 mm/year, accounting for 0.3% (Figure 11a,b). In the area obtained by land reclamation, there are 10,039 targets, and there are 376 targets with displacement velocity greater than −10 mm/year. More than half of the targets with displacement velocity higher than −10 mm/year are located in the area obtained by land reclamation. There are 44,464 targets in the PSI inversion results, of which 374 targets points with a displacement velocity over −10 mm/year, accounting for 0.8%. There are 1519 targets in the area obtained by land reclamation, and 185 targets with displacement velocity above −10 mm/year (Figure 11c,d). Similarly, more than half of the targets with an annual deformation rate greater than −10 mm/year are located in the land reclamation area. In the sub-region, the target with annual deformation rate greater than −10 mm/year by SBAS and PSI inversion result is mainly located in the area obtained by land reclamation. It is worth noting that PSI results will are sparser due to the stricter screening conditions of the PS points. The area where significant subsidence signals appear in the land reclamation area, combined with satellite optical image information, shows that areas with large surface deformations are mostly located in the area where the underground construction project in progress or recently completed areas. The following three regions that are representative of the region are selected for analysis.

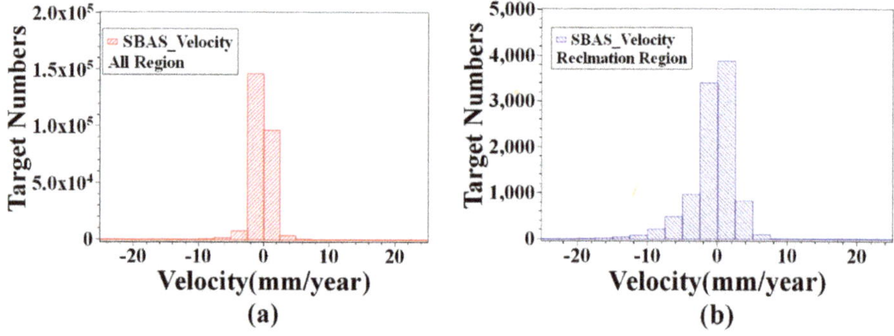

Figure 11. *Cont.*

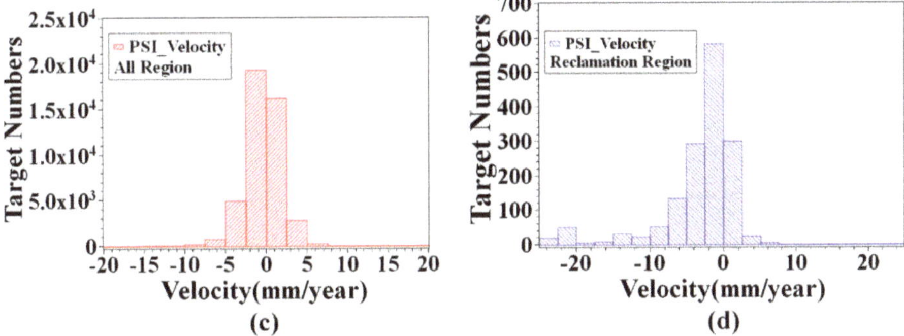

Figure 11. SBAS and PSI inversion result histogram in the sub-region and the land reclamation region.

Seen in Figure 12, the Qianhai Bay of Shenzhen acquired more land for urban construction by various phases of land reclamation projects from 1999 to 2016. The area where the subsidence signal appears in the SBAS results distributed in the land reclamation area that was generated after 1999. It is worth noting that the results of InSAR in these areas are missing because of the lack of effective PS targets in areas where land reclamation was completed in recent years. Moreover, because the conditions for PS searching are stricter than the calibration of ordinary coherent target points, the result point density of PSI inversion in the study area was lower than the SBAS results (Table 2). The three metro lines located in the land reclamation area are Line 1, Line 5, and Line 11. We found that there are a series of areas that appeared as subsidence signals in the area along the subway.

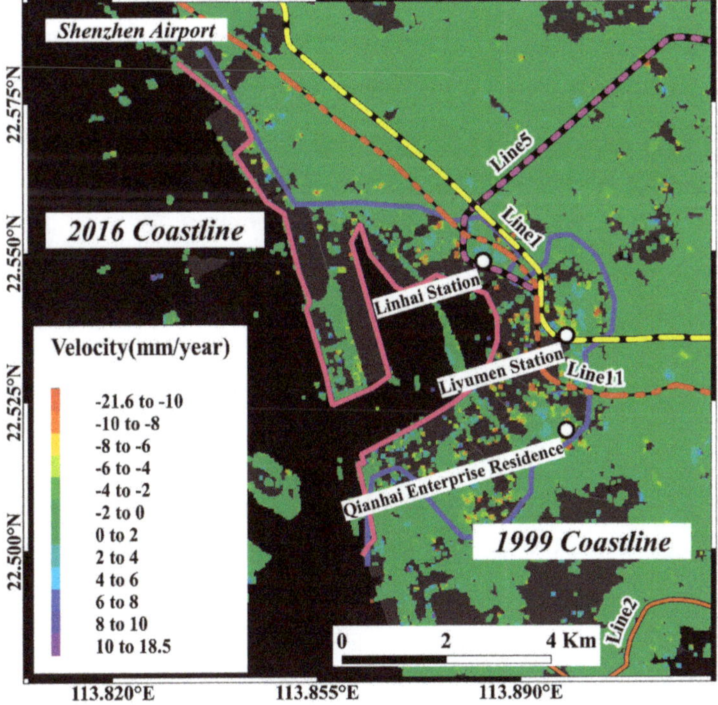

Figure 12. SBAS result of Qianhai bay. The blue line represents the coastline of 1999, and the pink line represents the coastline of 2016.

Table 2. Inversion results between Permanent Scatterer InSAR (PSI) and Small BAseline Subset InSAR (SBAS).

	Land Reclamation Region (PSI)	All Region (PSI)	Land Reclamation Region (SBAS)	All Region (SBAS)
Target Points	1519	44,464	10,039	255,433
Velocity >−5 mm/y Points	610	6016	1827	9702
Velocity >−10 mm/y Points	185	374	376	702

Among them, there is a significant settlement signal on the north side of the Liyumen station of Shenzhen Metro Line 1 (Figure 13). Extracting the time series analysis of pixel deformation in time series InSAR inversion results, we can find that, after November 2016, a trend of continuous subsidence occurred when the cumulative settlement reached 300 mm. On the sample point #5, #6, which is relatively far from the subway line, exhibits a relatively gradual change trend (Figure 14). At the same time, the PSI inversion results show that the PSI #1 sampling points with SBAS #7 sampling points coincide with a relatively uniform subsidence level and subsidence trend.

The upper cover of the Linhai station of Shenzhen Metro Line 5, which is also located in Qianhai District, also showed a subsidence zone (Figure 15). The north side of the upper cover of the station in the cross-validation results of the time series InSAR inversion result show a consistent settlement trend throughout the monitoring period. Among them, #6 which is closest to the subway line shows an accelerated subsidence trend after November 2016. Since the PSI results are sparse, there are only sampling points #1, #4, #6 that have both PSI and SBAS inversion results (Figure 16). At these three sampling points, the R^2 between SBAS results and PSI results can reach 0.6 or higher (Figure 16b,d,f).

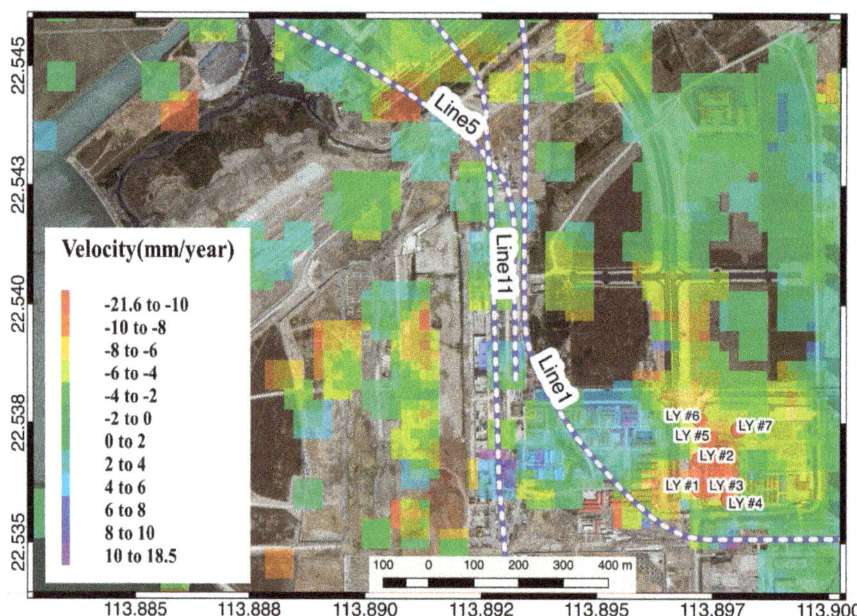

Figure 13. Time series InSAR inversion result of Liyumen station.

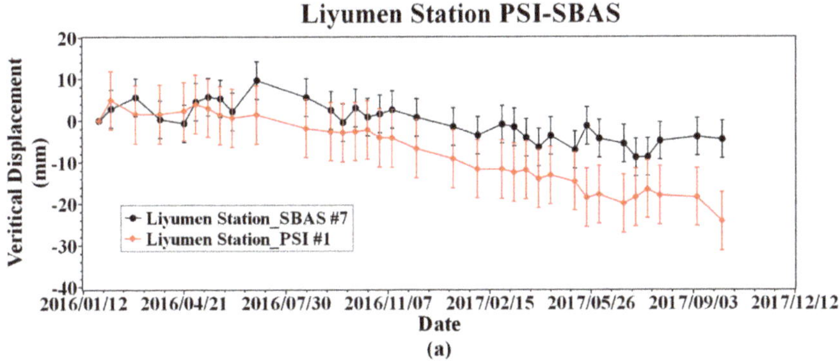

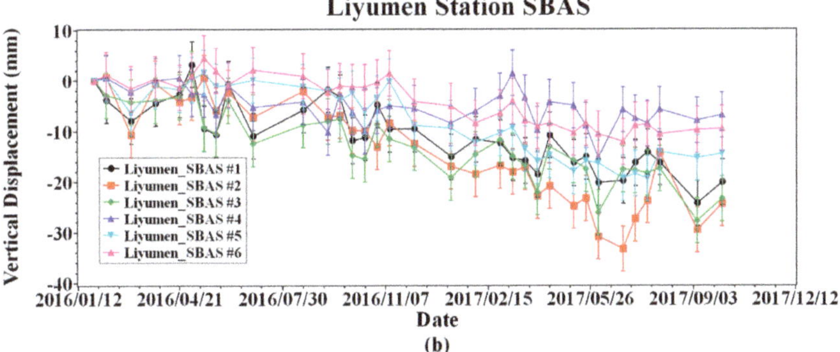

Figure 14. Time-series of sampling point in the Liyumen Station region.

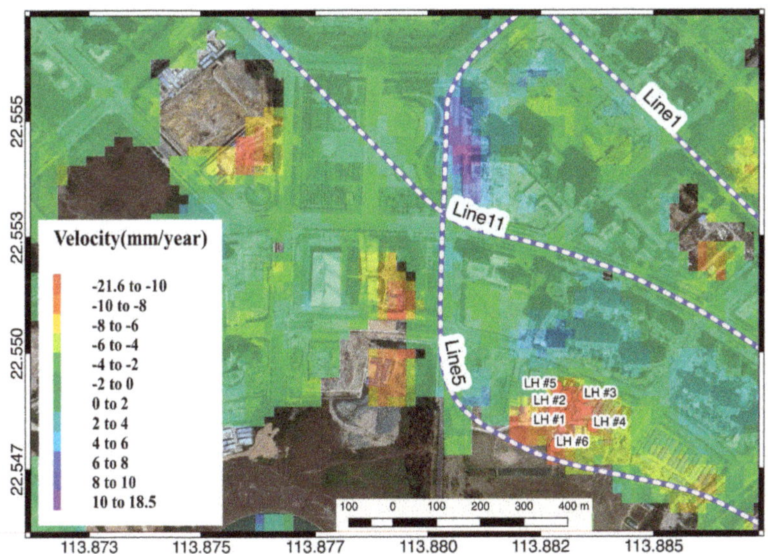

Figure 15. Time series InSAR inversion result of Linhai station.

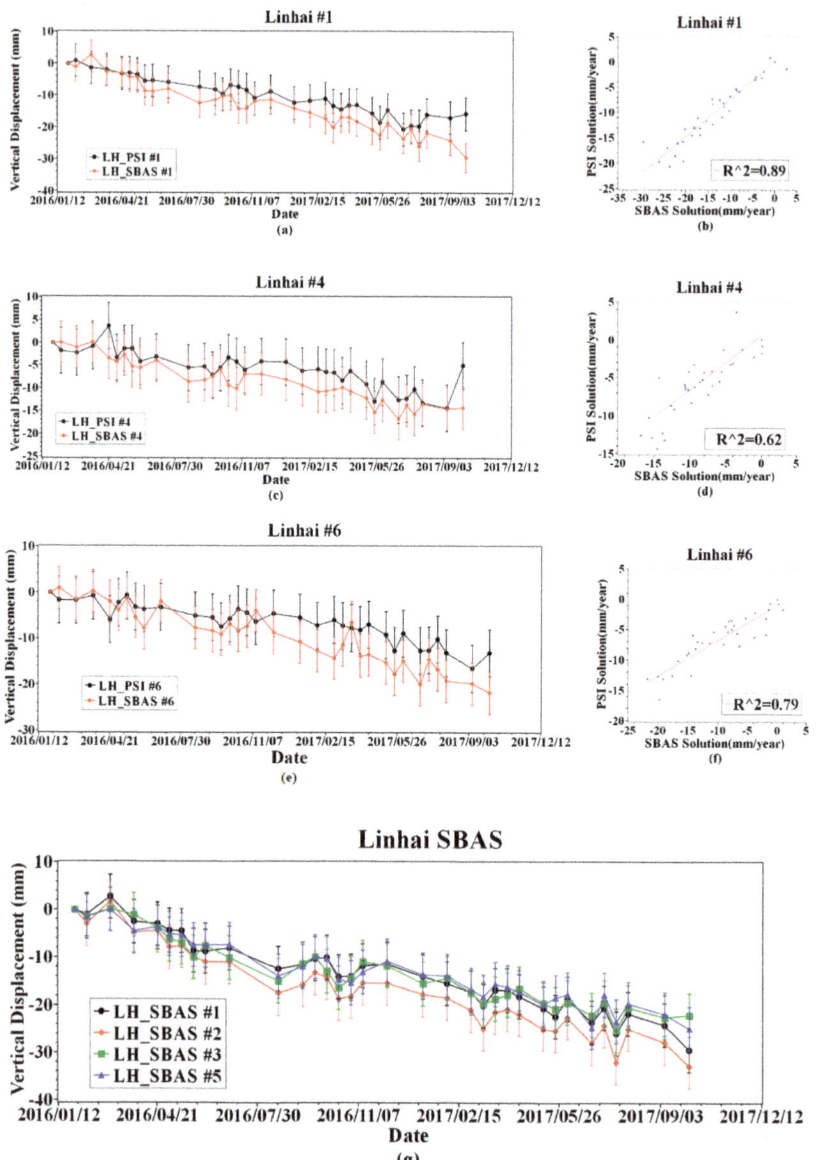

Figure 16. Time-series of sampling point in the Linhai Station region result.

The Qianhai Enterprise Residence, completed in 2014, showed subsidence during the entire monitoring period, and the accumulated settlement reached 40 mm (Figure 17). The Qianhai Enterprise Residence is located in the weak formation area. In the results of two different InSAR inversion methods, the main subsidence area occurred at the south side of the building. The coefficient of determination, R^2 between the PSI result and the SBAS result on the sampling point #1, #2, #3 (Figure 18b,d,f) can reach over 0.79, which shows a high consistency between the PSI and the SBAS results. In the whole monitoring period, the land deformation rate of the settlement trend always tended to be moderate and, considering the completion time, indicates that the area is possible in a self-consolidation situation during the monitoring time. However, due to the active activity of infrastructure activities in the Qianhai area, the landform changes faster which means that the radar reflection signals provided by these type of ground targets cannot maintain high coherence between the master image and the slave image. When using PSI for inversion, fewer PS points were searched. It makes PSI not as good as the SBAS method in terms of spatial resolution in this application scenario.

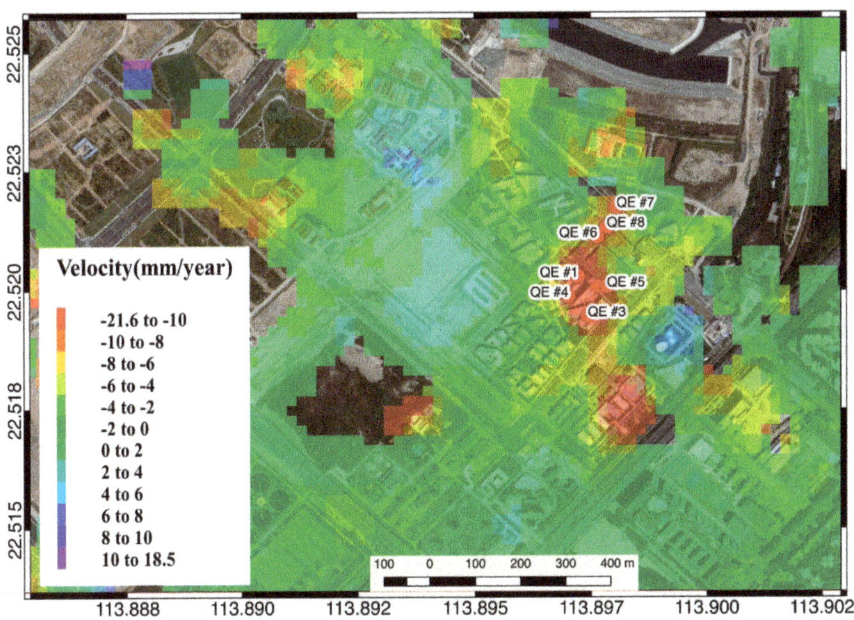

Figure 17. Time series InSAR inversion result of Shenzhen Qianhai Enterprise Residence.

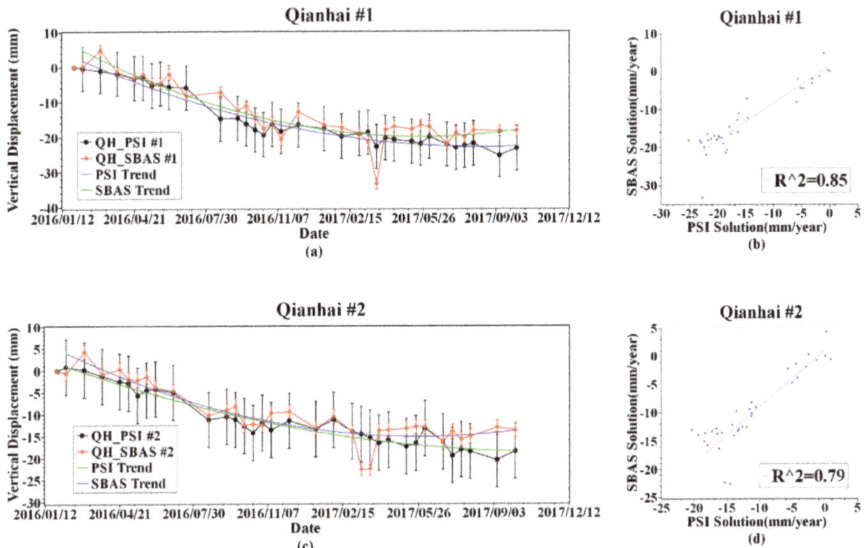

Figure 18. *Cont.*

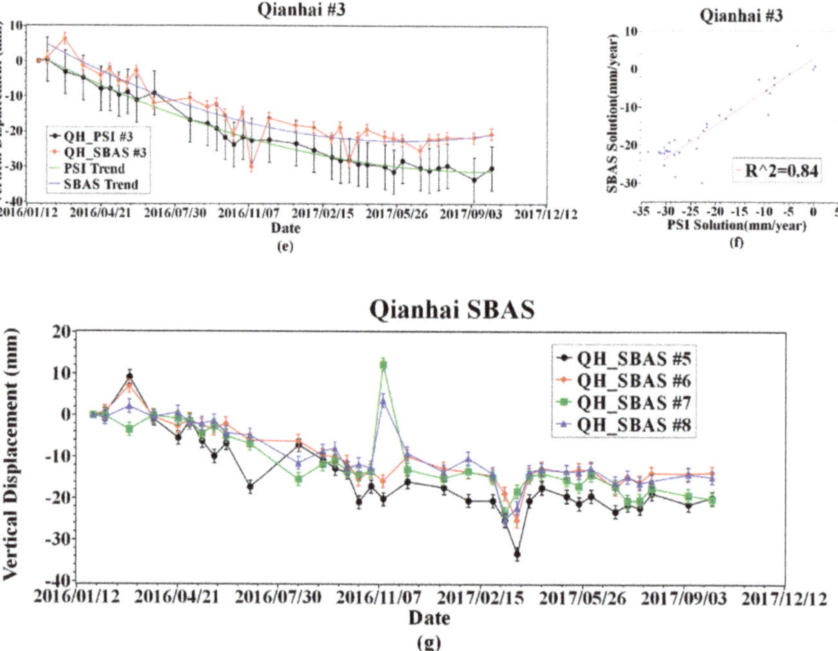

Figure 18. Time-series of sampling point in the Shenzhen Qianhai Enterprise Residence.

6. Conclusions

In this research, 34 frames of Sentinel-1A SAR images were used for SBAS and PSI processing, and GNSS observation data were used to perform surface deformation observation of the study area. From the results of data processing, three different methods have the ability to monitor the surface displacement, no matter the station or regional. The results show that the surface deformation obtained by the inversion of SBAS and PSI is highly consistent in magnitude and trend. Both PSI and SBAS show that the subsidence area in the study area was mainly concentrated in Qianhai District, which has recently completed land reclamation activities. In the reclamation area, especially along the underground traffic facilities [33] are the area that the subsidence phenomenon mainly concentrated. However, the TS-InSAR technology, which we mainly focus on, can locate and continuously monitor the areas with subsidence phenomena without prior information through periodic SAR image data. The application of TS-InSAR, especially SBAS technology, for surface deformation inversion has a higher spatial resolution than GNSS technology, hence it has advantages in the subsidence area detection. Since InSAR technology uses time and spatial filtering to remove the APS of each interferogram [1], the results of InSAR inversion are more stable than GNSS calculations. Although the GNSS method can obtain a three-dimensional variation of the station, the GNSS solution results are less accurate in the vertical direction. In the results, the implementation effect in monitoring the vertical displacement is not as good as the InSAR. However, GNSS technology has significant advantages in horizontal displacement monitoring, making it possible to apply GNSS technology to the region where the deformation signal appears in the InSAR result to obtain the horizontal change information of the deformation region. In the InSAR inversion results, land subsidence phenomena occur in areas with frequent human activities, although the current InSAR inversion model in this research does not distinguish well between the foundation trench excavation and the urban surface subsidence signal. From the result, we found that the land reclamation area of Qianhai District began to carry out infrastructure construction within ten years of completion of land reclamation work which with a higher risk. However, due to the extensive development of underground projects, such as subways, and the extensive geological distribution of land reclamation projects and underground caves, it has also increased the risk of the occurrence of surface deformation.

In terms of data processing, we expect to use GNSS at MIT/GLOBal Kalman (GAMIT/GLOBK) software for a higher accuracy baseline solution. In the process of collecting GNSS observation data, it is also possible to add non-synchronous observation data to constitute the asynchronous loop to enhance the reliability of the GNSS static network. With the use of two orbital SAR images in the same region for InSAR data processing, it is possible to acquire three-dimensional variations of coherent targets using InSAR technology.

Author Contributions: All three authors contributed to this work. J.C. elaborated the SAR data, implemented the methodology, results interpretation and finished the manuscript. X.Z. provided the validation and analyses of GNSS processing. B.H. designed the research program and supervised the research and provided valuable suggestions for the revision.

Funding: This work was supported by the Science and Technology Plan Project in Fujian Province (grant No. 2017Y3004) and by the National Natural Science Foundation of China (Grant No. 41204012 and 41674006).

Acknowledgments: We wish to thank the European Space Agency (ESA) for arranging the Sentinel-1 data and Japanese Aerospace Agency (JAXA) for providing ALOS-2 and "ALOS World 3D(AW3D)-30m" data. We thank the anonymous editors for their influential suggestions.

Conflicts of Interest: The authors declare that there is no conflict of interest regarding the publication of this paper.

References

1. Berardino, P.; Fornaro, G.; Lanari, R.; Sansosti, E. A new algorithm for surface deformation monitoring based on small baseline differential SAR interferograms. *IEEE Trans. Geosci. Remote Sens.* **2002**, *40*, 2375–2383. [CrossRef]
2. Perissin, D.; Wang, T. Time-Series InSAR Applications Over Urban Areas in China. *IEEE J. Sel. Top. Appl. Earth Obs. Remote Sens.* **2011**, *4*, 92–100. [CrossRef]
3. Herrera, G.; Gutiérrez, F.; García-Davalillo, J.C.; Guerrero, J.; Notti, D.; Galve, J.P.; Fernández-Merodo, J.A.; Cooksley, G. Multi-sensor advanced DInSAR monitoring of very slow landslides: The Tena Valley case study (Central Spanish Pyrenees). *Remote Sens. Environ.* **2013**, *128*, 31–43. [CrossRef]
4. Nagler, T.; Rott, H.; Kamelger, A. Analysis of Landslides in Alpine Areas by Means of SAR Interferometry. In Proceedings of the IEEE International Geoscience and Remote Sensing Symposium, Toronto, ON, Canada, 24–28 June 2002.
5. Wang, T.; Perissin, D.; Liao, M.; Rocca, F. Deformation Monitoring by Long Term D-InSAR Analysis in Three Gorges Area, China. In Proceedings of the IEEE International Geoscience and Remote Sensing Symposium, Boston, MA, USA, 8–11 July 2008.
6. Ferretti, A.; Savio, G.; Barzaghi, R.; Borghi, A.; Musazzi, S.; Novali, F.; Prati, C.; Rocca, F. Submillimeter accuracy of InSAR time series: Experimental validation. *IEEE Trans. Geosci. Remote Sens.* **2007**, *45*, 1142–1153. [CrossRef]
7. Hooper, A. A multi-temporal InSAR method incorporating both persistent scatterer and small baseline approaches. *Geophys. Res. Lett.* **2008**, *35*, L16302. [CrossRef]
8. Qu, W.; Zhang, B.; Lu, Z.; Kim, J.W.; Zhang, Q.; Gao, Y.; Hao, M.; Zhu, W.; Qu, F. Source Parameter Estimation of the 2009 Ms6.0 Yao'an Earthquake, Southern China, Using InSAR Observations. *Remote Sens.* **2019**, *11*, 462. [CrossRef]
9. Yang, Y.H.; Hu, J.C.; Tung, H.; Tsai, M.C.; Chen, Q.; Xu, Q.; Zhang, Y.J.; Zhao, J.J.; Liu, G.X.; Xiong, J.N.; et al. Co-Seismic and Postseismic Fault Models of the 2018 Mw 6.4 Hualien Earthquake Occurred in the Junction of Collision and Subduction Boundaries Offshore Eastern Taiwan. *Remote Sens.* **2018**, *10*, 1372. [CrossRef]
10. Béjar-Pizarro, M.; Álvarez Gómez, J.A.; Staller, A.; Luna, M.P.; Pérez-López, R.; Monserrat, O.; Chunga, K.; Lima, A.; Galve, J.P.; Martínez Díaz, J.J.; et al. InSAR-Based Mapping to Support Decision-Making after an Earthquake. *Remote Sens.* **2018**, *10*, 899. [CrossRef]

11. Xia, Y.; Michel, G.W.; Reigber, C.; Klotz, J.; Kaufmann, H. Seismic Unloading and Loading in Northern Central Chile as Observed by Differential Synthetic Aperture Radar Interferometry (D-INSAR) and GPS. *Int. J. Remote Sens.* **2003**, *24*, 4375–4391. [CrossRef]

12. Hooper, A.; Zebker, H.; Segall, P.; Kampes, B. A new method for measuring deformation on volcanoes and other natural terrains using InSAR persistent scatterers. *Geophys. Res. Lett.* **2004**, *31*, L23611. [CrossRef]

13. Wang, T.; Perssin, D.; Rocca, F.; Liao, M. Three Gorges Dam stability monitoring with time series InSAR analysis. *Sci. China Ser. D Earth Sci.* **2011**, *54*, 720–732. [CrossRef]

14. Zhang, Y.; Gong, H.; Li, X.; Liu, T.; Yang, W.; Chen, B.; Li, A.; Su, Y. InSAR Analysis of Land Subsidence Caused by Ground Water Exploitation in Changping, Beijing, China. In Proceedings of the IEEE International Geoscience and Remote Sensing Symposium, Boston, MA, USA, 8–11 July 2008.

15. Zerbini, S.; Richter, B.; Rocca, F.; Van Dam, T.; Matonti, F. A Combination of Space and Terrestrial Geodetic Techniques to Monitor Land Subsidence: Case Study, the Southeastern Po Plain, Italy. *J. Geophys. Res.* **2007**, *112*, B05401. [CrossRef]

16. Zhang, Y.; Liu, Y.; Jin, M.; Jing, Y.; Liu, Y.; Liu, Y.; Sun, W.; Wei, J.; Chen, Y. Monitoring Land Subsidence in Wuhan City (China) using the SBAS-InSAR Method with Radarsat-2 Imagery Data. *Sensors* **2019**, *19*, 743. [CrossRef]

17. Qin, X.; Yang, M.; Zhang, L.; Yang, T.; Liao, M. Health Diagnosis of Major Transportation Infrastructures in Shanghai Metropolis Using High-Resolution Persistent Scatterer Interferometry. *Sensors* **2017**, *17*, 2770. [CrossRef]

18. North, M.; Farewell, T.; Hallett, S.; Bertelle, A. Monitoring the Response of Roads and Railways to Seasonal Soil Movement with Persistent Scatterers Interferometry over Six UK Sites. *Remote Sens.* **2017**, *9*, 922. [CrossRef]

19. Xu, B.; Feng, G.; Li, Z.; Wang, Q.; Wang, C.; Xie, R. Coastal Subsidence Monitoring Associated with Land Land reclamation Using the Point Target Based SBAS-InSAR Method: A Case Study of Shenzhen, China. *Remote Sens.* **2016**, *8*, 652. [CrossRef]

20. Ferretti, A.; Prati, C.; Rocca, F. Permanent scatterers in SAR interferometry. *IEEE Trans. Geosci. Remote Sens.* **2001**, *39*, 8–20. [CrossRef]

21. Perissin, D.; Rocca, F. High accuracy urban DEM using permanent scatterers. *IEEE Trans. Geosci. Remote Sens.* **2006**, *44*, 3338–3347. [CrossRef]

22. Legco.gov.hk. *Legislative Council, Background brief on the development of the Three-Runway System at the Hong Kong International Airport*; Legco.gov.hk: Hong Kong, China, 19 November 2016.

23. Hu, L.; Jiao, J. Modeling the influences of land reclamation on groundwater systems: A case study in Shekou peninsula, Shenzhen, China. *Eng. Geol.* **2010**, *114*, 144–153. [CrossRef]

24. Yang, J.; Chen, J.; Le, X.; Zhang, Q. Density-oriented versus development-oriented transit investment: Decoding metro station location selection in Shenzhen. *Transp. Policy* **2016**, *51*, 93–102. [CrossRef]

25. Lai, L.W.; Lu, W.W.; Lorne, F.T. A catallactic framework of government land reclamation: The case of Hong Kong and Shenzhen. *Habitat Int.* **2014**, *44*, 62–71. [CrossRef]

26. Chen, K.; Jiao, J.J. Metal concentrations and mobility in marine sediment and groundwater in coastal reclamation areas: A case study in Shenzhen, China. *Environ. Pollut.* **2008**, *151*, 576–584. [CrossRef]

27. Eldhuset, K.; Weydahl, D.J. Using Stereo SAR and Insar by Combining the COSMOSkymed and the Tandem-X Mission Satellites for Estimation of Absolute Height. *Int. J. Remote Sens.* **2013**, *34*, 8463–8474. [CrossRef]

28. Jiang, T.C. Ameliorative Minimum Cost Flow Algorithm for Phase Unwrapping. *Procedia Environ. Sci.* **2011**, *10*, 2560–2566.

29. Yu, H.; Lan, Y.; Lee, H.; Cao, N. 2-D Phase Unwrapping using Minimum Infinity-Norm. *IEEE Geosci. Remote Sens. Lett.* **2018**, *15*, 1887–1891. [CrossRef]

30. Yu, H.; Lan, Y.; Yuan, Z.; Xu, J.; Lee, H. Phase Unwrapping in InSAR. *IEEE Geosci. Remote Sens. Mag.* **2019**, *7*, 40–58. [CrossRef]

31. Poggi, G.; Ragozini, A.P.R.; Servadei, D. A Bayesian approach for SAR interferometric phase restoration. *Proc. IEEE Int. Geosci. Remote Sens. Symp.* **2000**, *7*, 3202–3205.

32. Cao, N.; Yu, H.; Lee, H. A multi-baseline InSAR phase unwrapping method using designed optimal baselines obtained by motion compensation algorithm. *IEEE Geosci. Remote Sens. Lett.* **2018**, *15*, 1219–1223. [CrossRef]

33. Vallone, P.; Crosetto, M.; Giammarinaro, M.S.; Agudo, M.; Biescas, E. Integrated Analysis of Differential SAR Interferometry and Geological Data to Highlight Ground Deformations Occurring in Caltanissetta City (Central Sicily, Italy). *Eng. Geol.* **2008**, *98*, 144–155. [CrossRef]

Article

Polarimetric Stationarity Omnibus Test (PSOT) for Selecting Persistent Scatterer Candidates with Quad-Polarimetric SAR Datasets

Xingjun Luo [1,2,3], **Changcheng Wang** [1,2,3,*] **and Peng Shen** [1]

1   School of Geosciences and Info-Physics, Central South University, Changsha 410083, China; luoxingjun@csu.edu.cn (X.L.); shen-peng@csu.edu.cn (P.S.)

2   Key Laboratory of Metallogenic Prediction of Nonferrous Metals and Geological Environment Monitoring Ministry of Education, Changsha 410083, China

3   Hunan Key Laboratory of Nonferrous Resources and Geological Hazards Exploration, Changsha 410083, China

*   Correspondence: wangchangcheng@csu.edu.cn; Tel.: +86-731-8883-6931

Received: 29 December 2019; Accepted: 6 March 2020; Published: 11 March 2020

Abstract: In the traditional single polarimetric persistent scatterers interferometric (PSI) technology, the amplitude dispersion index (ADI) is usually used to select persistent scatterer candidates (PSC). Obviously, based on single polarimetric information, it is difficult to use the statistical characteristics for comprehensively describing the temporal stability of scatterers, which leads to a decrease in persistent scatterer (PS) density. Considering that the temporal polarimetric stationarity of PS, the paper is based on complex Wishart distribution and proposes the polarimetric stationarity omnibus test (PSOT) for identifying PSC. The nonstationary pixels can be removed by the preset significance threshold, which reduces the subsequent processing error and the calculation cost. Then, the exhaustive search polarimetric optimization (ESPO) method is selected for improving the phase quality of PSCs while suppressing the sidelobe of the strong scatterer effectively. For validating the effectiveness of the proposed method, we select a time-series quad-polarimetric ALOS PALSAR-1 images in an urban area as experimental data and mainly perform five group experiments for detailed analysis, including the PSOT+ESPO, ADI+ESPO, ADI+HH, ADI+HV, and ADI+VV. The results show that the proposed PSOT+ESPO method has a better performance on both PSC selection and interferometric phase optimization aspects than that of other methods. Specifically, compared to the last four methods, both the PSCs and PSs identified by the proposed PSOT+ESPO are more concentrated in the high-coherence region. The PSs with the standard deviation (STD) less than 5mm in the PSOT+ESPO method account for 94% of all PSs, which is greater than that of the ADI+ESPO, ADI+HH, ADI+HV, and ADI+VV methods, respectively.

Keywords: persistent scatterers; polarimetric optimization; deformation monitoring

---

1. Introduction

Interferometric synthetic aperture radar (InSAR) technology is one of the most popular geodetic techniques with the advantages of high precision, high resolution and all-weather work. Since Gabriel first used differential InSAR (DInSAR) to obtain deformation information of farmland in 1989 [1], researchers have successively improved DInSAR technology, such as improvement of the interferometric phase, and separation of multiple-phase signals, including orbital, atmospheric and residual topographic phases [2–5], etc. However, the accuracy of DInSAR is still affected by factors such as temporal and spatial decorrelation and atmosphere delay.

Therefore, to overcome the mentioned problems above, Time-series InSAR (TS-InSAR) technology based on DInSAR technology has gradually developed [6] and mainly includes the persistent scatterers interferometric (PSI) [7] and the small baseline approaches (SBAS) [8]. The PSI can identify the targets, persistent scatterers (PS), with stable scattering characteristics on the ground and monitor the surface deformation based on the reliable phase and amplitude information. PSI is widely used in volcanoes, earthquakes, urban subsidence, and landslides [9–14]. Especially in urban areas, artificial buildings in urban areas can be considered as ideal persistent scatterers (PS) [15], and the PSI technology is based on the PSs to explore the impact of human or natural activities on cities [16].

However, how to identify PS pixels more steadily is still a widely concerned problem, and it restricts the application of the PSI technology. More reliable PSs mean that the noise can be suppressed more effectively, and the accuracy of deformation solution can be improved in the subsequent processing. Some scholars have done some work in improving PS density. Shanker et al. proposed a method of maximum likelihood ratio to find more PS, and the average phases of PS pixels clearly show the slip along the Hayward fault [17]. Foroughnia et al. proposed a novel iterative PSI method (IPSI) to increase the PS points, which are lost in the PS-InSAR technique due to unwrapping failure [18]. Xiang et al. fully take advantage of the signal amplitude and phase information in the monitored scene and propose a combined PS selection (CPSS) method [19]. Gheorghe et al. Combine the ascending and descending SAR images to improve the monitoring density [20]. With the development of tomography SAR technology, Budillon et al. address the complementarity of the two techniques, and in particular it assesses the increase of measurement density that can be achieved by adding the double scatterers from SAR tomography to the persistent scatterer interferometry measurements [21]. These methods mentioned above are based on single-pol image for the improvement and application of PSI technology.

With the number of multi-polarization satellites increasing, it is necessary to introduce polarimetric observation for more effectively selecting PS and improving phase quality [22–26]. Polarimetric optimization becomes a direction of using polarimetric images to monitor deformation. The dual-polarimetric SAR dataset supported by Sentinel-1A and TerraSAR-X has been proved to be able to obtain more PSs with higher coherence [22,27,28]. It has been demonstrated that the polarimetric information introduction can increase the PS density. Higher PS density can help to detect more local deformation information and construct a more robust unwrapping network for removing the atmospheric phase [29]. To obtain PSs with stable phase, persistent scatterer candidates (PSC) need to be extracted firstly from full scenes. The effective identification of PSC efficiently extracts most high-coherence pixels while reducing the subsequent processing error and the calculation cost. Usually, the traditional PSI technology uses amplitude dispersion index (ADI) as the quality indicator to identify PSC [30]. However, in traditional PSI technology, the single-pol statistical characteristics are not comprehensively used for describing the temporal stationarity of scatterers.

In addition, in the urban area, the side lobes of high-intensity scatterers will interfere with nearby scatterers and even cover up low-intensity scatterers [31]. These problems are challenges when using PSI technology. In [24,32,33], Navarro-Sanchez et al. propose a general framework for PSI technology based on the exhaustive search polarimetric optimization (ESPO) algorithm [34]. The ESPO method optimizes the phase quality while suppressing noise and sidelobe, and allows some scatterers that are only sensitive to specific polarizations to be detected. However, the computational cost of the ESPO algorithm increases exponentially with the improvement of accuracy, which also limits the use of this method.

In this paper, a PSC selection method based on quad-polarimetric SAR image is proposed. Considering the temporal stationarity of PS, the paper is based on complex Wishart distribution and proposes the polarimetric stationarity omnibus test (PSOT) for identifying PSCs [35]. The proposed PSOT method can greatly reduce the amount of calculations of the ESPO algorithm and increase the number of PSs. The selected PSCs are optimized by ESPO, and it can improve phase quality and reduce noise and sidelobe. Finally, deformation velocity can be estimated by the PSI technology. The hypothesis and method are verified on real SAR images, and all experimental results are evaluated quantitatively.

The rest of this article is as follows. Section 2 presents the basic theory and flowchart of the proposed method. Section 3 describes the study area and the polarimetric SAR datasets for experiments. Section 4 is the experimental results and detailed analysis. Finally, Section 5 gives the conclusion.

2. Method

2.1. Persistent Scatterer Candidates Identification Based on Polarimetric Stationarity Omnibus Test (PSOT)

In an ideal situation, for a stable scatterer on the ground, the polarimetric information in time-series remains unchanged. In this paper, a time-series Polarimetric Stationarity Omnibus Test (PSOT) is proposed to evaluate whether the polarimetric information of scatterers changes in time. For the quad-polarimetric sensor, the scattering matrix S can be obtained:

$$S = \begin{bmatrix} S_{HH} & S_{HV} \\ S_{VH} & S_{VV} \end{bmatrix} \tag{1}$$

In a monostatic radar system, the scattering target satisfies the reciprocity and the Sinclair matrix is restricted to a symmetric matrix, i.e., $S_{HV} = S_{VH}$. Therefore, 3-dimension Pauli feature vector $\underline{k}$ is:

$$\underline{k} = \frac{1}{\sqrt{2}}[S_{HH} + S_{VV} \; S_{HH} - S_{VV} \; S_{HV} + S_{VH}]^T \tag{2}$$

The polarimetric coherence matrix T can be obtained by the cross product between $\underline{k}$ and its conjugate transpose [34]

$$T = \underline{k}_i \cdot \underline{k}_i^\dagger \tag{3}$$

where † denotes the conjugate transpose.

We supposed that there are k quad-polarimetric observations in time-series, let $\Sigma_i = T_3^i$, $(2 \leq i \leq k)$, and the $X_i = n\hat{\Sigma}_i$. The Σ_i (and the X_i) are p by p (p = 3 for quad-polarization image), following the complex Wishart distribution, i.e., $X_i \sim W_T(p, n, \Sigma_i)$. Further, $X = \sum_{i=1}^k X_i \sim W_T(p, nk, \Sigma)$.

It is supposed that the polarimetric information of the ground scatterers in k observations is unchanged, i.e., the polarimetric stationarity hypothesis. The complex coherency matrix is stationary under the polarimetric stationarity hypothesis. In order to evaluate whether all the complex covariance matrixes are equal when $k \geq 2$, the null hypothesis is tested [22]:

$$H_0 : \Sigma_1 = \Sigma_2 = \cdots = \Sigma_k \tag{4}$$

For all cases, the statistic Q can be constructed:

$$Q = \left\{ k^{3k} \frac{\prod_{i=1}^k |X_i|}{|X|^k} \right\}^n \tag{5}$$

where $|\cdot|$ denotes the determinant of the matrix. If the hypothesis is true ("under H0" in statistical parlance), $Q = 1$ and it means the polarimetric stationarity.

For the logarithm of the test statistic we get [36]:

$$\ln Q = n\left\{ 3k \ln k + \sum_{i=1}^k \ln|X_i| - k\ln|X| \right\} \tag{6}$$

Setting

$$\begin{cases} f = 9(k-1) \\ \rho = 1 - \frac{17}{18(k-1)}\left(\frac{k}{n} - \frac{1}{nk}\right) \\ \omega_2 = \frac{3}{\rho^2}\left(\frac{k}{n^2} - \frac{1}{(nk)^2}\right) - \frac{9(k-1)}{4}\left(1 - \frac{1}{\rho}\right)^2 \end{cases} \tag{7}$$

The probability of finding a smaller value of $-2\rho \ln Q$ is ($z = -2\rho \ln q_{obs}$)

$$P\{-2\rho \ln Q \leq z\} \cong P\{\chi^2(f) \leq z\} + \omega_2\left[P\{\chi^2(f+4) \leq z\} - P\{\chi^2(f) \leq z\}\right] \tag{8}$$

$P\{-2\rho \ln Q \leq -2\rho \ln q_{obs}\} = P\{Q \geq q_{obs}\}$ is the change probability, $1 - P\{-2\rho \ln Q \leq -2\rho \ln q_{obs}\} = P\{Q < q_{obs}\}$ is the no-change probability [37,38].

It is worth noting that when the number of looks of the original image is smaller than the matrix dimension, the coherency matrix is singular and no longer obeys the complex Wishart distribution. In order to maintain the spatial resolution of the original image and the accuracy of deformation extraction, a direct way to solve this problem is to adjust the non-diagonal elements for forcing the polarimetric coherence matrix to be a full rank matrix [39,40]. Therefore, the forced polarimetric coherence matrix T' can be expressed as:

$$\begin{cases} \forall i = j, \ T'_{i,j} = T_{i,j} \\ \forall i \neq j, \ T'_{i,j} = cT_{i,j}, \ c = \sqrt[3]{\min(n/q, 1)} \end{cases} \tag{9}$$

where $T_{i,j}$ represents the elements of row i and column j in the original coherence matrix; $T'_{i,j}$ represents the elements of the forced coherency matrix; n is the Equivalent Number of Looks (ENL); q is the dimension of matrix T; $q = 3$ for the quad-polarimetric image. When the forced coherence matrix is full rank, the equivalent number of looks of the corresponding data n is considered to be 3, which is equal to the matrix dimension.

2.2. Polarimetric Optimization of PSC Using Exhaustive Search Polarimetric Optimization Method

Usually, traditional single-pol PSI technology is based on a co-polarized polarization (i.e., HH or VV) image to monitor the ground deformation, because the image quality in co-polarized polarization is better than that in cross-polarized polarization (i.e., HV). With the support of quad-polarimetric SAR images, it is possible to find an optimal polarization in the quad-polarimetric signal space [32], whose phase quality is better than that of the co-polarized polarization (i.e., HH or VV) image. Navarro-Sanchez et al. proposes the ESPO algorithm for polarimetric optimization in PSI technology, which can find the optimal polarization to improve phase quality [32].

The quad-polarimetric observation information of the scattering target can be denoted by the complex scattering vector $\underline{k}$. In order to obtain the interferometric phase and coherence of quad-polarimetric images, we need to convert $\underline{k}$ to μ using unitary complex vector ω [23]:

$$\mu = \omega^\dagger \underline{k} \tag{10}$$

where $\dagger$ denotes the conjugate transpose; μ denotes a scalar complex scattering coefficient, which is equal to single look complex (SLC) image. Therefore, we can apply the existing PSI techniques to μ. For quad-polarimetric image, ω can be parameterized by the four parameters of $\omega(\alpha, \beta, \delta, \psi)$, which depend on the geometry and electromagnetic properties of the scatterers:

$$\omega = \begin{bmatrix} \cos(\alpha) \\ \sin(\alpha)\cos(\beta)e^{j\delta} \\ \sin(\alpha)\sin(\beta)e^{j\psi} \end{bmatrix}, \quad \begin{cases} 0 \leq \alpha \leq \pi/2 \\ 0 \leq \beta \leq \pi/2 \\ -\pi \leq \delta < \pi \\ -\pi \leq \psi < \pi \end{cases} \tag{11}$$

The PSI technology generally uses amplitude dispersion index (ADI) D_a to identify PSC. In this regard, the proposed ESPO method purposefully takes the minimum D_a of the pixel in the time-series as the goal of polarimetric optimization. D_a can be expressed as [22]:

$$D_a = \frac{\sigma_\alpha}{\overline{\alpha}} = \frac{1}{|\omega^\dagger \underline{k}| \sqrt{k-1}} \sqrt{\sum_{i=1}^{k}\left(|\omega^\dagger \underline{k}_i| - \overline{|\omega^\dagger \underline{k}|}\right)^2} \tag{12}$$

where k denotes the number of images, the upper line denotes the average value, and $|*|$ denotes the absolute value. In the PSC selection process of PSI, a pixel whose amplitude deviation is less than the preset threshold can be selected as a PSC. Here we use the exhaustive search polarimetric optimization (ESPO) algorithm to search for the ω to minimize D_a [33].

Figure 1 is the algorithm flow of this paper. The quad-polarimetric image obtained by the satellite can be denoted as scattering vector $\underline{k}$. The complex coherency matrix T, which contains all polarimetric scattering information of ground object, can be obtained by the cross product between $\underline{k}$ and its conjugate transpose. Then the significance level of each pixel is computed with the proposed PSOT method. Next, the PSCs obeying the polarimetric stationarity hypothesis are selected by setting the threshold T_{ots}. The optimal SLC image μ can be obtained by the ESPO method, and then $k-1$ differential interferograms can be obtained. In this paper, StaMPS technology is used for the subsequent processing of PSI on the selected PSCs. PS can be selected by controlling the threshold values (i.e., T_{n-max}, T_{n-std} and T_γ), and downsampling are carried out. Spatial-correlation errors (i.e., DEM, atmosphere and orbit error) are estimated and removed after 3D phase unwrapping. Finally, the deformation velocity can be estimated.

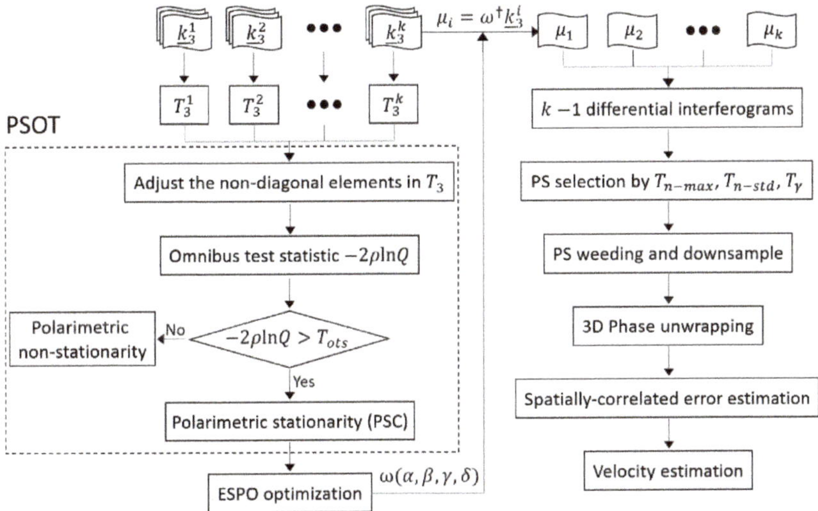

Figure 1. The flowchart of the proposed method. T_{ots}: significance threshold of polarimetric stationarity hypothesis test; T_{n-max}: threshold for the maximum noise allowed for a pixel; T_{n-std}: threshold for noise standard deviation; T_γ: threshold for temporal coherence.

3. Datasets

In order to verify the effectiveness of the proposed method, 13 scene quad-polarimetric ALOS PALSAR-1 images covering the San Fernando Valley CA are used. The coverage of the image is shown in Figure 2a, the black rectangle shows the spatial area of the original image, and the red rectangle shows the study area. This paper mainly selects the urban area as the study area, and the main scattering mechanism is double-bounce scattering (Figure 2b).

The quad-polarimetric PALSAR-1 datasets acquisition time is between June 8, 2006, and August 1, 2009. The temporal and spatial baselines of the dataset are listed in Table 1. The image of April 26, 2007 is selected as the master image. The estimated ENL of the original image is 0.7296.

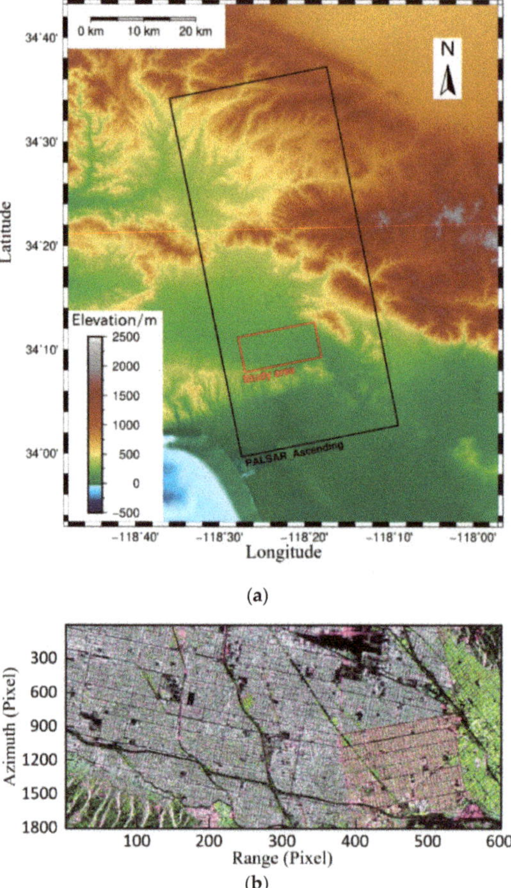

Figure 2. The scope and scattering characteristics of the study area. (**a**) Scope of the synthetic aperture radar (SAR) image: The black rectangle shows the spatial range of the original image, and the red rectangle shows study area; (**b**) Composite RGB image of the study area—red: $|S_{HH} - S_{VV}|^2$, green: $|S_{HV} + S_{VH}|^2$, blue: $|S_{HH} + S_{VV}|^2$.

Table 1. Temporal and spatial baselines of ALOS PALSAR-1 quad-polarimetric image.

Date	Perpendicular Baseline (m)	Temporal Baseline (Days)
20060608	−1129.9990	−322.00199
20060908	749.6921	−230.00109
20070311	1092.7781	−46.00006
20070426	0.0000	0.00000
20070727	1338.2859	91.99989
20071027	2264.7495	183.99958
20080127	2754.1354	275.99899
20080729	518.1668	459.99757
20080913	−1637.5801	505.99824
20081029	−1356.7558	551.99885
20090129	−655.3229	643.99985
20090316	−61.6506	690.00021
20090801	−204.9118	828.00092

4. Discussion

4.1. Selection of Significance Threshold

By evaluating the polarimetric stationarity of the pixels, the significance level of the hypothesis test of all the pixels in the study area is obtained in Figure 3. At airports, highways, vegetated areas, and the foreshortening area, the possibility of maintaining stable polarimetric characteristics is relatively low (high significance). The nonstationarity of the airport and highway is caused by the randomness and weakness of the echo signal. The vegetation growth and volume scattering variation make the vegetation area instable. The nonstationarity of the foreshortening area is caused by the backscatter signal aliasing.

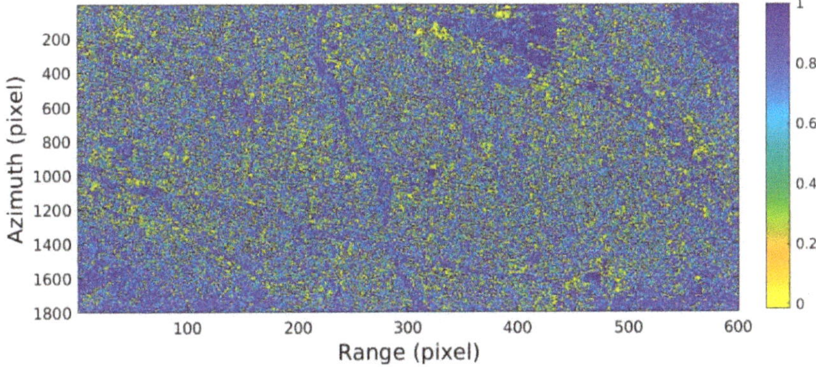

Figure 3. Significance level of polarimetric stationarity in the experimental area.

Except for the regions with a significantly higher significance level (significance level > 0.6), the remaining areas maintained a certain degree of stable polarimetric characteristics (significance level ≤ 0.4). The main surface cover types of these areas are urban buildings, and the main scattering mechanism is double-bounce scattering. Different significance thresholds are used to identify PSC, and the number and statistical characteristics of selected PSCs (and PSs) are also different. We conducted further experiments using HH polarization for identifying PSC, and then the PSC was post-processed using StaMPS to obtain PS. In StaMPS, the noise pixel can be removed by the downsampling parameter (merge_resample_size), but at the same time, the number of PS pixels will inevitably be reduced. The resolution of azimuth and ground range direction of the experimental data used in this paper is 3.54 m and 24.1 m, respectively. Therefore, we set the downsampling parameter to 25 m.

For studying the influence of increased significance threshold on the temporal coherence distribution, different significance (≤ 0.6) are selected for the PSC selection experiment, and the corresponding temporal coherence distribution of PSC and PS in HH polarization are obtained. Figure 4a shows that with the decrease of significance threshold, PSC can be greatly reduced. Pixels in the low-coherence range (< 0.85) are more sensitive to the change of significance threshold, which also shows that the PSOT can extract PSC effectively. In Figure 4b, the decrease of significance threshold mainly affects PS with temporal coherence between 0.6–1. The change of significance threshold has little effect on the number of PS. A higher threshold can extract more PS. Even if the significance threshold is set to 0.01, which is very strict, the number of PS (HH polarization) can still stay at 28355.

In Figure 5, within the range of significance threshold greater than 0.3, the ratio of Nps (the number of PS) to Npsc (the number of PSC) are approximately linear. Therefore, it can be considered that the pixels above the significance threshold of 0.3 have no significant polarimetric stationarity characteristics. When the significance threshold is less than 0.3, Npsc and Nps drastically decrease, but the ratio (Nps/Npsc) rapidly increases. It can be considered that the selected PS and PSC with a significance less than 0.3 under the polarimetric stationarity hypothesis have polarimetric stationarity. When the significance is 0.3, the proportion of PS in the total pixels (1080000) is 3.69%, and the proportion of PS in the PSC is 14.78%. Therefore, the range of significance threshold is 0.1–0.3.

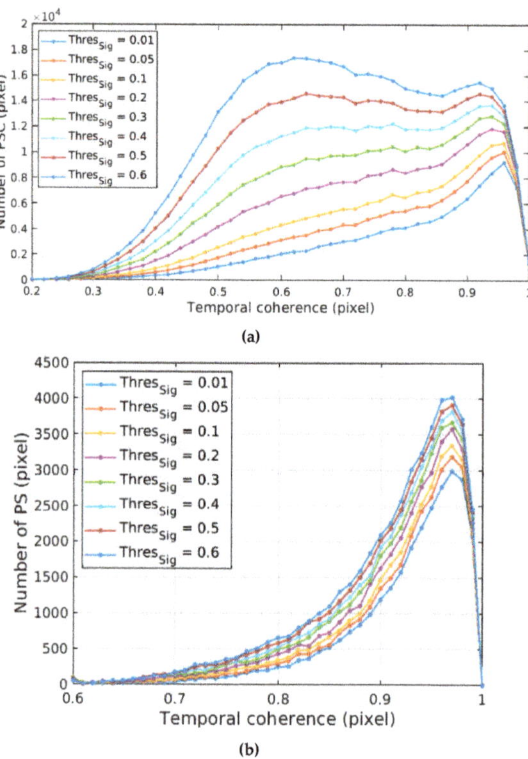

Figure 4. Temporal coherence distribution of persistent scatterer candidates (PSCs) and persistent scatterers (PSs) in HH polarization under different significance thresholds. (**a**) PSC; (**b**) PS.

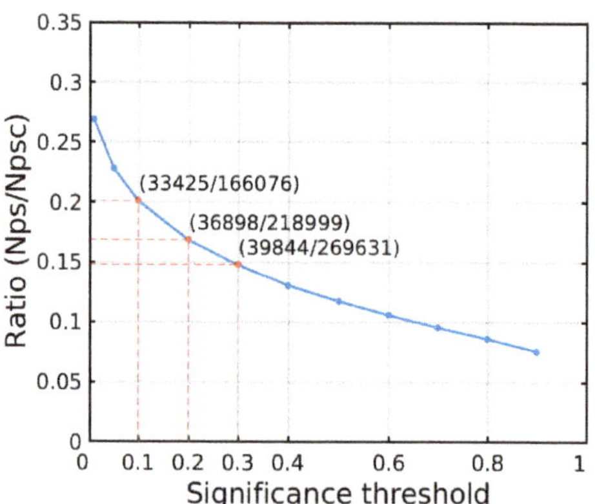

Figure 5. Changes of the ratio of Nps to Npsc under different significance thresholds. Nps: the number of PS, Npsc: the number of PSC. The tags in the figure correspond to the number of PS and PSC under different significance thresholds, namely (Nps/Npsc).

The temporal coherence is an important indicator for evaluating the phase quality of both PSC and PS. In order to evaluate the bias of coherence estimation, the paper makes a detailed analysis on the statistical characteristics of temporal coherence estimation, as shown in Figure 6. Touzi et al. illustrated the deviation between estimated coherence and true coherence for statistically independent samples [41]:

$$d = \frac{\Gamma(L)\Gamma\left(\frac{3}{2}\right)\left(1 - D^2\right)^L}{\Gamma\left(L + \frac{1}{2}\right)}\ {}_3F_2\left(\frac{3}{2}, L, L; L + \frac{1}{2}, 1; D^2\right) \tag{13}$$

where D is the true degree of coherence, d is the estimated coherence, Γ denotes the gamma function, ${}_3F_2$ denotes the hypergeometric function, and L is the number of statistically independent samples. Then the phase standard deviation under different coherence was obtained [36]:

$$\sigma_\varphi = \sqrt{\frac{1 - |D|^2}{2L|D|^2}} \tag{14}$$

The following is to quantitatively analyze the estimated bias and the phase standard deviation condition of the temporal coherence with different coherence value and number of interferograms. Figure 6 shows that when the number of samples is 12, the deviation between estimated coherence and true coherence is very small in the high-coherence region (the main feature of PS). When the coherence is 0.8, the deviation of coherence is −0.00391, and the variance of estimated coherence is 0.0228, which shows that the estimated coherence can evaluate the performance of the proposed method.

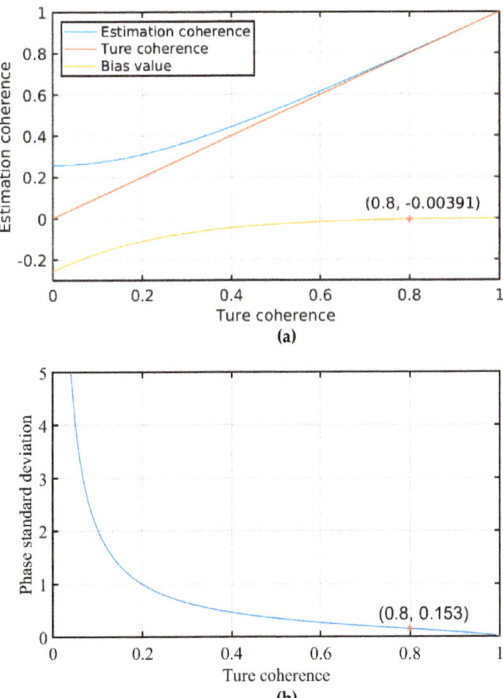

Figure 6. Statistical characteristics of temporal coherence estimation with different coherence value and 12 interferograms. (**a**) Estimated coherence, true coherence and the bias value; (**b**) variance of estimation coherence.

In order to compare the effect of the PSOT on different polarizations, the thresholds are set to 0.1, 0.2 and 0.3 respectively to select the points in ESPO, HH, HV and VV. The PSC (PS) coherence distribution of different polarizations at the same threshold can reflect the performance of the ESPO.

Figure 7 shows that the temporal coherence of PSC identified by the PSOT method to the ESPO polarization is the lowest. For HH and VV polarizations, the PSOT selects more high-coherence pixels (> 0.7) than VV polarizations, and selects fewer low-coherence pixels (< 0.7). The ESPO selects less PSC in the coherence range of 0.6–0.85 than other polarizations, and selects the most PSC in the coherence range of greater than 0.85, which makes the distribution of PS more concentrated in the high-coherence region.

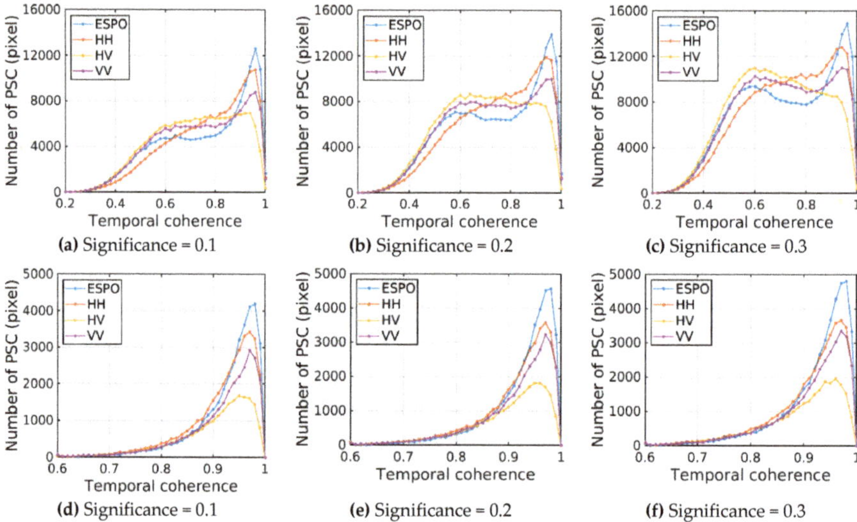

Figure 7. Coherence distribution is selected for different polarizations (ESPO, HH, HV, VV) under different significance thresholds; above is the coherence distribution of PSC, below is the coherence distribution of PS; (**a**,**d**) significance = 0.1; (**b**,**e**) significance = 0.2; (**c**,**f**) significance = 0.3.

4.2. Comparison of Different PSC Selection Methods

In order to compare the difference between this method (PSOT+ESPO) and the traditional method, we compared the performance of amplitude dispersion index (ADI) in selecting PSC for different polarizations (ESPO, HH, HV, VV). The ADI and the PSOT describe the stability of scatterer from different angles, so the threshold value is not comparable. However, the purpose of different PSC selection methods is to select the "optimal" PSC set, so it is more reasonable to compare the distribution of time coherence in the same number of PSC sets. Based on PSOT, different significance thresholds (0.1, 0.2, 0.3) are used to select PSC, and then the same number of PSCs for different polarizations are selected by the ADI to compare. The following table shows the actual thresholds of PSCs selected by different methods.

Figure 8 shows that the effect of the ADI+ESPO is only better than that of the ADI+HV, which shows that, with the ESPO method based on ADI index, it is easy to select low-coherence PSC in the aspect of selecting PSC by mistake. The traditional co-polar HH and VV can effectively select high-coherence PSC. The main reason is that the co-polar polarization contains most of the scattering energy of the ground objects. It is worth noting that when the PSOT is used to select PS, the best performance can be achieved by using the ESPO method to optimize polarization, which is more concentrated in the region with coherence greater than 0.9. This shows the effectiveness of the PSOT in PSC selection, and effectively makes up for the shortcomings of the ESPO method. Moreover, under different confidence thresholds, the PSOT has the same effect and good robustness.

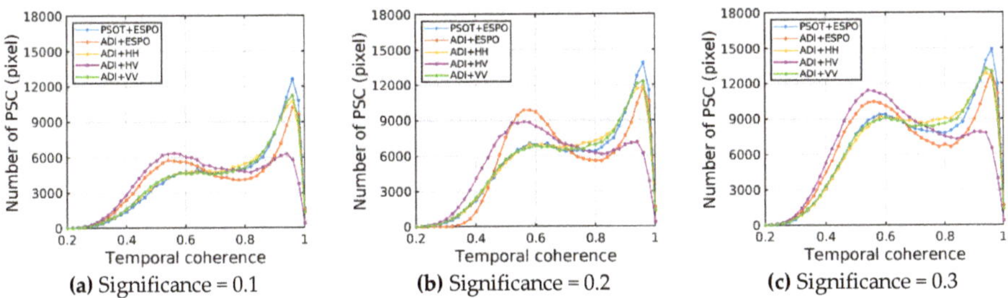

(a) Significance = 0.1 **(b)** Significance = 0.2 **(c)** Significance = 0.3

Figure 8. PSC temporal coherence distribution is selected for different methods under different significance thresholds; (**a**), significance = 0.1; (**b**), significance = 0.2; (**c**), significance = 0.3.

To further compare the performance of different methods, it is necessary to compare the coherence distribution of PS. In this paper, the same PSI process is used to process the PSC obtained by different methods, and the coherence distribution of PS obtained by various methods (Figure 9).

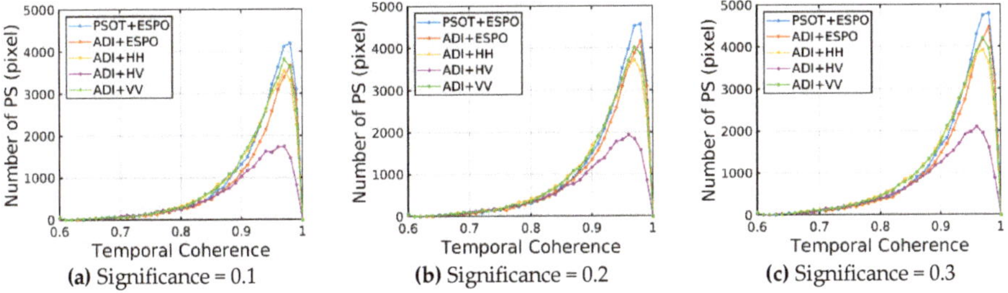

(a) Significance = 0.1 **(b)** Significance = 0.2 **(c)** Significance = 0.3

Figure 9. PS temporal coherence distribution is selected for different methods under different significance thresholds; (**a**) significance = 0.1; (**b**) significance = 0.2; (**c**) significance = 0.3.

After removing most of the low-coherence PSC, the coherence of PS is mainly distributed in the range above 0.8. The performance of the ADI+ESPO is equivalent to that of the ADI+HH/VV. This is because the ESPO method based on ADI is still the statistical information based on a single polarization, which is easy to greatly reduce the amplitude dispersion index of the scatterer (see the threshold value in Table 2). This will result in a larger proportion of points that are unstable being selected as PSC, so the amplitude dispersion index can not identify the PSCs of the ESPO very well. In this method, the quad-polarimetric information of the scatterer in time-series is considered to represent the stability of the scatterer, which avoids the limitation of a single measure. It can effectively make up for the shortcomings of the ADI+ESPO and select PSCs more effectively.

Table 2. The threshold value of different PSC selection methods and the number of selected PSC.

Method	Threshold 1	Threshold 2	Threshold 3
PSOT+ESPO	0.1	0.2	0.3
ADI+ESPO	0.1680	0.1773	0.1851
ADI+HH	0.3628	0.3822	0.3985
ADI+VV	0.3750	0.3934	0.4088
ADI+HV	0.3636	0.3830	0.3993
Number of PSC	166081	219006	269638

The performance of the ESPO has been discussed in [22,23,34]. In the experiment in this paper, we also found that the EPSO method can suppress the side lobe effect to a certain extent, and the texture of the surface coverage is clearer. We optimized the polarization of the quad-polarization image with an accuracy of 6°. To evaluate the effect of the ESPO, we compared it with three single polarization (HV and VH are equivalent) respectively. In order to compare the ability of different polarization to retain details of ground targets, representative buildings and parks were selected for analysis. The effects of different polarization will be analyzed in terms of intensity details and phase quality.

Figure 10a,b and d shows that ESPO inhibits the sidelobe effect to some extent. In Figure 10, the building with sidelobe is Oak Tree Aviation Services LLC, Burbank Glendale Pasadena (BUR) airport and the direction of the building is parallel to the LOS (line of sight) direction (Figure 10e), therefore the sidelobe effect is very serious in the co-polarized polarization (Figure 10b,d). However, the signal of BUR is weak in cross-polarized polarization (Figure 10c). ESPO not only suppressed the sidelobe effect, preserved the scattering information of the building, but also solved the problem that it was difficult to observe the building in the cross-polarized polarization. For Pierce Brothers Valhalla in Figure 10f, ESPO preserves scattering information of the park path with clear details and noise suppression. The noise suppression effect of ESPO is also well reflected in the odd scattering regions (airport runway, etc.).

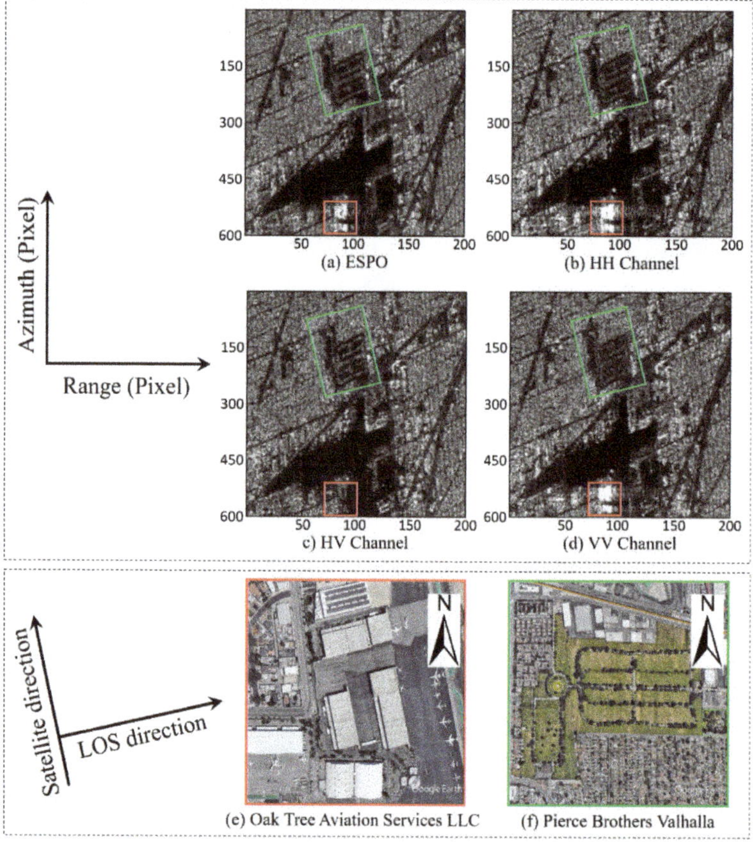

Figure 10. Intensity images of ESPO, HH, VV and HV polarizations, and Google Earth images of two ground objects; the red box is Oak Tree Aviation Services LLC and the green box is Pierce Brothers Valhalla. (**a**) ESPO; (**b**) HH; (**c**) HV; (**d**) VV; (**e**) Google Earth image of Oak Tree Aviation Services LLC; and (**f**) Google earth image of Pierce Brothers Valhalla. The strong scatterers in the red box show the effect of ESPO on sidelobe suppression, and the park in the green box shows the effect of ESPO on noise suppression and detail retention.

5. Analysis of Deformation Results

After verifying the performance of PSOT in Section 4, combining the constraint degree of significance threshold on the number of PSs, we select the median value (0.2) of the suggested significance threshold interval ([0.1 0.3]) for PSC selection. Then the ESPO method is carried out on the selected PSC, and the PS and time-series deformation is obtained by StaMPS (Figure 11a). In order to compare the effect of deformation monitored with different method, the ADI+ESPO (or HH, HV, VV) uses the same post-processing.

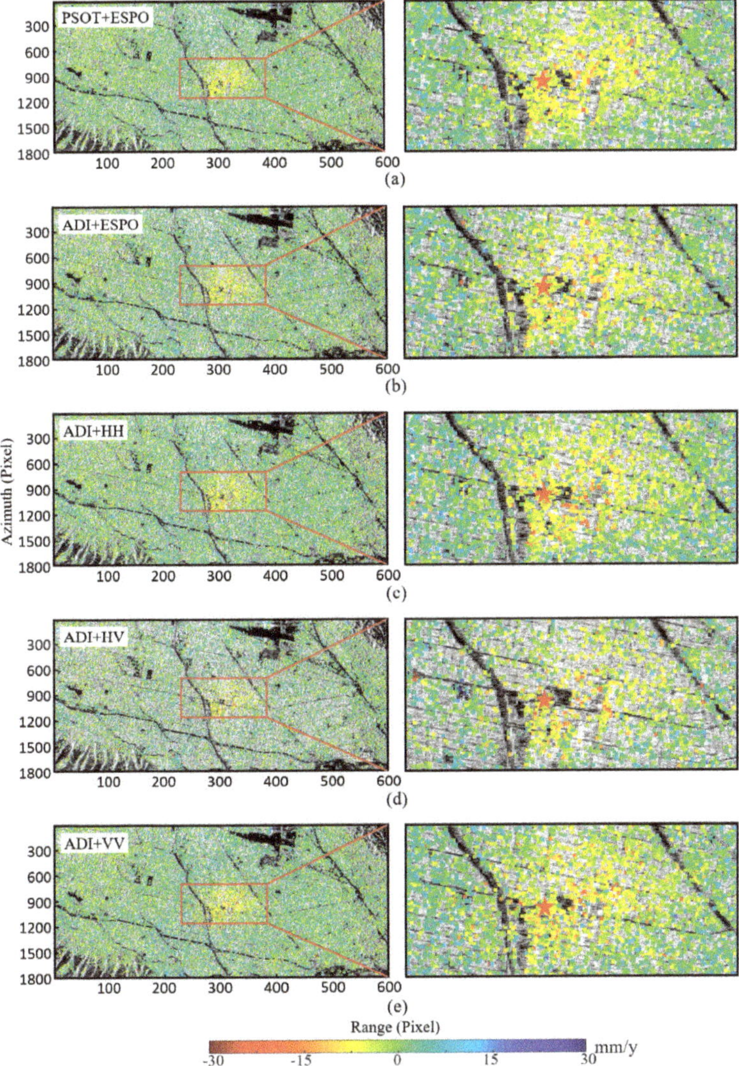

Figure 11. Mean deformation rate obtained by different methods (**a**) PSOT+ESPO; (**b**) ADI+ESPO; (**c**) ADI+HH; (**d**) ADI+HV; (**e**) ADI+VV; PSOT and ADI are the methods to identify PSC. the left shows the deformation results of the experimental area, the right is a zoomed-in view of deformation; the pentagram is position of the North Hollywood Station.

Figure 11a,d shows that the selected PS density of the PSOT+ESPO is higher than that of the ADI+HV, and there is less noise. Figure 11a–c,e looks similar, and the effect of deformation needs to be observed locally. It can be seen that Figure 11c,e has more noise pixels. The PSOT+ESPO and the ADI+ESPO have fewer noisy pixels. In order to quantitatively analyze the deformation results, we compared the standard deviation (STD) distribution of the time-series deformation of the five methods (Figure 12). PSs with larger STD indicate either larger errors (due to atmosphere or unwrapping errors) or deformation that is non-linear.

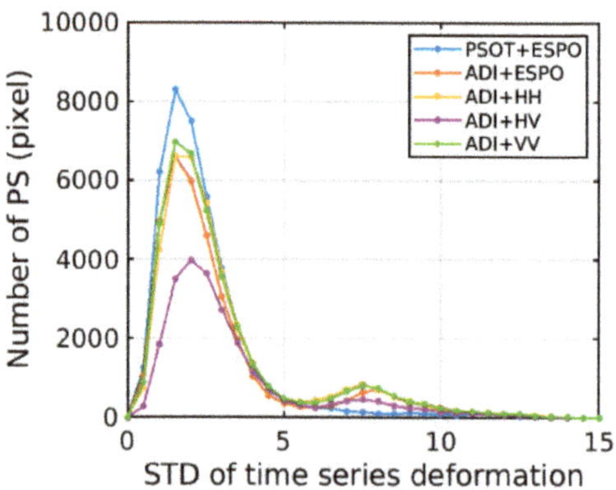

Figure 12. STD distribution of time-series deformation results obtained by different methods. STD is calculated by the residual time series after subtracting the linear component, units of STD is mm.

Figure 12 shows that the STD distribution of time-series deformation obtained by the PSOT+ESPO is significantly better than the other method. But the ratio of the ADI+ESPO in error points is higher than ADI+HH/VV, which also indicates that there is a problem of misselection. The PSOT method proposed in this paper can select the ESPO points more effectively, and the error points are far less than the ADI+ESPO/HH/VV. Specifically, in Table 3, we can see that only 2393 pixels of the deformation obtained by the PSOT+ESPO have STD greater than 5. For the specific deformation time-series, select the corresponding points of the pentagram in Figure 11 for analysis.

Table 3. The mean intensity of each polarization and the polarization-optimized image.

Method	PSOT+ESPO	ADI+ESPO	ADI+HH	ADI+HV	ADI+VV
Number of PS	39620	35185	38408	23923	39015
Number of PS (STD > 5)	2393	5127	6263	3956	6113

Figure 13 shows that the time-series deformation of different methods. Assuming that the deformation is linear. Deformation rate estimation of ADI+HV is different from other methods, and its STD is the largest, so the reliability of this method is the lowest. Figure 13a shows that the time-series deformation with PSOT+ESPO has a higher linear fitting degree (STD = 1.638 mm) and its deformation rate estimation is similar to ADI+VV.

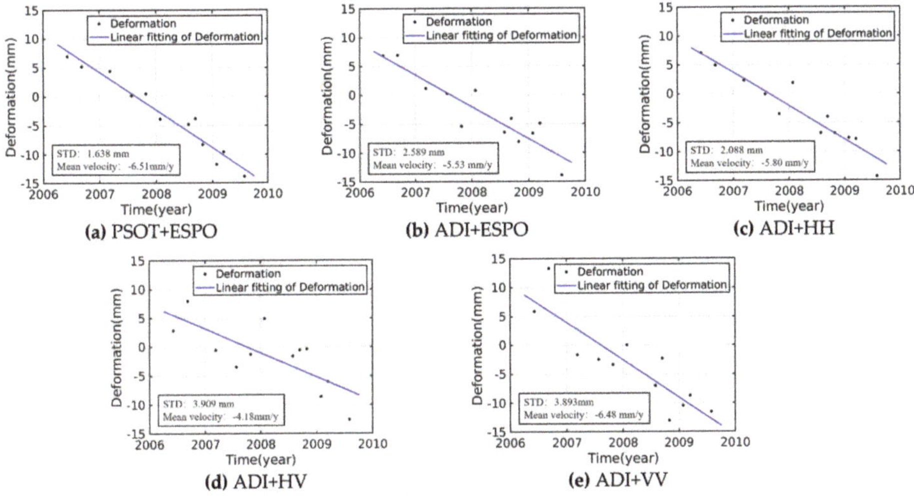

Figure 13. Time-series deformation of the pentagram in Figure 11; (**a**) PSOT+ESPO; (**b**) ADI+ESPO; (**c**) ADI+HH; (**d**) ADI+HV; (**e**) ADI+VV. STD is calculated by the residual time-series after subtracting the linear component.

In order to compare the efficiency of different methods, the paper records the time consumption of different methods (Table 4). The ADI+HH/HV/VV methods do not require polarization optimization, so there is no ESPO time cost. Time consumption for PSI (StaMPS) includes coherence estimation, error estimation, phase unwinding, deformation estimation and so on. All the programs of these methods are executed under the condition of single-core processor without parallel processing.

Table 4. Calculation time of the proposed PSC selection and non-selection of PSC (CPU: AMD Ryzen 5 2600 Six-Core Processor × 12, Memory: 64G DDR4, Operation system: Ubuntu 16.04 LTS 64bit).

Time Consumption (h)	PSOT+ESPO	ADI+ESPO	ADI+HH	ADI+HV	ADI+VV
PSOT	0.080	—	—	—	—
ESPO	8.195	40.230	—	—	—
PSI(StaMPS)	0.155	0.154	0.173	0.140	0.171
Total	8.430	40.384	0.173	0.140	0.171

The time consumption shown in the figure can be seen that the processing time for different methods to select the same number of PSC in PSI is equivalent. The PSOT+ESPO method proposed in this paper (the significance threshold is 0.2) is 79% less than the traditional ESPO for deformation monitoring. In addition, the calculation amount of the ESPO varies exponentially with accuracy. In this paper, we use accuracy of 6°. Reducing the accuracy of ESPO can also reduce the amount of calculation.

The deformation area monitored in this paper is about 3 km^2 and the center is North Hollywood Station. The deformation is funnel-shaped, and the deformation rate is approximately linear. During the monitoring period, there was no major earthquake damage in the deformation area, and most of the buildings were built early. North Hollywood Station was also unable to cause the deformation of such a large area as 3 km^2. Therefore, it is speculated that this is the inelastic deformation caused by groundwater exploitation.

6. Conclusions

The PSI is an important means for InSAR to monitor surface deformation. Improving the density and phase quality of PS are key problem in PSI. The ADI in the traditional single-pol PSI is difficult to describe the statistical characteristics of scatterers, so it is not robust when selecting PSC. In this paper, we propose a PSOT method to identify the polarimetric stationary scatterers as the PSCs. Experimental results show that the phase quality of PSCs identified by the PSOT+ESPO is higher than that by the ADI+ESPO method and traditional single-pol PSI. Through error analysis, the proposed PSOT+ESPO method can obtain the maximum number of PS, and the deformation estimation is more robust. Specifically, when the significance threshold is 0.2, 219006 PSCs were selected. After error analysis, 39620 PS remained, accounting for 3.67% of the total pixels (1080000).

Experiments show that the ESPO method can not only achieve the polarimetric optimization of the PSC interferometric phase, but also suppress the sidelobe of the strong scatterer effectively and make the details of the ground object clearer. The ESPO improves the coherence of the scatterers by optimizing the amplitude dispersion index, which will select some incorrect PSC when using the ADI method. The PSOT method based on the polarimetric SAR image can avoid the unsteadiness of PSC selected by the ESPO using ADI. Therefore, The PSOT combined with ESPO can identify PSC more accurately and improve the phase quality.

Author Contributions: X.L. performed the experiments, wrote the paper; C.W. contributed the ideas, analyzed the experimental results and revised the paper; P.S. analyzed the experimental results and revised the paper. All authors have read and agreed to the published version of the manuscript.

Funding: This work was funded by the National Key Research and Development Program of China (No. 2018YFC1505103), the National Natural Science Foundation of China (No. 41671356) and the Fundamental Research Fund for the Central Universities of Central South University (No. 2019zzts639).

Acknowledgments: The authors would like to thank the Japanese Space Agency (JAXA) for providing the ALOS PALSAR-1 SAR image.

Conflicts of Interest: The authors declare no conflict of interest.

References

1. Gabriel, A.K.; Goldstein, R.M.; Zebker, H.A. Mapping small elevation changes over large areas: Differential radar interferometry. *J. Geophys. Res.* **1989**, *94*, 9183–9191. [CrossRef]

2. Shirzaei, M.; Walter, T.R. Estimating the Effect of Satellite Orbital Error Using Wavelet-Based Robust Regression Applied to InSAR Deformation Data. *IEEE Trans. Geosci. Remote Sens.* **2011**, *49*, 4600–4605. [CrossRef]

3. Wei, J.; Li, Z.; Hu, J.; Feng, G.; Duan, M. Anisotropy of atmospheric delay in InSAR and its effect on InSAR atmospheric correction. *J. Geod.* **2019**, *93*, 241–265. [CrossRef]

4. Iglesias, R.; Monells, D.; López-martínez, C.; Mallorqui, J.J.; Fabregas, X.; Aguasca, A. Polarimetric Optimization of Temporal Sublook Coherence for DInSAR Applications. *IEEE Geosci. Remote Sens. Lett.* **2015**, *12*, 87–91. [CrossRef]

5. Schneider, R.Z.; Papathanassiou, K. Pol-DinSAR: Polarimetric SAR Differential Interferometry Using Coherent Scatterers. In Proceedings of the 2007 IEEE International Geoscience and Remote Sensing Symposium, Fort Worth, TX, USA, 23–28 July 2007; pp. 196–199.

6. Li, Z.; Fielding, E.J.; Cross, P. Integration of InSAR Time-Series Analysis and Water-Vapor Correction for Mapping Postseismic Motion After the 2003 Bam (Iran) Earthquake. *IEEE Trans. Geosci. Remote Sens.* **2009**, *47*, 3220–3230.

7. Ferretti, A.; Prati, C.; Rocca, F. Permanent Scatterers in SAR Interferometry. *IEEE Trans. Geosci. Remote Sens.* **2001**, *39*, 8–20. [CrossRef]

8. Berardino, P.; Fornaro, G.; Lanari, R.; Sansosti, E. A new algorithm for surface deformation monitoring based on small baseline differential SAR interferograms. *IEEE Trans. Geosci. Remote Sens.* **2002**, *40*, 2375–2383. [CrossRef]

9. Wu, B.; Tong, L.; Chen, Y.; He, L. Improved SNR Optimum Method in POLDINSAR Coherence Optimization. *IEEE Geosci. Remote Sens. Lett.* **2016**, *13*, 982–986. [CrossRef]

10. Salehi, M.; Mohammadzadeh, A.; Maghsoudi, Y. Multitemporal multidimensional speckle filtering of PolSAR images. In Proceedings of the 2016 IEEE International Geoscience and Remote Sensing Symposium (IGARSS), Beijing, China, 10–15 July 2016.

11. Kampes, B.M. *Radar Interferometry Persistent ScattererTechnique*; Springer: Berlin/Heidelberg, Germany, 2006.

12. Costantini, M.; Falco, S.; Malvarosa, F.; Minati, F. A new method for identification and analysis of persistent scatterers in series of sar images. *Int. Geosci. Remote Sens. Symp.* **2008**, *2*, 449–452.

13. Liu, X.; Xu, W. Logarithmic Model Joint Inversion Method for Coseismic and Postseismic Slip: Application to the 2017 Mw 7.3 Sarpol Zahāb Earthquake, Iran. *J. Geophys. Res. Solid Earth* **2019**, *124*, 12034–12052. [CrossRef]

14. Wang, C.; Cai, J.; Li, Z.; Mao, X.; Feng, G.; Wang, Q. Kinematic Parameter Inversion of the Slumgullion Landslide Using the Time Series Offset Tracking Method With UAVSAR Data. *J. Geophys. Res. Solid Earth* **2018**, *123*, 8110–8124. [CrossRef]

15. Di Martino, G.; Iodice, A.; Poreh, D.; Riccio, D.; Ruello, G. Physical models for evaluating the interferometric coherence of potential persistent scatterers. In Proceedings of the 2017 IEEE International Geoscience and Remote Sensing Symposium (IGARSS), Fort Worth, TX, USA, 23–28 July 2017; pp. 3163–3166.

16. Schneider, R.Z.; Papathanassiou, K.P.; Hajnsek, I.; Moreira, A. Polarimetric and Interferometric Characterization of Coherent Scatterers in Urban Areas. *IEEE Trans. Geosci. Remote Sens.* **2006**, *44*, 971–984. [CrossRef]

17. Shanker, P.; Zebker, H. Persistent scatterer selection using maximum likelihood estimation. *Geophys. Res. Lett.* **2007**, *34*, 2–5. [CrossRef]

18. Foroughnia, F.; Nemati, S.; Maghsoudi, Y.; Perissin, D. An iterative PS-InSAR method for the analysis of large spatio-temporal baseline data stacks for land subsidence estimation. *Int. J. Appl. Earth Obs. Geoinf.* **2019**, *74*, 248–258. [CrossRef]

19. Xiang, X.; Chen, J.; Wang, H.; Pei, L.; Wu, Z. PS selection method for and application to GB-SAR monitoring of dam deformation. *Adv. Civ. Eng.* **2019**, *2019*, 8320351. [CrossRef]

20. Gheorghe, M.; Armaş, I.; Dumitru, P.; Călin, A.; Bădescu, O.; Necşoiu, M. Monitoring subway construction using Sentinel-1 data: A case study in Bucharest, Romania. *Int. J. Remote Sens.* **2020**, *41*, 2644–2663. [CrossRef]

21. Budillon, A.; Crosetto, M.; Johnsy, A.C.; Monserrat, O.; Krishnakumar, V.; Schirinzi, G. Comparison of persistent scatterer interferometry and SAR tomography using Sentinel-1 in urban environment. *Remote Sens.* **2018**, *10*, 1986. [CrossRef]

22. Sadeghi, Z.; Valadan Zoej, M.J.; Hooper, A.; Lopez-Sanchez, J.M. A new polarimetric persistent scatterer interferometry method using temporal coherence optimization. *IEEE Trans. Geosci. Remote Sens.* **2018**, *56*, 6547–6555. [CrossRef]

23. Mullissa, A.G.; Perissin, D.; Tolpekin, V.A.; Stein, A. Polarimetry-Based Distributed Scatterer Processing Method for PSI Applications. *IEEE Trans. Geosci. Remote Sens.* **2018**, *56*, 3371–3382. [CrossRef]

24. Navarro-Sanchez, V.D.; Lopez-Sanchez, J.M.; Ferro-Famil, L. Polarimetric approaches for persistent scatterers interferometry. *IEEE Trans. Geosci. Remote Sens.* **2014**, *52*, 1667–1676. [CrossRef]

25. Xiao, S.P.; Chen, S.W.; Chang, Y.L.; Li, Y.Z.; Sato, M. Polarimetric coherence optimization and its application for manmade target extraction in PolSAR data. *IEICE Trans. Electron.* **2014**, *E97-C*, 566–574. [CrossRef]

26. Ishitsuka, K.; Matsuoka, T.; Tamura, M. Persistent Scatterer Selection Incorporating Polarimetric SAR Interferograms Based on Maximum Likelihood Theory. *IEEE Trans. Geosci. Remote Sens.* **2016**, *55*, 38–50. [CrossRef]

27. Esmaeili, M.; Motagh, M. Improved Persistent Scatterer analysis using Amplitude Dispersion Index optimization of dual polarimetry data. *ISPRS J. Photogramm. Remote Sens.* **2016**, *117*, 108–114. [CrossRef]

28. Azadnejad, S.; Maghsoudi, Y.; Perissin, D. Evaluation of polarimetric capabilities of dual polarized Sentinel-1 and TerraSAR-X data to improve the PSInSAR algorithm using amplitude dispersion index optimization. *Int. J. Appl. Earth Obs. Geoinf.* **2020**, *84*, 101950. [CrossRef]

29. Hooper, A.; Zebker, H.; Segall, P.; Kampes, B. A new method for measuring deformation on volcanoes and other natural terrains using InSAR persistent scatterers. *Geophys. Res. Lett.* **2004**, *31*, 1–5. [CrossRef]

30. Mora, O.; Mallorqui, J.J.; Duro, J.; Broquetas, A. Long-term subsidence monitoring of urban areas using differential interferometric SAR techniques. *Int. Geosci. Remote Sens. Symp.* **2001**, *3*, 1104–1106.

31. Huang, Y.; Ferro-Famil, L. 3-D characterization of buildings in a dense urban environment using L-band pol-insar data with irregular baselines. *Int. Geosci. Remote Sens. Symp.* **2009**, *3*, 29–32.

32. Navarro-sanchez, V.D.; Lopez-sanchez, J.M.; Vicente-guijalba, F. A Contribution of Polarimetry to Satellite Differential SAR Interferometry: Increasing the Number of Pixel Candidates. *IEEE Geosci. Remote Sens. Lett.* **2010**, *7*, 276–280. [CrossRef]

33. Navarro-sanchez, V.D.; Lopez-sanchez, J.M. Improvement of Persistent-Scatterer Interferometry Performance by Means of a Polarimetric Optimization. *IEEE Geosci. Remote Sens. Lett.* **2012**, *9*, 609–613. [CrossRef]

34. Navarro-Sanchez, V.D.; Lopez-Sanchez, J.M. Spatial adaptive speckle filtering driven by temporal polarimetric statistics and its application to PSI. *IEEE Trans. Geosci. Remote Sens.* **2014**, *52*, 4548–4557. [CrossRef]

35. Salehi, M.; Mohammadzadeh, A.; Maghsoudi, Y. Adaptive Speckle Filtering for Time Series of Polarimetric SAR Images. *IEEE J. Sel. Top. Appl. EARTH Obs. Remote Sens.* **2011**, *5*, 567–576.

36. Lee, J.; Pottier, E. *Polarimetric Radar Imaging From Basics To Applications*; CRC Press: Taylor & Francis, UK, 2009.

37. Nielsen, A.; Conradsen, K.; Skriver, H. Omnibus test for change detection in a time sequence of polarimetric SAR data. In Proceedings of the 2016 IEEE International Geoscience and Remote Sensing Symposium (IGARSS), Beijing, China, 10–15 July 2016; pp. 3398–3401.

38. Wang, C.; Shen, P.; Li, X.; Zhu, J.; Li, Z. A Novel Vessel Velocity Estimation Method Using Dual-Platform TerraSAR-X and TanDEM-X Full Polarimetric SAR Data in Pursuit Monostatic Mode. *IEEE Trans. Geosci. Remote Sens.* **2019**, *57*, 6130–6144. [CrossRef]

39. Deledalle, C.; Denis, L.; Tupin, F.; Reigber, A.; Jäger, M. NL-SAR: A unified nonlocal framework for resolution-preserving (Pol)(In) SAR denoising. *IEEE Trans. Geosci. Remote Sens.* **2015**, *53*, 2021–2038. [CrossRef]

40. Shen, P.; Wang, C.; Gao, H.; Zhu, J. An adaptive nonlocal mean filter for PolSAR data with shape-adaptive patches matching. *Sensors* **2018**, *18*, 2215. [CrossRef]

41. Touzi, R.; Lopes, A.; Bruniquel, J.; Vachon, P.W. Coherence Estimation For Sar Imagery. *IEEE Trans. Geosci. Remote Sens.* **1999**, *37*, 135–149. [CrossRef]

Article

Coherent Markov Random Field-Based Unreliable DSM Areas Segmentation and Hierarchical Adaptive Surface Fitting for InSAR DEM Reconstruction

Qian Qian [1,2,*], Bingnan Wang [1,*], Xiaoning Hu [1,2] and Maosheng Xiang [1,2]

[1] National Key Laboratory of Microwave Imaging Technology, Aerospace Information Research Institute, Chinese Academy of Sciences, Beijing 100190, China; huxiaoning16@mails.ucas.ac.cn (X.H.); xms_iecas@163.com (M.X.)

[2] School of Electronic, Electrical and Communication Engineering, University of Chinese Academy of Sciences, Beijing 100190, China

* Correspondence: qianqian_19940216@163.com (Q.Q.); wbn@mail.ie.ac.cn (B.W.)

Received: 31 December 2019; Accepted: 3 March 2020; Published: 4 March 2020

Abstract: A digital elevation model (DEM) can be obtained by removing ground objects, such as buildings, in a digital surface model (DSM) generated by the interferometric synthetic aperture radar (InSAR) system. However, the imaging mechanism will cause unreliable DSM areas such as layover and shadow in the building areas, which seriously affect the elevation accuracy of the DEM generated from the DSM. Driven by above problem, this paper proposed a novel DEM reconstruction method. Coherent Markov random field (CMRF) was first used to segment unreliable DSM areas. With the help of coherence coefficients and residue information provided by the InSAR system, CMRF has shown better segmentation results than traditional traditional Markov random field (MRF) which only use fixed parameters to determine the neighborhood energy. Based on segmentation results, the hierarchical adaptive surface fitting (with gradually changing the grid size and adaptive threshold) was set up to locate the non-ground points. The adaptive surface fitting was superior to the surface fitting-based method with fixed grid size and threshold of height differences. Finally, interpolation based on an inverse distance weighted (IDW) algorithm combining coherence coefficient was performed to reconstruct a DEM. The airborne InSAR data from the Institute of Electronics, Chinese Academy of Sciences has been researched, and the experimental results show that our method can filter out buildings and identify natural terrain effectively while retaining most of the terrain features.

Keywords: coherence coefficient; DEM; DSM; hierarchical adaptive surface fitting; InSAR; markov random field; residue

---

1. Introduction

An interferometric synthetic aperture radar (InSAR) has the ability of acquiring a large-area and high-precision digital surface model (DSM) in all-times and all weather. The information of a digital elevation model (DEM) is required for many applications, therefore it is necessary to reconstruct a DEM from a DSM by removing the above-ground objects such as buildings. The DEM reconstruction is involved in photogrammetry [1,2], laser detection and ranging (LiDAR) [3–6], or InSAR [7–10]. Many methods have been proposed in this subject especially in the field of LiDAR [6], however reconstruction research based on InSARs is relatively rare. The main reason is that the accuracy of an InSAR DSM is lower than that of LiDAR due to the unique side looking imaging mechanism of synthetic aperture radar (SAR). For example, in an InSAR DSM, there can be a lot of layover and shadow areas in the building scene and the interferometric phase inversion of these areas are not reliable, which may generate a lot of points with incorrect extreme elevations in an InSAR DSM. Therefore, the reconstruction of an InSAR DEM is more challenging than that of LiDAR data.

Wang and Mercer [7] proposed an InSAR DEM reconstruction method based on image pyramid. Each level needs to be reconstructed in this algorithm, thus the error in the middle level will affect the next level, which is prone to error accumulation. Jiang [8] combined the slope information and the image pyramid method to filter non-ground points by calculating the slope between the candidate points. Zhang and Tao [9] proposed a surface-fitting-based method of an InSAR DEM reconstruction. The DEM is generated from InSAR DSM by extracting candidate ground points in a fixed-size grid, adjusting points with a distance of more than the given threshold from fitted surface, and and using ground points for interpolatio. These methods assume that the point with the minimum elevation in the fixed-size grid is the ground point, without considering the unreliable DSM points with the large spike noise belonging to layovers and shadows in the InSAR building areas. When the local minimum points fall into these unreliable DSM areas, extreme points are selected as the ground points, causing significant errors in the DEM reconstruction. Therefore, to avoid the adverse effects of these areas on ground points selection, it would make sense to segment the unreliable DSM area before selecting the ground point. At the same time, the selection of grid size and threshold in surface fitting may also significantly affect the reconstruction of the DEM. When the grid size is too large, some ground details will be lost, and the terrain will be smoothed. When the grid size is too small, the local minimum point will fall into the building, resulting in reduced DEM reconstruction accuracy.

Unreliable DSM areas mainly include the layover and shadow in a building scene, which can be segmented by the intensity of pixel gray because of their different brightness in SAR images. Due to the existence of speckle noise and complex texture characteristics of ground objects, the segmentation results are not satisfactory in the general image segmentation algorithm. To improve segmentation performance, the spatial relationship is usually considered. Markov random field (MRF) is recognized in the field of image segmentation due to its ability to utilize spatial context information [11], and it has been widely applied in SAR image segmentation [12–14]. In a traditional MRF, the ability of the neighborhood energy to describe the spatial correlation is insufficient, and the fixed parameter causes the neighborhood pixels to have the same impact on the central pixel. Moreover, the context information is not fully utilized [15,16], therefore the segmentation result is prone to misclassification points. In this paper, considering the potential of interferometric information and the coherence coefficient and residue information are incorporated into the traditional MRF model for improving segmentation performance.

Based on the above discussion, this paper proposed a DEM reconstruction method based on unreliable DSM area segmentation and hierarchical adaptive surface fitting. The contributions of this paper can be summarized as follows:

(1) In order to avoid the influence of the extreme points in the unreliable DSM areas when performing DEM reconstruction, segmentation based on the intensity of pixel gray levels in the InSAR amplitude image (which is helpful for the selection of ground points) was firstly used to identify the unreliable DSM areas for improving the performance of the subsequent DEM reconstruction.

(2) In order to improve the segmentation performance, we considered the potential of InSAR data information, such that this paper combined the coherence coefficient and residue information of interferometric phase with the neighborhood energy of the MRF, and the full use of contextual relationship was achieved by using the interferometric information between neighboring pixels.

(3) In the general surface fitting-based method, the fixed grid size and threshold will affect the filtering accuracy. Therefore, a new idea of progressively reducing the grid size and setting the adaptive threshold is proposed. It can realize the step-by-step filtering of ground points and the preservation of terrain detail information. At the same time, inverse distance weighted (IDW) interpolation with coherence coefficient is performed for completing the reconstruction of the DEM.

The rest of the paper is organized as follows. In Section 2, details of the proposed method are described. The experimental results and discussion are in Section 3, and Section 4 is the conclusions.

2. Proposed Method

2.1. Unreliable DSM Areas Segmentation with Coherent Markov Random Field (CMRF) Method

2.1.1. Image Segmentation Based on a MRF Model

A MRF model regards an image as a points set S, and the segmentation label X is a random field corresponding to S. The spatial relationship between neighboring pixels is constructed by defining neighborhood cliques $\eta = \{\eta_{ij}: (i, j) \in S, \eta_{ij} \in S\}$. According to Bayesian theory, we need to find the estimate of segmentation label X_{MAP} that maximizes the posterior probability distribution:

$$\hat{X}_{MAP} = \mathrm{argmax}P(X|Y) = \mathrm{argmax}\frac{P(Y|X)P(X)}{P(Y)} = \mathrm{argmax}P(Y|X)P(X) \tag{1}$$

where X is the segmentation label, and Y is the observation image. According to the equivalence of MRF and Gibbs Random Field (GRF), which can be proved by the Hammersley-Clifford theorem and the Gibbs theorem, the posterior probability distribution can be represented as:

$$P(X = x|Y = y) = Z^{-1}\exp(-U(x|y)) \tag{2}$$

where U is the energy function; and Z denotes the normalizing constant. From Equation (2), it can be seen that maximizing the posterior probability $P(X|Y)$ means minimizing energy function $U(x|y)$. Moreover $U(x|y)$ which is called posterior energy in this letter can be decomposed into Equation (3)

$$\begin{aligned} U(x|y) \quad &= U(y|x) + U(x) \\ &= -\sum_{s}\ln p(y_s|x_s) + \sum_{c \in V_s} V_c(x) \end{aligned} \tag{3}$$

where V_s is a set of all neighborhood cliques; $U(y|x)$ denotes the likelihood energy which represents the contribution of the pixel itself to the energy; and $U(x)$ denotes the neighborhood energy. $V_c(x)$ is expressed as Equation (4) [17]:

$$V_c(x) = \begin{cases} 0 & x_i = x_j \\ \beta & x_i \neq x_j \end{cases} \tag{4}$$

where x_i is the segmentation label of pixel i; x_j is the segmentation label of pixel j which is neighboring pixel of i; and β is a parameter to control the contribution between $U(y|x)$ and $U(x)$, which is usually determined by experience.

As shown in Equation (3), the likelihood energy is related to the likelihood function of pixels. According to the imaging structure and pixel gray of the building scene in the SAR image, the following three classes are determined, and the unreliable DSM areas include the layover and shadow areas.

(1) Layover areas: The characteristics of this area are scattered signals of targets at different positions overlapping at the same distance resolution unit, causing high brightness in the SAR image.
(2) Shadow areas: This area is characterized by an extremely low backscattered signal strength, which is caused by steep terrain or occlusion by towering targets.
(3) Background areas: The other areas which don't belong to the layover or shadow in the scene are grouped into the background, which mainly includes roofs, trees, and bare ground.

A Fisher distribution model is used to describe the probability distribution of building scenes in high-resolution SAR images by Tison [18], and it can be described as follows:

$$p_{Fisher}(u) = \frac{\Gamma(L+M)}{\Gamma(L)\Gamma(M)} \frac{L}{M\mu} \frac{\left(\frac{L}{M\mu}u\right)^{L-1}}{\left(1 + \frac{L}{M\mu}u\right)^{L+M}}, L > 0, M > 0 \tag{5}$$

where L and M represent the shape parameters; μ denotes the weight parameter; and Γ is the Gamma function.

After selecting areas of different classes defined above as the supervising information, we can estimate the parameters as follows:

$$M = \frac{4R_1 - 3R_2 - 1}{2R_1 - R_2 - 1} \tag{6}$$

$$L = \frac{2(R_1 - R_2)}{-R_1 + 2R_2 - R_1 R_2} \tag{7}$$

$$\mu = m_1 \frac{2(R_1 - R_2)}{4R_1 - 3R_2 - 1} \tag{8}$$

where $R_1 = m_2/(m_1 * m_1)$, $R_2 = m_3/(m_1 * m_2)$, and m_1, m_2, m_3 are the statistical histogram central moments of corresponding orders.

Therefore, according to the estimated Fisher probability distribution corresponding to Equation (5) and the neighborhood energy shown in Equation (4), the class label can be obtained by the following formula:

$$\begin{aligned} \hat{X} &= \text{argmin} U(x|y) \\ &= \text{argmin}\left(-\sum_s \ln p(y_s|x_s) + \sum_{c \in V_s} V_c(x)\right) \end{aligned} \tag{9}$$

These class labels are firstly obtained by the initial segmentation, and then the labels are updated iteratively. The neighborhood energy is related to the class labels of the neighboring pixels, and the likelihood energy is determined by the probability distribution function of the pixel values. The pixel value and neighboring label are used to calculate the posterior energy of a single pixel, and the label with the minimum energy value is used as the segmentation result. Finally, iterative solution is performed until the energy is stable.

2.1.2. CMRF Segmentation

In the traditional MRF model, when the center pixel label and the neighborhood pixel label are the same, the neighborhood energy is a certain value, and when the labels are different, it is zero. This results in the adjacent pixels having the same effect on the center pixel [15], therefore it cannot fully utilized the contextual information. Driven by this problem, this paper redefined the neighborhood energy model of MRF based on the coherence coefficient and residue information to make full use of the contextual interferometric information.

The coherence coefficient is used to evaluate the quality of the InSAR interferogram, which is defined as follows [19]:

$$\gamma = \frac{\left|E[s_1 s_2^*]\right|}{\sqrt{E\left[|s_1|^2\right] E\left[|s_2|^2\right]}} \tag{10}$$

where s_1 and s_2 are the interferometric complex image pair; and E represents mathematical expectation. The interferometric coherence is an elemental parameter for InSAR applications, which is estimated by comparing the radar echo across several nearby radar images pixels [20]. The related coherent change detection (CCD) [21], maximum-likelihood (ML) CCD [22], and ML-polarimetric InSAR-CCD (ML-PolInSAR-CCD) [23] are important applications of satellite earth observation. The coherence coefficient is related to the characteristics of the scatterers. For example, pixels which belong to shadow area tend to have low a coherence coefficient because the scattering signal in these areas is dominated by noise, while the coherence coefficient in other areas is usually higher than shadow. This property can be used to distinguish different classes [24]. Meanwhile, the coherence coefficient usually shows consistency and uniformity in areas with pixels belonging to the same category, which can be used to further improve the performance of image segmentation. This paper defines a coherence coefficient distance that measures the difference in coherence between the central pixel and the neighboring pixels, and it is expressed as follows:

$$D = |\gamma_i - \gamma_j| \tag{11}$$

where γ_i is the coherence coefficient of the pixel i; and γ_j is the coherence coefficient of pixel j, which is the neighboring pixel of i.

Furthermore, the residue information of the interferometric phase is also helpful for SAR image segmentation. Under ideal conditions, the absolute value of the phase gradient should be less than π. However, due to the existence of low scattering areas such as shadow, smooth roads, and water, etc., the absolute value of the wrapped phase gradient may be greater than π. This is called the phase discontinuity point and is known as residue [25]. The residue distribution in the interferometric phase image is obtained according to the following formula:

$$\begin{aligned}
\psi_1 &= W(\varphi_{i,j+1} - \varphi_{i,j}) \\
\psi_2 &= W(\varphi_{i+1,j+1} - \varphi_{i,j+1}) \\
\psi_3 &= W(\varphi_{i+1,j} - \varphi_{i+1,j+1}) \\
\psi_4 &= W(\varphi_{i,j} - \varphi_{i+1,j}) \\
R &= \psi_1 + \psi_2 + \psi_3 + \psi_4
\end{aligned} \tag{12}$$

where $\varphi_{i,j}$ represents the wrapped phase at the pixel (x, y); and W represents the wrapped phase operator. When $R > 0$ it is a positive residue, otherwise it is a negative residue, and $R = 0$ is the normal point. The residues are caused by phase discontinuity in low-scattering areas such as shadows. If both points are residues, they are likely to be divided into shadows, thus residue information can be helpful for segmenting InSAR amplitude images.

Considering the effects of coherence coefficient and residue information, if the coherence coefficient distance is small and both points are residues, the possibility of being divided into the same class is greater, and vice versa.

More specifically, when the class labels are the same between the center pixel and the neighboring pixel, a small coherence coefficient distance should mean low neighborhood energy, which may increase the probability of being identified as the same class for the two pixels. Meanwhile, if the center pixel and the neighboring pixel are both residues, the corresponding neighborhood energy should be lower than the energy that the two points are not both residues, and it is more likely to be classified into the same label. When the class labels are different between the center pixel and the neighboring pixel, the opposite is true. Based on the above analysis, the improved neighborhood energy form is as follows:

$$V_{c-CMRF}(x) = \begin{cases}
(1 - e^{-\alpha D})\beta & x_i = x_j \ \ r(x_i) \neq r(x_j) \\
(1 - e^{-\mu\alpha D})\beta & x_i = x_j \ \ r(x_i) = r(x_j) \\
(e^{-\alpha D} - 1)\beta & x_i \neq x_j \ \ r(x_i) \neq r(x_j) \\
(e^{-\mu\alpha D} - 1)\beta & x_i \neq x_j \ \ r(x_i) = r(x_j)
\end{cases} \tag{13}$$

where α is a constant greater than zero and it is used to control the shape of the curve; μ is the weighting coefficient of the residue information; and $r(x_i) = r(x_j)$ means that both x_i and x_j are residues, and $r(x_i) \neq r(x_j)$ means the opposite. Figure 1 shows a curve of the neighborhood energy as a function of coherence coefficient distance and residue information.

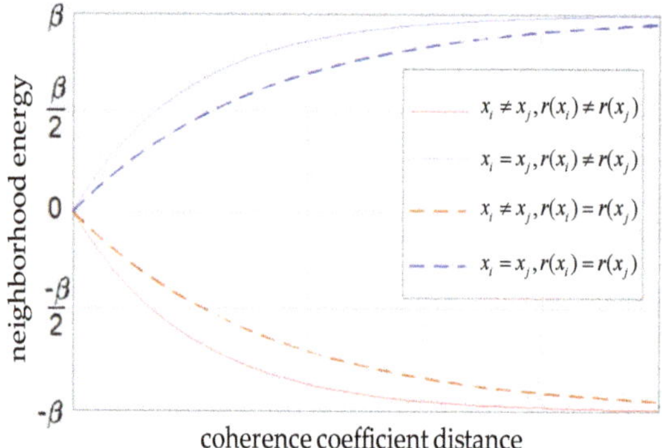

Figure 1. Curve of the neighborhood energy changing with the coherence coefficient distance and residue information.

Therefore, Equation (3) is represented as Equation (14) in the CMRF model:

$$U_{CMRF}(x|y) = -\sum_s \ln p(y_s|x_s) + \sum_{c \in V_s} V_{c-CMRF}(x) \tag{14}$$

Equation (9) is represented as Equation (15):

$$\begin{aligned}\hat{X} &= \mathrm{argmin} U_{CMRF}(x|y) \\ &= \mathrm{argmin}(-\sum_s \ln p(y_s|x_s) + \sum_{c \in V_s} V_{c-CMRF}(x))\end{aligned} \tag{15}$$

2.2. DEM Reconstruction Based on Hierarchical Adaptive Surface Fitting

2.2.1. Reconstruction Method Based on Surface Fitting

After removing the points of the unreliable DSM areas, the lowest points of the grids which don't belong to the unreliable DSM areas are used for surface fitting to realize DEM reconstruction. Zhang [9] took the local minimum points in a given grid as the candidate ground points, which were further optimized by surface fitting. Assuming that the terrain surface is a complex spatial surface, and it can be approximated by a quadric surface, as shown in the following equation:

$$z = a_0 + a_1 x + a_2 y + a_3 x^2 + a_4 y^2 + a_5 xy \tag{16}$$

where z represents the value of DEM; and x, y represent the horizontal and vertical coordinates of the candidate ground points, respectively. According to the least squares method, the parameters of the surface equation can be determined by the following equation:

$$A = (M^T P M)^{-1}(M^T P Z) \tag{17}$$

where $A = [a_0, a_1, \cdots, a_5]^T$, $Z = [z_1, z_2, \cdots, z_n]^T$. M, and P are described as follow:

$$M = \begin{bmatrix} 1 & x_1 & y_1 & x_1{}^2 & x_1 y_1 & y_1{}^2 \\ 1 & x_2 & y_2 & x_2{}^2 & x_2 y_2 & y_2{}^2 \\ \vdots & \vdots & \vdots & \vdots & \vdots & \vdots \\ 1 & x_n & y_n & x_n{}^2 & x_n y_n & y_n{}^2 \end{bmatrix} \tag{18}$$

$$P = \begin{bmatrix} p_1 & & & 0 \\ & p_2 & & \\ & & \ddots & \\ 0 & & & p_n \end{bmatrix} \tag{19}$$

where $p_1, p_2, \cdots, p_n$ are the weights of the corresponding points. This paper considers that all points have the same effect on surface fitting, therefore $p_1 = p_2 \cdots = p_n = 1$, and n is the number of points used for fitting. If the difference between the actual elevation and the fitted elevation is greater than the given threshold, the point is filtered out; otherwise, the original value remains unchanged.

Considering the continuity of the terrain, this paper added the neighborhood grids, and the fitting surface of each grid is obtained based on the minimum points which exclude the detected unreliable DSM points of the 3×3 neighborhood grids, as shown in Figure 2. The left part represents the original DSM data, and the red grid is surrounded by its 3×3 neighborhood grids. The point in each grid in the right part is the local lowest point of the grid, which cannot be the detected DSM unreliable point. These points in the right part are fitted to the surface of the red grid by Equation (16). The surface fitting using the minimum points of the neighborhood grids can maintain the characteristics of the terrain as much as possible.

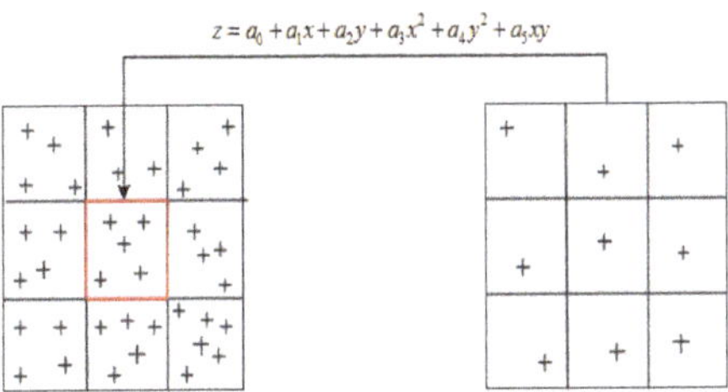

Figure 2. Fitting process with minimum points in neighborhood grids.

In the surface fitting-based method, the choice of grid size is important. As shown in Figure 3, when the grid size is set to a large value such as l_1, the lowest point will not fall near the ridge, thus it is difficult to completely retain the true terrain at the ridge during subsequent surface fitting. When the grid size is set to a small value such as l_2, the lowest point will fall on the roof of the building, and the fitted terrain will deviate from the real terrain, resulting in incomplete filtering of the buildings. At the same time, the threshold in the filtering process is not changed adaptively, which will affect the reconstruction result.

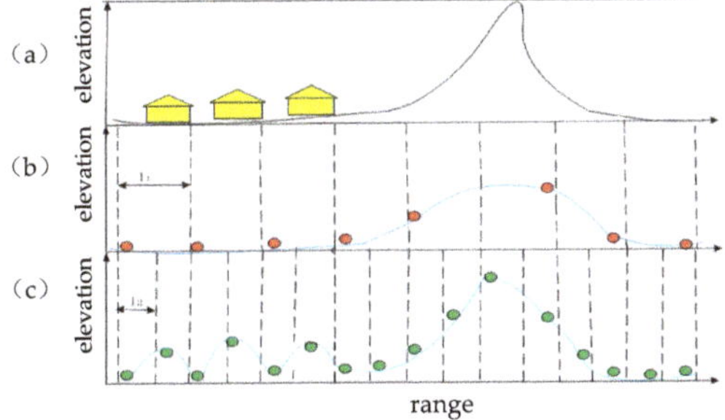

Figure 3. Ground points selection for a steep terrain area with buildings. In part (**a**), buildings and terrain are shown in different colors. In parts (**b**) and (**c**), dashed lines define the grid cells for ground points selection; the red and blue circles represent the lowest points, and the blue lines represent the initial terrain constructed with the lowest points.

2.2.2. Hierarchical Adaptive Surface Fitting

In order to solve the above problem, this paper proposed a hierarchical adaptive surface fitting method. Inspired by Zhang [9], the performance of the algorithm is improved by the following process.

(1) Hierarchical surface fitting: In the first iteration, the DSM data is first divided evenly by relatively large-sized grids, and then the minimum elevation points in each grid that are not the unreliable DSM areas are used as candidate ground points. The candidate ground points are compared with the surface obtained by fitting the candidate ground points in the 3×3 neighborhood grids. If the difference between the elevation of the candidate ground point and the fitted surface is greater than the threshold, the candidate point will be marked as non-ground points. Due to the large mesh size in the first iteration, it cannot represent the true topographic relief well, and the threshold should be set relatively loosely, filtering out buildings with large elevation values. In order to further locate potential non-ground points, we continuously reduce the size of the mesh and repeated the above steps until the mesh size is less than the preset minimum. Figure 4 shows a schematic diagram of the hierarchical surface fitting process.

(2) Determination of adaptive threshold: As mentioned above, considering the influence of grid size and elevation variance, this paper proposed a method for adaptively determining the threshold, which is shown in the Equation (20). The basic idea is that smaller grid size and variance of elevation difference usually correspond to a more reliable fitting result, which means that the threshold should be relatively strict. Conversely, with the increase of grid size and variance, its ability to represent real terrain is weakened, indicating that the fitted terrain has large deviations and the threshold should be relatively loose.

$$T = \mu_1 \times l + \mu_2 \times \sigma^2 \tag{20}$$

where l represents the grid size; and σ^2 represents the variance of elevation difference. μ_1 and μ_2 represent the weights of the grid size and variance of elevation difference, respectively.

(3) Interpolation with Coherence-Coefficient-Based IDW: After the ground points have been acquired by hierarchical surface fitting, the next step is to perform the interpolation with discrete ground points. In this study, the IDW algorithm was selected to interpolate the ground DEM, and it determines the weighting coefficient of ground points based on the distance between the known ground point and the interpolation point. This algorithm searches for ground points within the initial area, and if the number of ground points meets the set threshold, the search is stopped and then the weight of the searched ground points is calculated and interpolation is performed; otherwise the search radius is increased and the search is continued until the condition is satisfied. Figure 5 shows the algorithm execution diagram. When calculating the elevation of the red box, which is the point to be interpolated, search for ground points around it. If the number of black boxes representing the ground points reaches the set threshold, the distance between each ground point and the point to be interpolated is calculated, and then the weight ω_{i-IDW} is obtained by Equation (21).

$$\omega_{i-IDW} = \frac{\frac{1}{d_i^2}}{\sum\limits_{n=1}^{N} \frac{1}{d_n^2}} \tag{21}$$

where d_i is the distance between the ground point i and the point to be interpolated; and N is the number of points participating in the calculation. Finally, the product of the weight and the elevation of ground point is summed to obtain the elevation of the point to be interpolated.

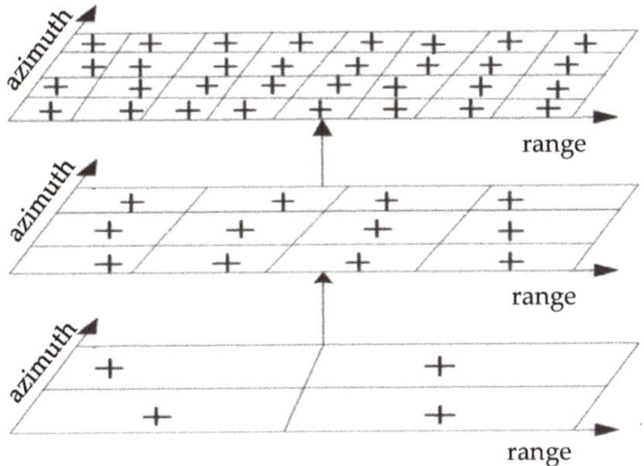

Figure 4. Hierarchical surface fitting with decreasing grid.

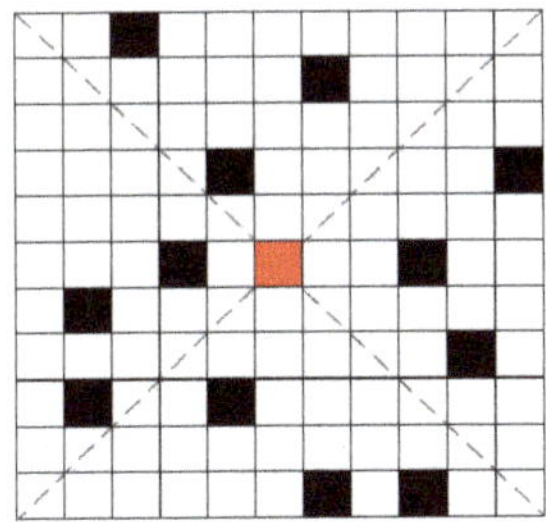

Figure 5. Inverse distance weighted (IDW) algorithm. Black boxes represent ground points, and red boxes represent points to be interpolated.

Considering the influence of the coherence coefficient, we combine the coherence coefficient and the inverse distance to improve the determination of the weight. The weight ω_i of the ground point i is expressed as follows:

$$\omega_i = \frac{\frac{q_i}{d_i^2}}{\sum_{n=1}^{N} \frac{q_n}{d_n^2}} \tag{22}$$

where q_i is the coherence coefficient of ground point i. The elevation of the point to be interpolated is estimated with the weighted sum:

$$h = \sum_{i=1}^{N} \omega_i h_i \tag{23}$$

where h represents the elevation of the point to be interpolated; and h_i represents the elevation of the ground point i.

As mentioned above, a DEM reconstruction method based on unreliable DSM area segmentation and hierarchical adaptive surface fitting was proposed in this method. As shown in Figure 6, in this method, an InSAR amplitude image is segmented initially, and the InSAR coherence coefficient and residue of interferometric phase are plugged into the neighborhood energy of the MRF model. Then we construct the likelihood energy and find the class labels that minimize the sum of the likelihood energy and the neighborhood energy as the segmentation result of the unreliable DSM areas. Next, the DSM is divided by a uniform grid and the minimum points of each neighborhood grids, which do not belong to the unreliable DSM area such as building layover and shadow, are used to fit a quadratic elevation surface. The difference between the true elevation and the fitted elevation is then calculated, and the points that are higher than the designed adaptive threshold are filtered out. Then the grid size changes step-by-step, iteratively filtering out the non-ground points. The surface fitting and filter is iterated in turn until the filter effect is not significantly different, or the filtering is stopped when the max number of iterations are met. Finally, the IDW interpolation combining the coherence coefficient is performed for completing the reconstruction of the DEM.

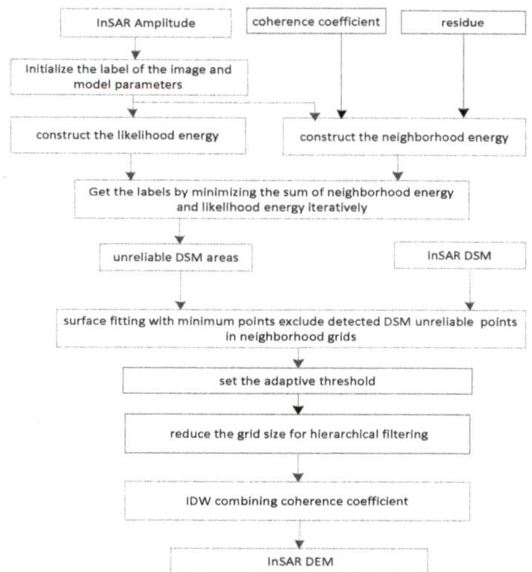

Figure 6. Flowchart of proposed digital elevation model (DEM) reconstruction method.

3. Results

3.1. Testing Data

In this paper, the InSAR data used for experimental verification was obtained by the Ku-band frequency modulation continuous wave (FMCW) InSAR system of the Institute of Electronics, Chinese Academy of Sciences in November 2015. The relative flight altitude of this experimental carrier aircraft was 1500 m, the incidence angle was 45 degrees, and the step size of DSM was 0.06 m. The experimental area was located in Jishan County, Yuncheng City, Shanxi Province, and belongws to hilly terrain where the buildings were densely distributed, and the terrain height was between 340 m and 420 m. The laser detection and ranging (LiDAR) bald earth DEM data from the same region was used as the reference DEM.

In this experimental data, three sites with buildings densely distributed were selected to evaluate the reconstruction results. Figure 7 shows the optical images of experimental areas.

3.2. The Segmentation Result of CMRF-Based Unreliable DSM Areas

According to Equations (6)–(8), the parameters of Fisher distribution were calculated in three areas, and the results are shown in Table 1. Thus, the likelihood energy could be obtained. Then the image was initially segmented, and its neighborhood energy could be calculated according to Equation (4). Finally, we found the class labels that minimize the sum of the likelihood energy and the neighborhood energy. This process needs to be iteratively calculated. An amplitude image of buildings was selected in the test sites for experiments, and the experimental results are shown in Figure 8.

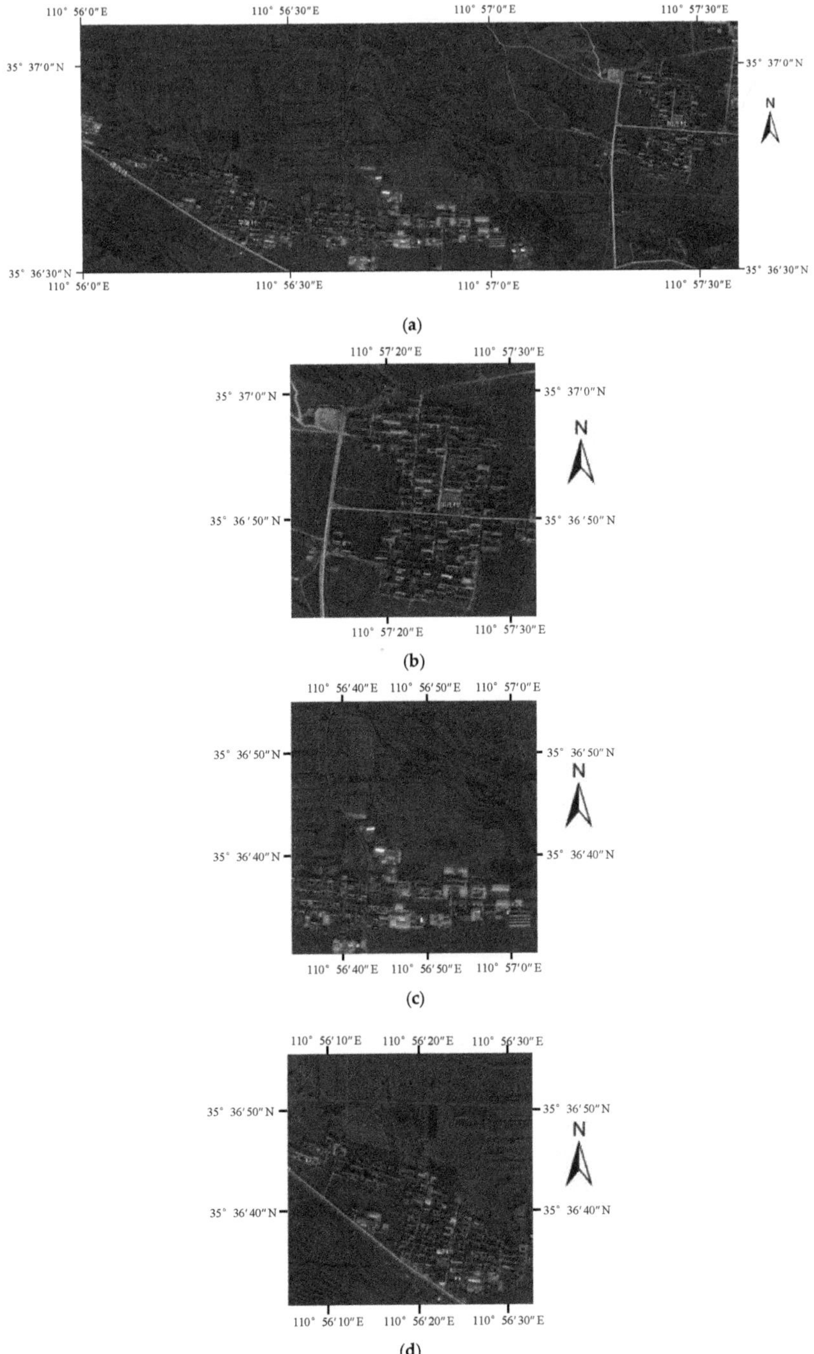

(a)

(b)

(c)

(d)

Figure 7. Optical image of experimental and evaluation areas. (**a**) Optical image of experimental areas. (**b**–**d**) Optical image of Site A to B.

Table 1. Estimations of Fisher distribution.

Class	M	L	μ
Layover	10.15	2.05	21.52
Shadow	12.31	5.21	3.72
Background	16.03	3.59	8.17

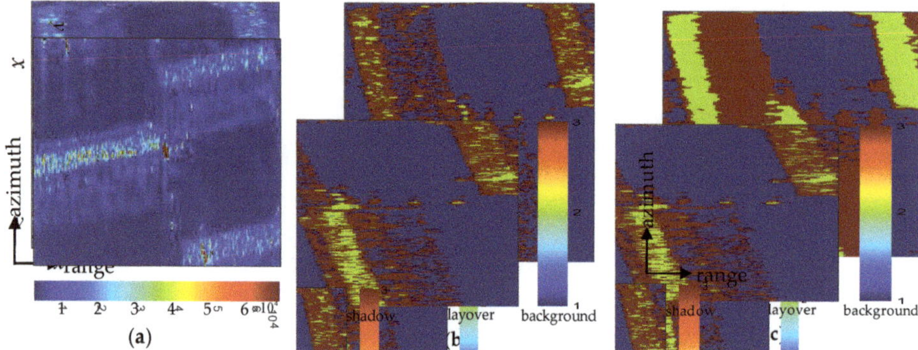

Figure 8. (**a**) The buildings in interferometric synthetic aperture radar (InSAR) amplitude image, and (**b**,**c**) the segmentation results based on traditional Markov random field (MRF) and coherent Markov random field (CMRF), respectively.

The building scene is shown in Figure 8a. Figure 8b,c are the segmentation results using traditional MRF and CMRF, respectively, where green represents layover and blue represents shadow, and red represents background areas.

It can be seen from Figure 8b that segmentation results generated by traditional MRF contain lots of holes and misclassifications. As shown in Figure 8c, the CMRF method detected most of the unreliable DSM areas and gave a better visual effect. The reason is that the introduction of coherence coefficient and residue can help the classifier make use of the interferometric information and better segment the InSAR amplitude image.

3.3. The DEM Reconstruction Result

In order to verify the effectiveness of hierarchical surface fitting, Figure 9 shows the first filtering result and the third filtering result. The ground and non-ground points of first filtering results are shown in Figure 9b,e,h, and the third filtering results are shown in Figure 9c,f,i, where blue represents the ground points and red indicates non-ground points. It can be seen from Figure 9b,e,h that some non-ground points are not detected in the first filtering, and some ground points are mistakenly classified as non-ground points, indicating that the grid size and the threshold of first filtering is too large. Therefore some non-ground points have not been filtered out, and the large grid size has lost the terrain detail information, causing some fluctuating ground points to be misidentified as non-ground points. The third fitting had detected more non-ground points than the first fitting, and the number of misjudging points was less, which means that buildings can be filtered out step-by-step while maintaining terrain features, indicating the effectiveness of hierarchical surface fitting.

To verify the effectiveness of the proposed method, experiments were performed using three test sites. In order to illustrate the necessity of unreliable DSM areas segmentation and hierarchical adaptive surface fitting, the methods compared in this paper were the original surface fitting, CMRF + surface fitting, and CMRF + hierarchical adaptive surface fitting. The experimental results are shown in Figure 10, and the comparison of altimetric profiles are shown in Figure 11.

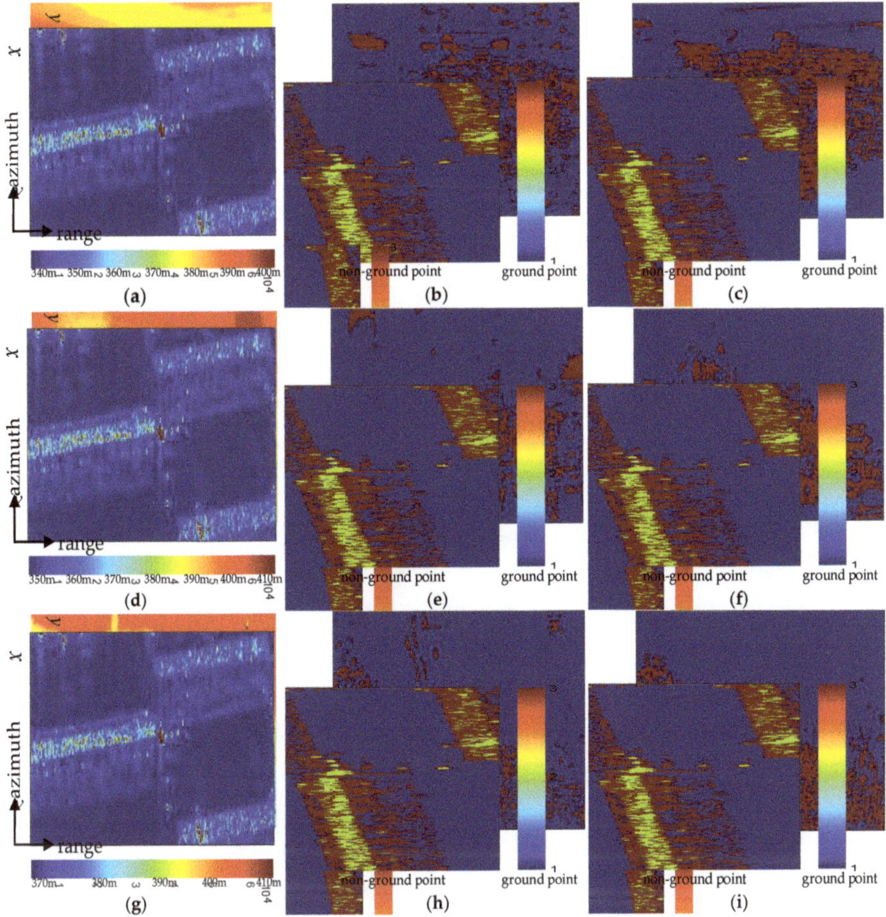

Figure 9. Hierarchical surface fitting results. (**a**), (**d**), and (**g**) are the InSAR digital surface model (DSM); (**b**), (**e**), and (**h**) are the first surface fitting results; and (**c**), (**f**), and (**i**) are the third surface fitting results.

Figure 10g–i show the DEM reconstruction results of the surface fitting method and the corresponding altimetric profiles are shown as a red line in Figure 11. It can be seen that the original surface fitting method had incorrect extreme values shown in the black rectangle, and the buildings were not completely filtered. This is because the unreliable DSM areas were not segmented in advance. Therefore some points of these areas were selected as ground points, and these points may be the extremely low points, or the higher points due to improper selection of grid size, thus causing the deviations in the interpolation result using ground points.

The results of the CMRF + surface fitting method are shown in Figure 10j–l, and the corresponding altimetric profiles are shown as a yellow line in Figure 11. Since the unreliable DSM areas were segmented first, and the lowest points in the grids were prevented from falling into these areas, the reconstruction results had fewer extreme values and the buildings were removed more thoroughly compared to surface fitting-based method, but there were still some buildings that had not been removed.

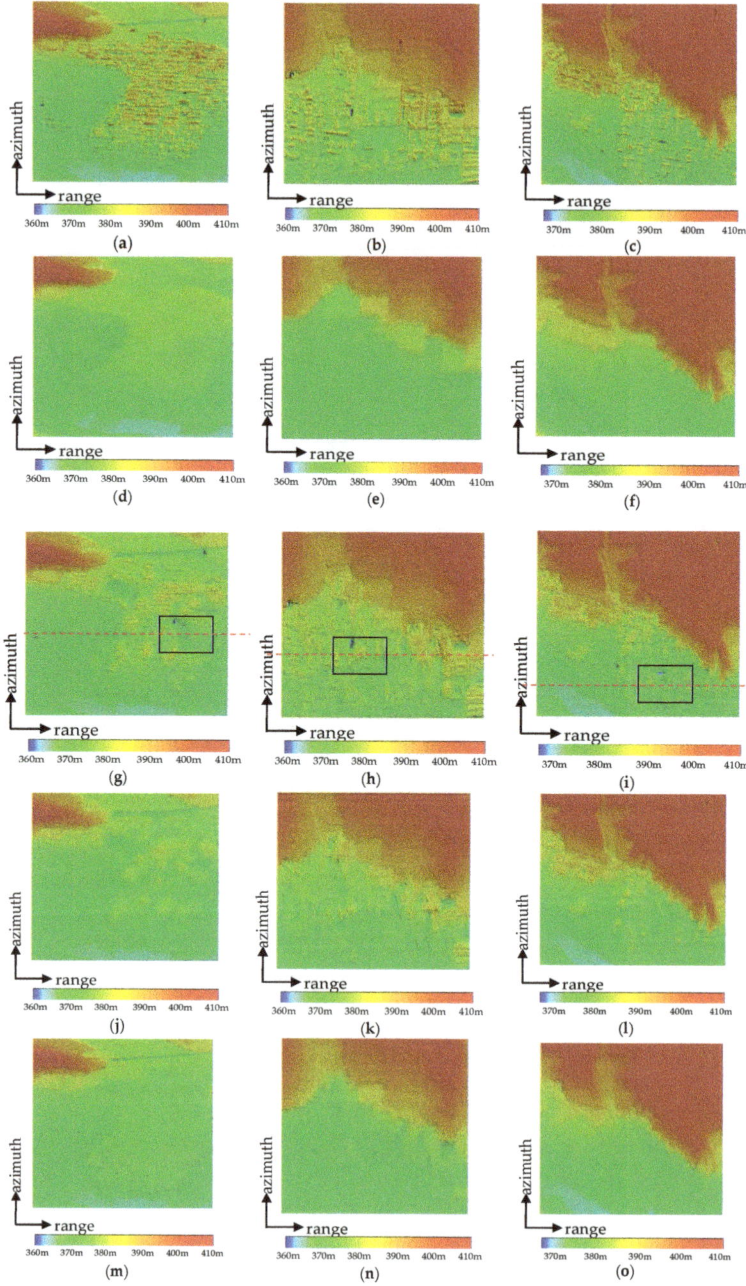

Figure 10. Visual comparison of before and after processing of the InSAR DSM, where (**a–c**) are the original DSMs, (**d–f**) are the reference DEMs obtained by laser detection and ranging (LiDAR), (**g–i**) are the reconstructed DEMs based on surface fitting, (**j–l**) are the reconstructed DEMs based on coherent Markov random field (CMRF)+surface fitting, and (**m–o**) are the reconstructed DEMs based on proposed method.

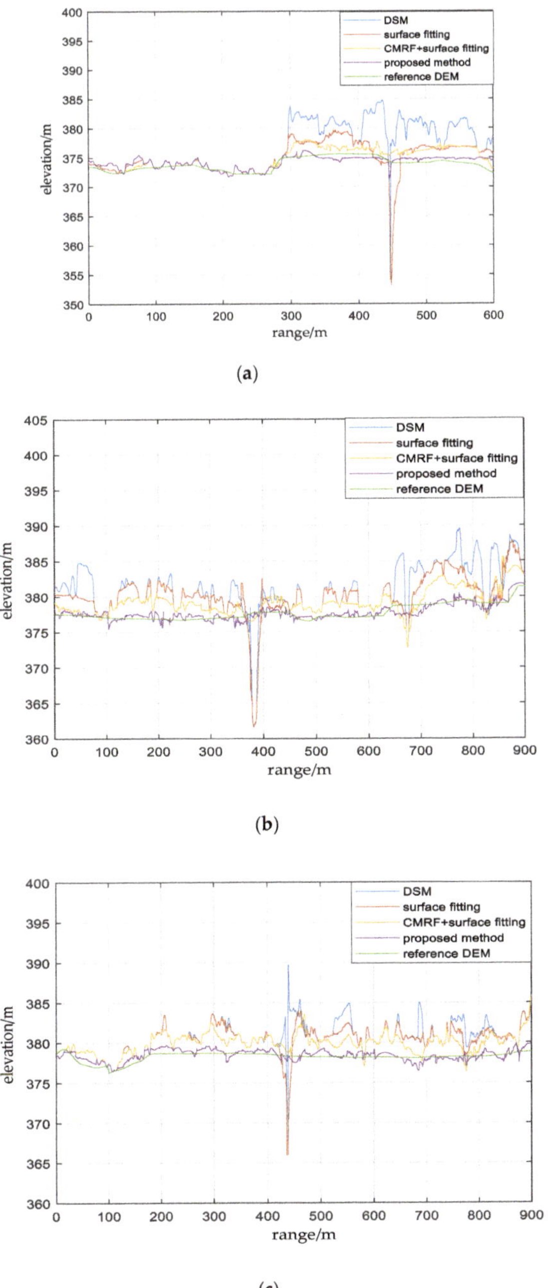

(a)

(b)

(c)

Figure 11. Altimetric profile comparison between the DSM, the reconstructed DEM based on different methods, and the reference DEM, where (**a**–**c**) are the altimetric profiles of experimental results corresponding to Site A–C. The profile position is at the red dashed line in Figure 10g–i.

The results of the proposed method in this paper are shown in Figure 10m–o, and the corresponding altimetric profiles are shown as a purple line in Figure 11. It was observed that the buildings had been completely filtered out and details of the undulating terrain had been retained. The reason is that the proposed method can gradually filter out the buildings and retain most of the ground points by keeping the grid size gradually smaller and setting the adaptive threshold, achieving the fine DEM reconstruction. Comparing the reconstruction results with the reference DEM of LiDAR in Figure 10d–f, the results of the proposed method are more accurate than other approaches. It confirms that the proposed method improves the performance of an InSAR DEM reconstruction.

3.4. Quantitative Evaluation

In order to quantify the performance of the proposed method, the three test sites were evaluated for accuracy, and the experimental methods including surface fitting, CMRF + surface fitting, and the methods proposed in this paper were used for evaluation. The reconstructed elevation was compared with the reference elevation to obtain the absolute elevation difference of each point, and finally the statistical values were calculated as the quantitative evaluation metrics of the DEM reconstruction result, such as the maximum difference (Max), the minimum difference (Min), and the root mean square error (RMSE) of the difference which can be expressed by Equation (24):

$$RMSE = \sqrt{\frac{\sum\limits_{i=1}^{n} (H_i - Z_i)^2}{n}} \tag{24}$$

where H_i is the elevation of reconstructed DEM for pixel i; Z_i is the corresponding elevation of reference DEM; and n is the number of the pixels involved in the calculation. The comparison results are shown in Table 2.

Table 2. Accuracy evaluation result.

Method	Min/m			Max/m			Root Mean Square Error (RMSE)/m		
	Site A	Site B	Site C	Site A	Site B	Site C	Site A	Site B	Site C
surface fitting	0.98	1.23	1.64	19.42	15.58	12.6	4.87	5.04	3.98
Coherent Markov Random Field (CMRF)++surface	0.81	1.01	1.33	3.12	3.79	5.18	2.32	2.76	2.84
the proposed	0.62	0.87	0.67	2.08	1.3	1.03	1.09	0.95	0.97

For the surface fitting algorithm, as shown in Figure 11, the maximum difference between the red line and the green line was the evaluation index Max, and the maximal Max in the three test sites was up to 19.42 m and the maximum RMSE was 5.04 m. This is because the unreliable DSM areas were not filtered in advance, which lead to the wrong selection of the extreme points as the ground points, as shown in the lowest point of the red line in Figure 11 and the gap between the reconstruction result and the real result is relatively large. In the CMRF + surface fitting, as shown by the yellow line in Figure 11, the Max was significantly reduced, showing the necessity of CMRF segmentation, while root mean square error (RMSE) was further reduced, and the reconstruction performance was improved. Compared with the above two methods, as shown in the comparison between the purple line and the green line in Figure 11, the CMRF + hierarchical adaptive surface fitting method proposed in this paper has obvious advantages in both indicators. The RMSEs of each test site were about 1 m and the Maxs were between 1 m and 2 m. The performance was significantly improved, which confirms the effectiveness of the proposed method.

4. Conclusions

In this paper, we proposed a new InSAR DEM reconstruction method in order to accurately extract a DEM from DSM generated by an InSAR system. The unreliable DSM areas were segmented in advance at the selection of ground point. Experiments show that the improved CMRF segmentation method was more accurate than the MRF method. Then, the hierarchical adaptive surface fitting can be used to mark ground points and non-ground points step-by-step, making the reconstruction result more accurate. The comparison results proved the superiority of the proposed algorithm qualitatively and quantitatively. However, there is still room for improvement. On the one hand, the hierarchical adaptive surface fitting can consider more interferometric phase information. On the other hand, the acceleration of the interpolation calculation may need further research.

Author Contributions: Q.Q. designed and performed the experiments; B.W. supervised the research and contributed to the article's organization; X.H. contributed to the interpretation of data; Q.Q. drafted the manuscript, which was revised by M.X. All authors have read and agreed to the published version of the manuscript.

Funding: This research is supported by the Development Department Preresearch Fund (Grant No.61404130308) and National Natural Science Foundation of China (Grant No.41971329).

Conflicts of Interest: The authors declare no conflicts of interest.

References

1. Perko, R.; Raggam, H.; Gutjahr, K.H.; Schardt, M. Advanced DTM generation from very high resolution satellite stereo images. In Proceedings of the PIA15+HRIGI15—Joint ISPRS Conference 2015, Munich, Germany, 25–27 March 2015; pp. 165–172.
2. Debella-Gilo, M. Bare-earth extraction and DTM generation from photogrammetric point clouds including the use of an existing lower-resolution DTM. *Int. J. Remote Sens.* **2015**, *37*, 3104–3124. [CrossRef]
3. Zhang, K.; Chen, S.C. A progressive morphological filter for removing nonground measurements from airborne lidar data. *IEEE Trans. Geosci. Remote Sens.* **2003**, *41*, 872–882. [CrossRef]
4. Zhao, X.; Guo, Q.; Su, Y.; Xue, B. Improved progressive TIN densification filtering algorithm for airborneLiDAR data in forested areas. *ISPRS J. Photogramm. Remote Sens.* **2016**, *117*, 79–91. [CrossRef]
5. Cai, S.; Zhang, W.; Liang, X.; Wan, P.; Qi, J.; Yu, S.; Yan, G.; Shao, J. Filtering Airborne LiDAR Data Through Complementary Cloth Simulation and Progressive TIN Densification Filters. *Remote Sens.* **2019**, *11*, 1037. [CrossRef]
6. Chen, Z.; Gao, B.; Devereux, B. State-of-the-Art: DTM Generation Using Airborne LIDAR Data. *Sensors* **2017**, *17*, 150. [CrossRef] [PubMed]
7. Wang, Y.; Mercer, B.; Tao, V.C.; Sharma, J.; Crawford, S. Automatic generation of bald earth digital elevation models from digital surface models created using airborne IFSAR. In Proceedings of the 2001 ASPRS Annual Conference, St. Louis, MO, USA, 23–27 April 2001; Available online: http://drmattnolan.org/kuparuk/ kupdem/library/asprs2001_intermap_e.pdf (accessed on 3 March 2020).
8. Jiang, L.; Xiang, M. Derivation of bald earth digital elevation models with X band airborne InSAR. In Proceedings of the 2009 2nd Asian-Pacific Conference on Synthetic Aperture Radar, Xi'an, Shanxi, China, 26–30 October 2009.
9. Zhang, Y.; Tao, C.V.; Mercer, J.B. An initial study on automatic reconstruction of ground DEMs from airborne IfSAR DSMs. *Photogramm. Eng. Remote Sens.* **2004**, *70*, 427–438. [CrossRef]
10. Geiß, C.; Wurm, M.; Breunig, M.; Felbier, A.; Taubenböck, H. Normalization of TanDEM-X DSM data in urban environments with morphological filters. *IEEE Trans. Geosci. Remote Sens.* **2015**, *53*, 4348–4362. [CrossRef]
11. Sun, L.; Wu, Z.; Liu, J.; Xiao, L.; Wei, Z. Supervised Spectral–Spatial Hyperspectral Image Classification with Weighted Markov Random Fields. *IEEE Trans. Geosci. Remote Sens.* **2015**, *53*, 1490–1503. [CrossRef]
12. Boudaren, M.E.Y.; Lin, A.; Pieczynski, W. Unsupervised Segmentation of SAR Images Using Gaussian Mixture-Hidden Evidential Markov Fields. *IEEE Geosci. Remote Sens. Lett.* **2016**, *13*, 1865–1869. [CrossRef]
13. Duan, Y.; Liu, F.; Jiao, L. Sketching model and higher order neighborhood Markov random field-based SAR image segmentation. *IEEE Geosci. Remote Sens. Lett.* **2016**, *13*, 1686–1690. [CrossRef]
14. Nazarinezhad, J.; Dehghani, M. A contextual-based segmentation of compact PolSAR images using Markov Random Field (MRF) model. *Int. J. Remote Sens.* **2019**, *40*, 985–1010. [CrossRef]
15. Zhang, H.; Shi, W.Z.; Wang, Y.J.; Hao, M.; Miao, Z.L. Spatial-Attraction-Based Markov Random Field Approach for Classification of High Spatial Resolution Multispectral Imagery. *IEEE Geosci. Remote Sens. Lett.* **2014**, *11*, 489–493. [CrossRef]

16. Wang, F.; Wu, Y.; Zhang, Q.; Zhao, W.; Li, M.; Liao, G. Unsupervised SAR image segmentation using higher order neighborhood-based triplet Markov fields model. *IEEE Trans. Geosci. Remote Sens.* **2013**, *52*, 5193–5205. [CrossRef]

17. Solberg, A.H.S.; Taxt, T.; Jain, A.K. A Markov random field model for classification of multisource satellite imagery. *IEEE Trans. Geosci. Remote Sens.* **1996**, *34*, 100–113. [CrossRef]

18. Tison, C.; Nicolas, J.M.; Tupin, F.; Maitre, H. A new statistical model for Markovian classification of urban areas in high-resolution SAR images. *IEEE Trans. Geosci. Remote Sens.* **2004**, *42*, 2046–2057. [CrossRef]

19. Touzi, R.; Lopes, A.; Bruniquel, J.; Vachon, P.W. Coherence estimation for SAR imagery. *IEEE Trans. Geosci. Remote Sens.* **1999**, *37*, 135–149. [CrossRef]

20. Zebker, H.A.; Chen, K. Accurate Estimation of Correlation in InSAR Observations. *IEEE Geosci. Remote Sens. Lett.* **2005**, *2*, 124–127. [CrossRef]

21. Cha, M.; Phillips, R.D.; Wolfe, P.J.; Richmond, C.D. Two-Stage Change Detection for Synthetic Aperture Radar. *IEEE Trans. Geosci. Remote Sens.* **2015**, *53*, 6547–6560. [CrossRef]

22. Wahl, D.E.; Yocky, D.A.; Jakowatz, C.V.; Simonson, K.M. A New Maximum-Likelihood Change Estimator for Two-Pass SAR Coherent Change Detection. *IEEE Trans. Geosci. Remote Sens.* **2016**, *54*, 2460–2469. [CrossRef]

23. Biondi, F. A new maximum likelihood polarimetric interferometric synthetic aperture radar coherence change detection (ML-PolInSAR-CCD). *Int. J. Remote Sens.* **2019**, *40*, 1–21. [CrossRef]

24. Askne, J.; Hagberg, J.O. Potential of interferometric SAR for classification of land surfaces. In Proceedings of the IEEE International Geoscience and Remote Sensing Symposium, Tokyo, Japan, 18–21 August 1993.

25. Dai, Z.; Zha, X. An accurate phase unwrapping algorithm based on reliability sorting and residue mask. *IEEE Geosci. Remote Sens. Lett.* **2011**, *9*, 219–223. [CrossRef]

Article

Mining-Induced Time-Series Deformation Investigation Based on SBAS-InSAR Technique: A Case Study of Drilling Water Solution Rock Salt Mine

Xiangbin Liu [1,2], **Xuemin Xing** [1,2,*], **Debao Wen** [3], **Lifu Chen** [1,4], **Zhihui Yuan** [1,4], **Bin Liu** [1,2] and **Jianbo Tan** [1,2]

[1] Laboratory of Radar Remote Sensing Applications, Changsha University of Science and Technology, Changsha 410014, China; liuxb0219@foxmail.com (X.L.); lifu_chen@139.com (L.C.); yuanzhihui@csust.edu.cn (Z.Y.); binliu@csust.edu.cn (B.L.); tanjianbo@imde.ac.cn (J.T.)

[2] School of Traffic and Transportation Engineering, Changsha University of Science and Technology, Changsha 410014, China

[3] School of Geographical Sciences, Guangzhou University, Guangzhou 510006, China; wdbwhigg@gzhu.edu.cn

[4] School of Electrical and Information Engineering, Changsha University of Science and Technology, Changsha 410014, China

* Correspondence: xuemin.xing@csust.edu.cn

Received: 9 October 2019; Accepted: 11 December 2019; Published: 13 December 2019

Abstract: Compared to traditional coal mines, the mining-induced dynamic deformation of drilling solution mining activities may result in even more serious damage to surface buildings and infrastructures due to the different exploitation mode. Therefore, long-term dynamic monitoring and analysis of rock salt mines is extremely important for preventing potential geological damages. In this work, the small baseline subset Interferometric Synthetic Aperture Radar (SBAS-InSAR) technique with Sentinel−1A imagery is utilized to monitor the ground surface deformation of a rock salt mining area. The time-series analysis is carried out to obtain the spatial–temporal characteristics of land subsidence caused by drilling solution mining activities. A typical rock salt mine in Changde, China is selected as the test site. Twenty-four scenes of Sentinel−1A image data acquired from June 2015 to January 2017 are used to obtain the time-series subsidence of the test mine. The temporal–spatial evolution of the derived settlement funnels is revealed. The time-series deformation on typical feature points has been analyzed. Experimental results show that the obtained drilling solution mining-induced subsidence has a spatial characteristic of multiplied peaks along the transversal direction. Temporally, the large-scale surface settlement for the rock salt mine area begins to appear in September 2016, with a time lag of 8 months, and shows an obvious seasonal fluctuation. The maximum cumulative subsidence is detected up to 199 mm. These subsiding characteristics are consistent with the connected groove mining method used in drilling water solution mines. To evaluate the reliability of the results, the SBAS-derived results are compared with the field-leveling measurements. The estimated root mean square error (RMSE) of ±11 mm indicates a high consistency.

Keywords: SBAS-InSAR; deformation; rock salt mine; drilling solution mining; time series

1. Introduction

The reserves of mirabilite deposits of China had been proved to be accumulated up to 1117.20 billion tons until the end of 2017 [1]. An omnidirectional advanced drilling solution mining is the dominate exploitation method for most mirabilite mines [2]. The connected groove mining method based on an oil pad is applied for most of the drilling solution mining activities. Figure 1 shows the schematic diagram of the connected groove mining method with two salt wells. It can be seen that each salt well is built on an oil pad with a dissolution cavity in the mining layer. The single well based on an oil pad is used to build grooves in the early stages of drilling solution mining (see Figure 1a), which can promote the side dissolution, control the upper dissolution, and speed up the connection of well groups. As the process of dissolution, the cavities derived by adjacent salt wells can be connected and merged in the mining layer (see Figure 1b). After the process of connection dissolution between different cavities, the fresh water at 40 °C is injected through one of the wells (demonstrated as well 1 in Figure 1). Subsequently, the mirabilite layer can be dissolved, and under the water injection residual pressure, the generated brine can be cramped out from the other well (demonstrated as well 2 in Figure 1) [2]. Due to the long time of the dissolution process and the certain supporting effect of the high-pressure injected fresh water to the roof, time lag and suddenness are the obvious characteristics of the ground deformation related to drilling solution mining activities. Compared to tunnel mining with the unidirectional propulsion of conventional coal mines, the depth of drilling solution mining is generally much deeper, and the thickness of the rock salt layer is even thicker. With a serious influence imposed by the water on the mechanical properties for the salt roof, the subsidence related to drilling solution mining will be even more severe and destructive [3]. Due to the omnidirectionality and uncertainty of drilling solution mining, the mechanical properties of the cavity may become unreasonable, which may induce the overburden or even serious collapse on the cavity [4,5]. Once the roof of cavity reaches the bottom ground, a sinkhole will be generated at the surface [6,7], which may induce potential damage to the nearby infrastructures (i.e., houses, roads, bridges, canals) [8]. Furthermore, the sustained mining of rock salt mines can easily lead to mechanical changes to the underground rock and water system. This may even cause brine pumping and land salinizing [9], which shows serious potential for environmental pollution [10]. Therefore, the long-term spatial–temporal deformation monitoring of rock salt mines is of practical significance to the prevention of mining-induced safety problems and the assurance of mining environmental protection.

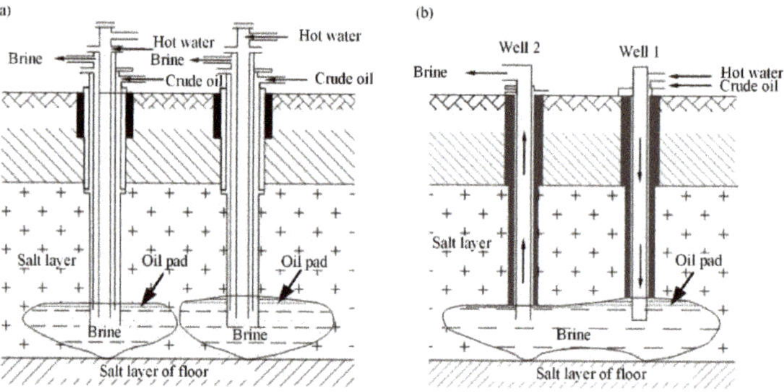

Figure 1. (a) Single well based on an oil pad before the process of dissolution connection, (b) The final connected groove based on an oil pad after the process of dissolution connection process (data from [2]).

Traditional geodetic monitoring methods, such as total station/prism, photogrammetry, leveling, and Global Navigation Satellite System (GNSS), have been widely applied in mining-induced deformation monitoring. Those methods are proven to be of high accuracy. However, due to the poor spatial–temporal resolution, those methods still have deficiencies in observing the overall ground surface subsidence of the mining area [11]. In addition, expensive labor force, and frequently in situ observation are necessary for the monitoring of mining area, which will consume an enormous amount of financial resources and inevitably aggravate the potential safety problems.

Interferometric Synthetic Aperture Radar (InSAR) offers a novel earth observation approach. It can provide wide spatial coverage, high imaging resolution, and non-intrusive surveying. Differential

InSAR (D-InSAR), as an extension of InSAR in terms of monitoring ground deformation, is mainly applied in ground deformation monitoring along the line of sight (LOS) of a radar satellite. The new monitoring approach is an important complement to the traditional geodetic surveying methods [12]. D-InSAR is widely applied to detect and monitor earthquake deformation [13], glacial shift [14], volcanic activity [15], and landslides [16], as well as man-made activities such as mining subsidence [17] and urban settlement caused by groundwater overdraft [18]. However, the unavoidable influences of the temporal and spatial decorrelation and atmospheric delay have brought restrictions on its application, especially on mining areas vulnerable to decorrelation. Small baseline subset InSAR (SBAS-InSAR) is an advanced InSAR technology proposed by Berardino [19], which utilizes the least squares (LS) and singular value decomposition (SVD) methods to obtain the deformation rates at the high coherence points based on the multi-scene of differential interferometric images. Although a large amount of successful cases using SBAS technology in coal mine areas have been published [20,21], the application in rock salt drilling solution mining has been rarely mentioned in previous studies.

The Sentinel−1 satellite, equipped with a C-band SAR sensor, is an Earth observation satellite launched by the European Space Agency's Copernicus Program in 2014. Sentinel−1A SAR data have the advantages of large global coverage and a short revisit period (12 days), which can be downloaded free of charge on the website (https://scihub.copernicus.eu/) [22]. Sentinel−1A SAR data have been widely and successfully applied in the monitoring of mining-induced subsidence [23,24]. In this work, a typical rock salt mine in Changde, China was selected as the test site. In order to verify the feasibility and reliability of the SBAS technique and Sentinel−1A imagery for the deformation monitoring of rock salt mines, we use SBAS and Sentinel−1A images to perform a case study. The time-series characteristics of the subsidence sequences related to drilling solution mining activities are revealed.

2. Methodology

Suppose $N + 1$ SAR images covering the same area are acquired in repeat orbits at different dates $(T_0, T_1, \ldots, T_N)$. Then, M interferometric pairs can be produced according to certain spatial–temporal baseline thresholds, where M satisfies the inequality $(N+1)/2 \leq M \leq N(N+1)/2$. Each of these interferometric pairs is generated by the two-orbit D-InSAR processing. In the processing, all images are registered and resampled to the same image first. Then, an external digital elevation model (DEM) is used to remove the topographic phase, and consequently, phase unwrapping is carried out for each interferometric pair. The unwrapped phase at pixel (x, r) in the i-th $(i = 1, 2, \ldots, N)$ interferogram can be written as [19].

$$\delta\varphi_i = \phi_B(x,r) - \phi_A(x,r) \approx \frac{4\pi}{\lambda}\Delta d + \frac{4\pi B_\perp}{\lambda r \sin\theta}\Delta h(x,r) + \Delta\varphi_{i,res(x,r)} \tag{1}$$

where λ, θ, and $B_\perp$ represent the SAR coordinate of the high coherence point, the radar wavelength, the radar incidence angle, and the perpendicular baseline of the two SAR acquisitions, respectively; $\Delta d = d(T_A, x, r) - d(T_B, x, r)$ is the time-series displacements along the LOS direction at date T_A and T_B respectively, with respect to the start time (i.e., $d(T_0, x, r) \equiv 0$); $\Delta\varphi_{i,res(x,r)}$ is the residual phase, including the phase noise, the atmospheric delay, and the high-pass (HP) deformation component; $\Delta h(x,r)$ represents the topographic error of the external DEM.

The deformation component Δd is of the main interest. The functional relationship between Δd and the deformation parameters can be written as [25]

$$\Delta d = \sum_{k=l+1}^{s} v_k(T_k - T_{k-1}) \tag{2}$$

where l and s define the index of the master image at time T_A and slave image at time T_B, respectively for the i-th interferometric pair. v_k defines the linear velocity for each temporal unit, which varies across different temporal units. According to Equations (1) and (2), we need to estimate N number of vs. as well as the unknown DEM error parameter Δh (in total $N + 1$ unknown parameters) in M generated functions. To solve the singular mathematical problem, the SVD algorithm and LS method are suggested here [26,27]. After the unknown parameter being estimated (v and Δh), integration over each temporal period is carried out to obtain the low-pass (LP) deformation component on all the high coherence points. Considering that the atmospheric delay phase component is a temporally random high frequency signal, it is spatially related to the low frequency signal. In contrast, the nonlinear deformation phase is a low-frequency signal both spatially and temporally [28,29]. Accordingly, in order to pick up the HP deformation component from the residual phase, a temporally high-pass filtering and a spatially low-pass filtering need to be applied. The final deformation on each coherent point will be obtained through summarizing both the LP deformation and HP deformation. The experimental flow of SBAS processing is shown in Figure 2.

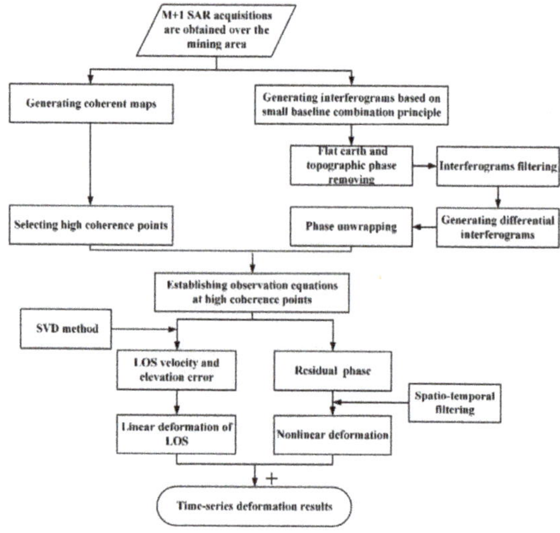

Figure 2. Experimental flow of small baseline subset (SBAS) algorithm.

In this work, the LOS deformation is converted to the vertical component in order to compare with the in situ leveling measurements (the horizontal displacement is omitted in the experiment, which will be discussed in Section 3.1), according to the following function [30]:

$$Def_{vertical} = Def_{LOS} / \cos \theta \tag{3}$$

where Def_{LOS} represents the LOS deformation and $Def_{vertical}$ represents the vertical component.

3. Experiments

3.1. Study Area and Geological Background

In this work, a typical water-soluble rock salt mine in Changde, Hunan Province is selected as our test site. Figure 3 shows the location of the test area. Figure 3a,b shows the corresponding study areas on a map of China. Figure 3c shows the optical images of the mine area. It can be seen from Figure 3c that the rock salt mine area is located in the Liyang Plain of Hunan Province, with a total area of 5.7 km^2. The red rectangle represents the spatial coverage of ascending Sentinel−1A images, whereas the white rectangle is the selected subset of interest in this work. Due to the location in the middle of the plain, the rock salt mine has a typical flat terrain characteristic, surrounded with dense pounds, natural water systems, and artificial channels. It is also located close to a wide area of surface rice fields. Since 2002, the long-term mining activities in this area have caused great damage to the surrounding environment and underground geological stratum (see Figure 4).

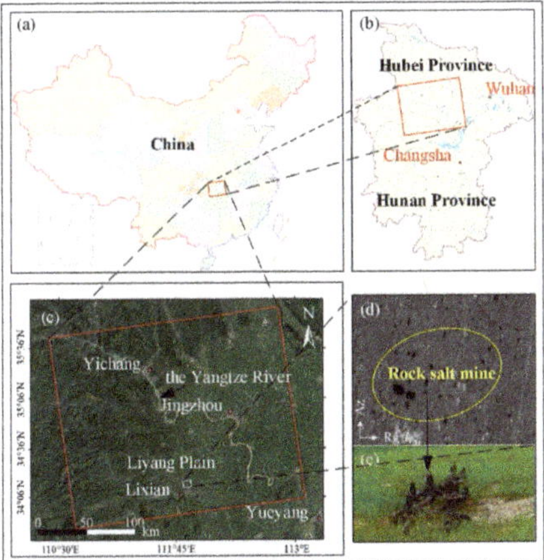

Figure 3. Study area overview. (**a,b**) Regional scale in China of the test mine. (**c**) The location of the study area. (**d**) Corresponding amplitude image of the area with the mining region of interest outlined in the white rectangle. (**e**) In situ picture of the drilling solution wellhead in (**d**).

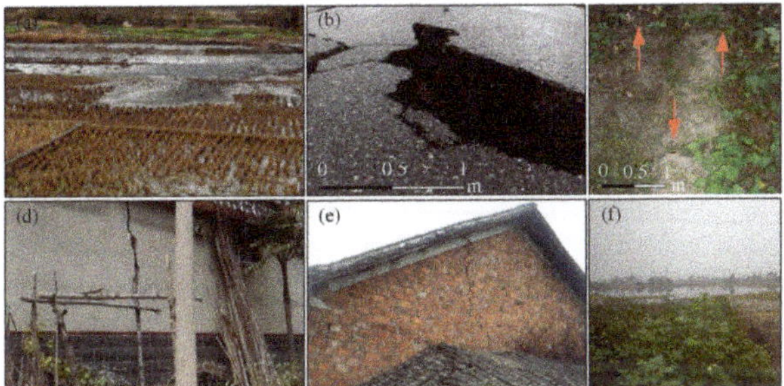

Figure 4. In situ pictures of ground ruptures in the rock salt mine. (**a**) The underground brine flowing into the cultivated land, inducing land salinization. (**b,c**) Accumulated deformation induced road surface cracks. (**d,e**) Cracks on resident houses (**f**) Gradually formed stagnant water area, with a settlement even deeper than 1.5 m.

Figure 5 illustrates the geological distribution of the test rock salt mine [31]. The strata encountered in drilling solution mining mainly includes the Quaternary System and Lower Tertiary System. The distribution of the strata from top to bottom is as follows: Holocene, Upper Pleistocene, Middle Pleistocene, and Lower Pleistocene. The total thickness of the strata is 77.95–138.55 m. The Lower Tertiary System consists of Eocene and Paleocene, with a total thickness of 562.96 m. The lithology of the Eocene Formation mainly includes mudstone, dolomite, siltstone, gypsum, glauberite, and thenardite. The extracted thenardite (Na_2SO_4, 62.76%–78.8%) and mirabilite of this mine are present in the salt-bearing section of the Xingouzui Formation of the Tertiary System (E_2x^3), with a cumulative thickness of 8.21–14.23 m. The fault structure of the mining area is mainly F10 fault, located in the south, with a stratum fault distance of 30–70 m and a 3–16 cm fracture zone. It is filled with fibrous gypsum cementation. After the south plate rises, F10 destroys the continuity of the seam in the south wing of the syncline and causes the minerals to dissolve. Therefore, F10 constitutes the natural boundary in the south of the mining area. A concealed fault F12, with a dip angle of 75°, belongs to the SEE (South East East) normal fault. The roof, floor, and interlayer of the ore bed, containing a small amount of anhydrite dolomitic mudstone, are mainly muddy dolomitic glauberite, which belongs to the weak layered rock mass. The rock, with poor stability, is easy to soften and collapse.

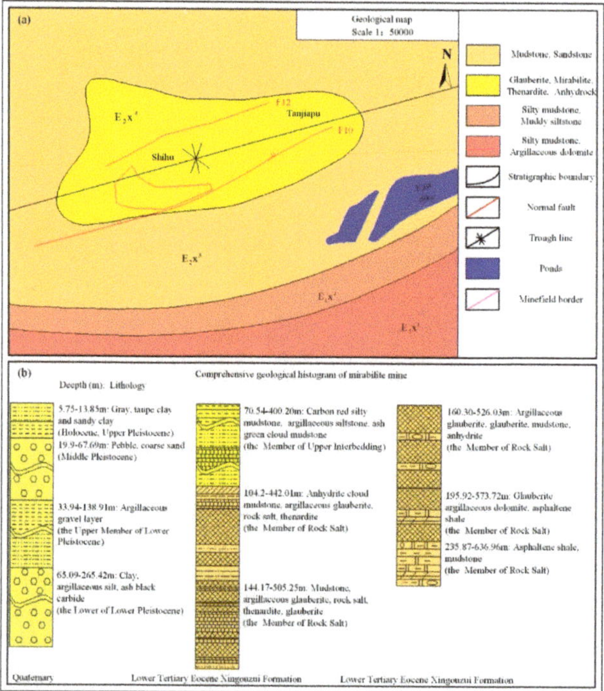

Figure 5. Geological map of the rock salt mine. (**a**) The plane geological distribution of the test rock salt mine (E_2X represents the Lower Tertiary Eocene Xingouzui Formation). (**b**) Vertical distribution of the comprehensive geological formations (data from [31]).

In order to prevent mining accidents and natural environment pollution caused by mining-induced roof collapse, hot water combined with the connected groove mining method based on an oil pad is utilized to extract thenardite in this salt mine. During the period from June 2016 to January 2017, the groove connection was completed in the test mine. Different cavities below different wells were mutually dissolved and connected (see Figure 1). The dissolution and transport channels of minerals were formed during this period. Since then, the stage of the upper dissolution started. The shape of the cavity started to change along the upper direction, which developed along the vertical deeper direction. During this stage, the side dissolution rate was significantly reduced, whereas the upper dissolution rate became twice as fast as the side dissolution, which performed as a significant ground subsidence along the vertical direction [32]. Therefore, the horizontal displacement is omitted in our experiment.

3.2. SAR Acquisitions and Data Processing

A total of 24 repeat-pass ascending Sentinel–1A images of the test rock salt mine area were collected. These acquisitions covered the period from 15 June 2015 to 30 December 2016. The parameters of these images are listed in Table 1. SARScape 5.2 and ENVI 5.3 were used to generate the unwrapped small baseline interferometric pairs. The subsequent procedures, including high coherence points identification and the LP–HP deformation component estimation, were carried out through MATLAB.

Table 1. List of the interferometric pairs and their parameters (Ascending).

	Drilling Water-Soluble Rock Salt Mine (Orbit No. 11)						
No.	Acquisition Date (yyyy/mm/dd)	Vertical Baseline (m)	Temporal Baseline (days)	No.	Acquisition Date (yyyy/mm/dd)	Vertical Baseline (m)	Temporal Baseline (days)
0	2015/06/15	26.89	−216	12	2016/05/16	−15.15	120
1	2015/07/09	88.17	−192	13	2016/07/03	−19.95	168
2	2015/08/02	1.87	−168	14	2016/08/20	22.27	216
3	2015/08/26	−36.04	−144	15	2016/09/25	−55.61	252
4	2015/09/19	−33.99	−120	16	2016/10/07	−21.78	264
5	2015/10/13	43.37	−96	17	2016/10/19	57.01	276
6	2015/12/24	121.57	−24	18	2016/10/31	54.31	288
7	**2016/01/17**	**0**	**0**	19	2016/11/12	42.52	300
8	2016/02/10	95.00	24	20	2016/11/24	0.96	312
9	2016/03/05	−23.73	48	21	2016/12/18	−20.47	336
10	2016/03/29	−48.43	72	22	2016/12/30	20.89	348
11	2016/04/22	39.75	96	23	2017/01/11	71.62	360

The thresholds for the spatial–temporal baseline of the interferometric combination were empirically set to 150 m and 360 days, respectively. In the two-pass D-InSAR processing, all the rest of the images were registered and resampled to the super master image. In order to remove the topographic phase, a 1-arc-second Shuttle Radar Topography Mission Digital Elevation Model (SRTM DEM, ~30 m spacing) provided by NASA was utilized. In addition, a Gaussian filter [33] was selected to suppress the phase noise. After the flat Earth phase removal and phase filtering processing, a polynomial fitting method was used to remove the orbital error; then, a relatively stable reference point was selected (see Figure 6a) and minimum cost flow [34] method was utilized to unwrap the wrapped interferometric deformation phases. Finally, a total of 58 unwrapped differential interferograms were generated. During the processing, based on a coherence threshold of 0.45 and amplitude dispersion threshold of 0.35, a total of 5559 high coherence candidates for the test mine were selected.

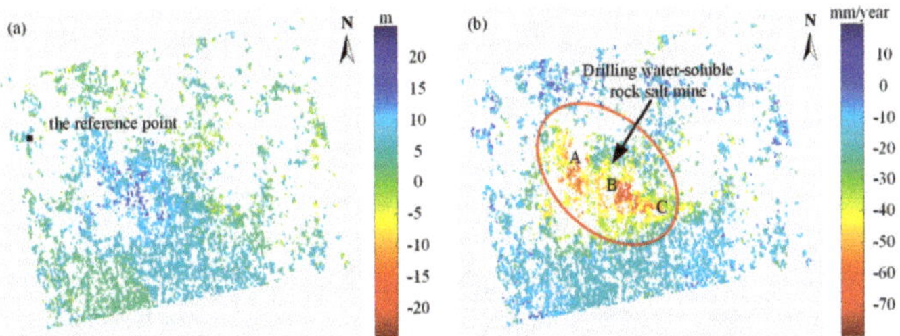

Figure 6. (**a**) Digital elevation models (DEM) errors (the solid black square represents the reference point for phase unwrapping). (**b**) Deformation rates.

4. Results and Discussion

4.1. Overall Deformation Results

The DEM error and the linear deformation rate of the high coherence points were obtained, which are shown in Figure 6. Through our quantitative analysis and statistics, the total number of the high coherence points, with the absolute DEM error values within the range of 0 m to 10 m, account for 86%, whereas the number within the range of 10 m to 20 m only account for 14% (blue to green color, as shown in Figure 6a). It is in good agreement with the accuracy of SRTM DEM data with 30-m resolution [35]. From Figure 6b, we can see an obvious subsidence bowl in the central region where the rock salt mine is located, with the color gradually changing from blue to red inwardly. According to our analysis, the subsidence rate of most coherence points is distributed within the range of 50 mm/year to 75 mm/year, with the maximum value of up to 109 mm/year.

According to [36], the accuracy of the retrieved topographic residuals is related to the thresholds of the perpendicular baseline and the quality of the differential interferogram. The accuracy of the DEM error will degrade the accuracy of the InSAR time-series deformation. Due to this, we controlled the spatial baseline threshold and selected the interferometric pairs strictly (the threshold was set as 150 m in our experiment). The external SRTM DEM had 30 m resolution, which is relatively high. Moreover, in order to show the correlation of the DEM error and the deformation, we conducted a simulated experiment. According to the phase contribution of DEM error, $\delta\varphi = \frac{4\pi}{\lambda}\frac{B_\perp}{R\sin\theta}\cdot\Delta h$, and the relationship between phase and the deformation velocity, $\delta\varphi = \frac{4\pi}{\lambda}v(T_B - T_A)$, the error of deformation velocity caused by a 20 m DEM error was only 5 mm/year. Consequently, compared to the large estimated subsidence (a maximum deformation velocity of 109 mm/year), the influence of DEM error on the deformation time series was ignored in our case study.

Figure 7 shows the overall time-series deformation of the rock salt mine. From the spatial characteristics of the color distribution, we can see that the obvious subsiding points were densely distributed in the center part of the images, where the rock salt mine was located (with a dark orange to red color in the subsidence bowl). The spatial distribution characteristics of the subsiding pixels in the mining area appeared as disperse zonal distribution in the northwest part and an overall sheet-like distribution in the central and southeast part. The reason for this phenomenon is that the drilling solution mining method based on connected well groups was utilized in the middle and southeast part, which induced a dissolution connection of different cavities underground; thus, the surface subsidence performed to be multi-subsiding bowls (see A, B, and C in Figure 6). Meanwhile, in the northwest of the area, the single well drilling solution method was adopted, which resulted in disperse zonal distribution characteristics.

As the temporal color variation shows in Figure 7, a temporal characteristic of seasonal fluctuation could be found (which will be analyzed quantitively in Section 4.2). For the period from 9 July 2015 to 10 February 2016, the subsidence velocity was relatively stable (which will be mentioned as the time lag in Section 4.2). From 5 March to 3 July 2016, a slow increase of subsiding occurred, while for the period from 3 July to 19 October 2016, a rapid subsiding dominated the deformation. The subsidence bowl started to appear on 5 March 2016. Since then, an obvious large subsiding velocity began to occur in the mining area. By 11 January 2017, the maximum subsidence in the bowl was accumulated to 199 mm.

By September 2016, large-scale subsidence began to occur. Subsidence bowls A, B, and C were gradually generated by rapid mining activities through multiple wells. Since B and C were connected by well groups, the caverns were interconnected underground. As the increasing of mineral exploitation, the volume of the caverns increased gradually, leading to more serious movements on the top edge of the chamber and closer distance between different caverns. Sequentially, funnels B and C would be merged into a large subsidence bowl.

Figure 7. Time-series deformation (with reference to 15 June 2015).

4.2. Discussions

As discussed in Section 4.1, the overall time-series deformational characteristics of the test rock salt mine follows spatially multi-distributed bowls and a temporal 8-month time lag, with a subsequent annual fluctuation. The reasons for the subsiding characteristics are supposed to be as follows:

(1) The process of the brine extraction was conducted by injecting solvent followed by rock salt dissolution, which takes a longer time than traditional coal mining activities; in addition, the depth of the drilling solution mining was deeper than that of common coal mines (the depth of wells in this study area was 200–500 m), which induced the lagging appearance of ground surface subsidence.

(2) The relationship between the solubility and the solvent temperature in Table 2 shows that the dissolution of mirabilite is significantly vulnerable to temperature [2,37]. The solubility under 30 °C is almost four times that of under 0–10 °C. This indicates that under the circumstance of high temperature in the warm season, the mineral dissolution was considerably rapid, inducing

a larger amount of brine extraction. On the contrary, for the cold season (the period of 24 December 2015 to 25 March 2016), the low temperature in winter suppressed the dissolution rate for the mirabilite.

(3)   The spatially multi-peak phenomenon was mainly due to the drilling solution mining method based on connected well groups and its comprehensively multi-direction advancing mode (which will be discussed in Section 4.2).

Table 2. Solubility of thenardite at different temperatures (g/100 g H_2O).

Mineral	Temperatures (°C)										
	0	**10**	**20**	**30**	**40**	**50**	**60**	**70**	**80**	**90**	**100**
Thenardite (Na_2SO_4)	5.0	9.0	19.4	40.8	48.8	46.7	45.3	44.1	43.7	42.9	42.5

In order to further reveal the characteristics of temporal deformation variation, five feature points (HCP1 to HCP5 shown in Figure 7) were selected for quantitive analysis. The extracted time-series deformation is illustrated in Figure 8a.

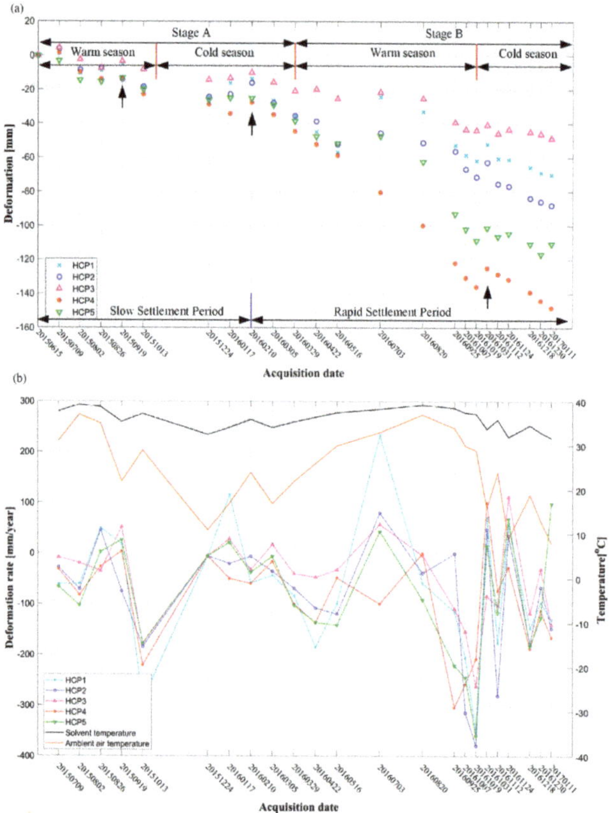

Figure 8. (**a**) Time-series deformation of feature points at drilling water-soluble rock salt mine (HCP1 to HCP5). (**b**) Correlation diagram of the deformation rate with the solvent temperature and the average air temperature.

From Figure 8a, we can see for the total period (15 June 2015 to 11 January 2017) that all the five feature points show similar temporal variations: a generally subsiding trend with an obvious seasonal fluctuation. HCP4 showed the most serious subsiding, with an accumulative subsidence of 148 mm until 11 January 2017, whereas HCP3 was relatively more stable, with the maximum deformation of 48 mm. For the cold season from 24 December 2015 to 10 February 2016 in stage A, and 19 October 2016 to 1 January 2017 in stage B, a relatively slow deformation trend occurred, with a small fluctuation of 11 mm and 12 mm, respectively. From 29 March 2016, a significant subsiding trend started. The seasonal fluctuation of the deformation in a rock salt mine was suggested to be mainly due to the dissolution rate of mirabilite and thenardite in water. The dissolution rate of thenardite was directly affected by the temperature of the solvent. In the production process of the rock salt mine, combined with the connected groove mining method based on an oil pad, hot water was suggested to be used as solvent to increase the dissolution. The hot water transported from the processing industry was pressurized by the injection pump, metered at the control station, and then directly injected into the well after distribution. According to our investigation, the temperature of the fresh water solvent injected into the well was about 40 °C. However, during the transportation from the processing industry and the injection process into the well through the injection pump, the solvent temperature was vulnerable to the air temperature. In summer, the temperature of fresh water could be well insulated. On the contrary, in winter, due to the decrease of external temperature, the temperature of fresh water was easy to decrease. Consequently, the high temperature in the warm season accelerated the dissolution of rock salt, which lead to the increase of subsidence. In contrast, the low temperature in cold seasons suppressed the process of water dissolution, inducing the slow or even uplifting trend of deformation. Three small uplifts for the five feature points can be found in the position pointed by the black arrows in Figure 8a, which showed good consistency with the measurements detected from Figure 7. The uplift phenomenon was mainly related to the aforementioned low temperature in cold seasons and the increase of rainfall (from 26 August to 19 September 2015, 19 October to 31 October 2016, and 17 January to 10 February 2016, as shown in Figure 9) [38].

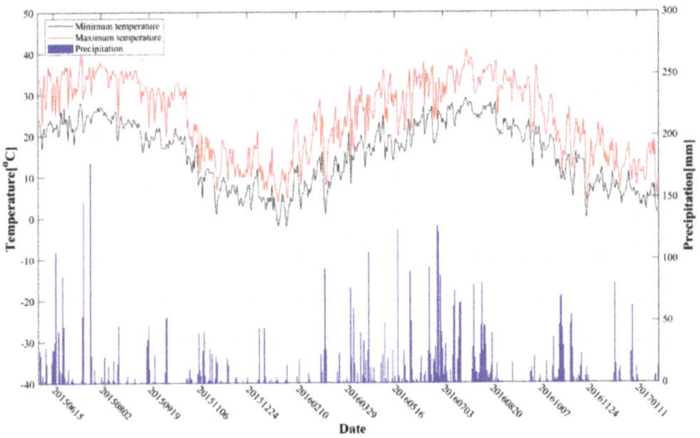

Figure 9. The temperature and precipitation in the study area (from 6 June 2016 to 31 January 2017).

In order to further prove the aforementioned hypothesis that the seasonal fluctuation of our obtained deformation was related to the temperature of solvent, we tried to obtain the temperature of the solvent during our observation period. Due to the limitation of the unavailable solvent temperature data, we used the principle of heat transfer and hydrodynamics introduced in [39] to derive the temperature of the solvent in our experiment. The temperature difference between the pipe inlet and outlet water can be written as:

$$\Delta t = t_g - t_o = \frac{k_g L \left(t_p - t_k \right)}{GC} \tag{4}$$

where t_g represent the temperature of the pipe inlet water, which was treated as a constant (40 °C in our experiment); t_o represents the pipeline outlet water temperature, which was the unknown temperature of the solvent injected into the cavity; k_g is the heat transfer coefficient of the pipeline, which could be indexed according to the pipe material; t_p is the average temperature, which could be calculated according to $t_p = \left(t_g + t_o\right)/2$; t_k is the ambient air temperature; L is the length of the pipeline, G is the mass flow of hot water (both L and G could be provided by the mining company); and C is the specific heat capacity of the hot water, which could be indexed from the standard industry document provided by the mining company.

Formula (4) can be transferred to

$$t_g - t_o = \frac{k_g L\left(\left(t_g + t_o\right)/2 - t_k\right)}{GC} \tag{5}$$

and then t_o can be estimated through Equation (5).

We added the obtained solvent temperature into the correlation analysis between the deformation velocities and the external air temperature, which is shown in Figure 8b. The five feature points on the graph are the time-series settlement points mentioned above. As can be seen from the figure, the temperature of the solvent was highly related to the external air temperature, and the linear deformation velocity also showed high correlation with the solvent temperature. In warm seasons, the subsidence rates of the mining area increased with the air temperature, whereas in cold seasons, the subsidence rates showed obvious dropping with the decrease of the temperature. This result proves the aforementioned hypothesis.

To further interpret the mechanics, we analyzed the accumulated number of coherence points in Figure 7. The statistical result is shown in Figure 10. It can be easily seen that the jumping happened at 10 February 2016, when the accumulated subsidence was above 30 mm, indicating a nearly 8-month period of stable surface condition. During the first 8 months, the subsidence was lower than 30 mm, which was mentioned as the time lag above. Since then, the number keeps increasing until 11 January 2017. From 5 March 2016, the increasing of the number of the high coherence points with subsidence above 60 mm began. Until January 2017, the accumulated number accounted for about 22%. From May 2016, the number of high coherence points, with subsidence above 90 mm and 120 mm, started increasing until January 2017, accounting for 8% and 2%, respectively. As mentioned above, the suggested reasons for the long time of the lagging phenomenon was mainly related to the rock salt dissolution delay and a much deeper mining depth in the process of drilling solution mining [3].

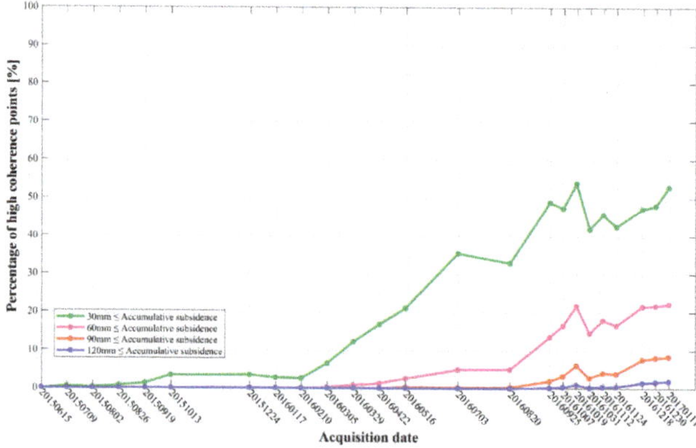

Figure 10. Percentage of accumulated number of coherence points with subsidence in the rock salt mine (with reference to 15 June 2015).

It can be seen from Figure 10 that the shape of the statistical curve performs to be a waveform curve. According to the principle of hydraulic transmission, in the process of drilling solution mining, the pressure generated by the new injected fresh water played a supporting role on the roof of the cavern; accordingly, the subsidence of the ground surface would be decreased [40]. This is also one of the reasons why the statistical curve is flat. With abundant precipitation in the rock salt mine area throughout 2016, the underground volume of the shallow aquifer was increased by the supplement of the nearby river network; thus, a small uplift of the ground surface appeared. This is another reason why the fluctuations occurred in the statistical curve. For example, from August to October 2016, high temperature and low rainfall dominated in the test area, as shown in Figure 9. Combined with the water evaporation in shallow aquifer, the ground surface subsidence was more serious. Accordingly, the cumulative number of pixels, with the subsidence greater than 30 mm, 60 mm, 90 mm, and 120 mm, increased rapidly, as shown in Figure 10. However, from 19 October 2016 to 31 October 2016, the external temperature decreased continuously. Combined with the large amount of precipitation, the slow dissolution rate of rock salt lead to the small uplift of the ground surface. Consequently, the cumulative number of ground surface subsiding pixels was significantly reduced for this period.

To further discuss and analyze the growing process of the typical subsidence bowls detected in the test mine, the profile analysis along the transversal and longitudinal directions (see the transversal line *l*1 and longitudinal lines *l*2, *l*3 and *l*4 in Figure 7) was carried out. The results are shown in Figure 11. We can see that obvious multi-peak phenomenon occurred along the transversal and longitudinal directions. According to our measurements, the peak subsidence along the *l*1 direction was 140 mm, 142 mm, 191 mm, 129 mm, and 128 mm on the fifth, ninth, 16th, 25th, and 29th pixels, whereas 129 mm and 113 mm at the third and 10th pixels along *l*2 direction. The maximum subsidence of 191 mm and 137 mm were detected at the 12th and 11th pixels along the *l*3 and *l*4 directions, respectively. The multi-peak phenomenon along the transversal and longitudinal directions was mainly related to the drilling solution mining method based on connected well groups and its comprehensively multi-direction advancing mode.

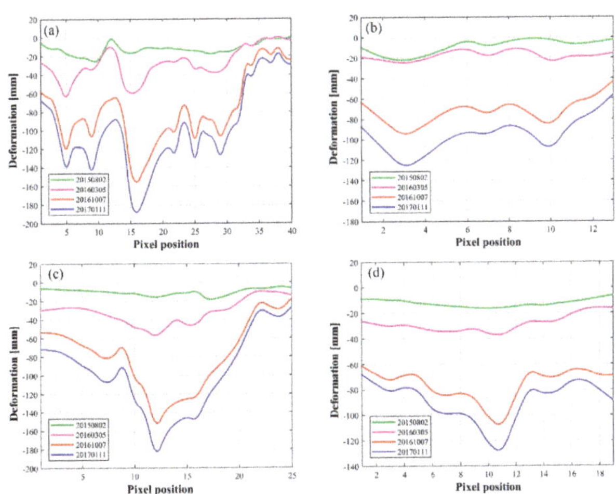

Figure 11. Profiles of the subsidence bowl in Figure 7. (**a**) Along the l1 direction, (**b**) along the l2 direction, (**c**) along the l3 direction, and (**d**) along the l4 direction.

4.3. Accuracy Assessment

In order to verify the reliability of the monitoring results obtained by SBAS technology in this work, an in situ leveling method was carried out to compare with the obtained InSAR measurements. The locations of leveling points (CP1 to CP10) are marked with red solid rectangles in Figure 12.

To perform an accurate comparison, we transferred the generated LOS deformation into vertical displacement according to Equation (3) and extracted the measurements that coincided temporally with our SAR acquisition dates.

Figure 12. Locations of the benchmarks in the rock salt mine (the corresponding amplitude images are shown in the red rectangle in the southeast corner).

Figure 13 shows the comparison results. Obviously, the leveling points of the mining area are continuously subsiding during the period of observation. The most serious subsidence occurred at CP3 in the rock salt mine, with a magnitude of 136 mm, which showed good consistency with the obtained SBAS measurements. According to our calculation, the final root mean square error (RMSE) of the rock salt mine is ±11 mm, accounting for 8% of the corresponding maximum deformation value. The result indicates that the SBAS results maintain a good consistency with that of the leveling measurements. It's also verified that SBAS-InSAR is feasible in the time-series deformation monitoring of rock salt mines.

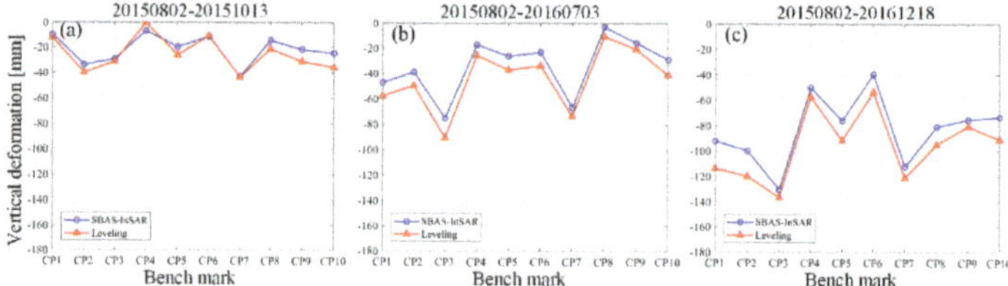

Figure 13. Times-series deformation results compared with leveling measurements on the benchmarks (the locations of CP1 to CP10 are shown in Figure 12). (**a**) from 2 August 2015 to 13 October 2015. (**b**) from 2 August 2015 to 3 July 2016. (**c**) from 2 August 2015 to 18 December 2016.

5. Conclusions

In this study, the SBAS-InSAR technique with Sentinel−1A imagery was used to obtain the spatial–temporal characteristics of the ground subsidence caused by drilling solution mining activities. To reveal the triggering mechanisms of the spatial–temporal ground subsidence, a typical rock salt mine in Hunan Province, China was detected, and its SBAS-derived time-series subsidence maps were obtained. The maximum cumulative subsidence was detected up to 199 mm.

The mechanical deformational characteristics of the rock salt mine were obtained through analyzing the time-series deformation maps, the temporal variations of selected feature points, the cumulative number of the coherence points, and the profiles of the subsidence bowls. Spatially, the distribution of the subsidence in the rock salt mine appeared as discrete strip-shaped in the northwest part and an overall sheet-like shaped distribution in the central and southeast part. Furthermore, the subsidence bowls were with multiple peaks along the transversal and longitudinal directions. This is related to the drilling solution mining method based on connected well groups, and its comprehensively multi-direction advancing mode. Temporally, the cumulative deformation variation curve of the rock salt mine showed a waveform characteristic, with a time lag of 8 months. The suggested reasons for this were that the pressure generated by the new injected fresh water played a supporting role on the roof of the cavern the large depth and thickness of the rock salt mine, and the process of rock salt dissolution induced the time delay in a combined manner. In addition, according to our measurements, the subsidence was greatly affected by the solvent temperature during the drilling solution mining process; thus, it showed obvious seasonal fluctuations. The reasons were supposed as the variations of the dissolution rate for mirabilite and thenardite. The high temperature in warm seasons accelerated the dissolution of rock salt, which led to the increase of subsidence. In contrast, the low temperature in cold seasons suppressed the process of water dissolution, inducing the slow or even uplift trend of the deformation.

Compared to the field leveling deformation measurements, the final accuracy was estimated to ±11 mm. The good consistency with the field measurements shows the feasibility and reliability of the SBAS technology and Sentinel−1 imagery in the application for the rock salt mine monitoring.

Author Contributions: X.X. designed the experiments; X.L. carried out the experiment.; X.X. and X.L. analyzed the experimental results; Z.Y. and L.C. analyzed the precipitation data; B.L. and J.T. helped collect and analyze the leveling measurement in the real data experiment; B.L., D.W. and J.T. contributed to the discussion of the results; X.X. and X.L. drafted the manuscript. All authors contributed to the study, reviewed, and approved the manuscript.

Funding: This work was supported by the National Natural Science Foundation of China [41701536, 61701047, 41201468, 41674040]; the Natural Science Foundation of Hunan Province [2017JJ3322, 2019JJ50639]; the Key Project of Education Department of Hunan Province [18A148, 16B004]; and the Graduate Student Research Innovation Fund of Hunan Province [CX2018B544].

Acknowledgments: The field leveling measurements were carried out by Liang Xiao, Qi Zhang, and Cong Liu, which provide abundant validation data for this paper. The Sentinel−1A dataset used in this paper was provided by the European Space Agency (ESA).

Conflicts of Interest: The authors declare no conflict of interest.

References

1. Ministry of Natural Resources, PRC. *China Mineral Resources Report*; Geological Publishing House: Beijing, China, 2018.
2. Wang, Q. *Drilling Solution Mining and Design*; Chemical Industry Press: Beijing, China, 2016.
3. Ren, S.; Jiang, D.; Yang, C.; Chen, J. *Study on the Mechanism and the Prediction Model of the Subsidence Caused by Rock Salt Solution Mining*; Chongqing University Press: Chongqing, China, 2012.
4. Johnson, K.S. Evaporite-karst problems and studies in the USA. *Environ. Geol.* **2008**, *53*, 937–943. [CrossRef]
5. Zhang, G.; Wang, Z.; Wang, L.; Chen, Y.; Wu, Y.; Ma, D.; Zhang, K. Mechanism of collapse sinkholes induced by solution mining of salt formations and measures for prediction and prevention. *Bull. Eng. Geol. Environ.* **2019**, *78*, 1401–1415. [CrossRef]
6. Kaufmann, G. Geophysical mapping of solution and collapse sinkholes. *J. Appl. Geophys.* **2014**, *111*, 271–288. [CrossRef]
7. Kaufmann, G.; Romanov, D.; Tippelt, T.; Vienken, T.; Werban, U.; Dietrich, P.; Mai, F.; Börner, F. Mapping and modelling of collapse sinkholes in soluble rock: The Münsterdorf Site, Northern Germany. *J. Appl. Geophys.* **2018**, *154*, 64–80. [CrossRef]
8. Buchignani, V.; Amato Avanzi, G.D.; Giannecchini, R.; Puccinelli, A. Evaporite karst and sinkholes: A synthesis on the case of Camaiore (Italy). *Environ. Geol.* **2008**, *53*, 1037–1044. [CrossRef]
9. Li, X.; Xu, B.; Tang, S. Research on ground subsidence mechanism of solution mining by drilling and control counter measures. *J. Saf. Sci. Technol.* **2009**, *5*, 131–134.
10. Gutiérrez, F.; Parise, M.; Waele, J.D.; Jourde, H. A review on natural and human-induced geohazards and impacts in karst. *Earth Sci. Rev.* **2014**, *138*, 61–88. [CrossRef]
11. Mura, J.C.; Gama, F.F.; Paradella, W.R.; Negrão, P.; Carneiro, S.; Oliveira, C.G.; Brandão, W.S. Monitoring the vulnerability of the dam and dikes in Germano iron mining area after the collapse of the tailings dam of Fundão (Mariana-mg, Brazil) using DInSAR techniques with TerraSAR-x data. *Remote Sens.* **2018**, *10*, 1507. [CrossRef]

12. Li, Z.W.; Yang, Z.F.; Zhu, J.J.; Hu, J.; Wang, Y.J.; Li, P.X.; Chen, G.L. Retrieving three-dimensional displacement fields of mining areas from a single InSAR pair. *J. Geod.* **2014**, *89*, 17–32. [CrossRef]

13. Massonnet, D.; Rossi, M.; Carmona, C.; Adragna, F.; Peltzer, G.; Feigl, K.; Rabaute, T. The displacement field of the Landers earthquake mapped by radar interferometry. *Nature* **1993**, *364*, 138–142. [CrossRef]

14. Necsoiu, M.; Onaca, A.; Wigginton, S.; Urdea, P. Rock glacier dynamics in Southern Carpathian Mountains from high-resolution optical and multi-temporal SAR satellite imagery. *Remote Sens. Environ.* **2016**, *177*, 21–36. [CrossRef]

15. Schaefer, L.N.; Lu, Z.; Oommen, T. Post-eruption deformation processes measured using ALOS-1 and UAVSAR InSAR at Pacaya Volcano, Guatemala. *Remote Sens.* **2016**, *8*, 73. [CrossRef]

16. Darvishi, M.; Schlögel, R.; Kofler, C.; Cuozzo, G.; Rutzinger, M.; Zieher, T.; Toschi, I.; Remondino, F.; Mejia-Aguilar, A.; Thiebes, B.; et al. Sentinel-1 and ground-based sensors for continuous monitoring of the Corvara Landslide (South Tyrol, Italy). *Remote Sens.* **2018**, *10*, 1781. [CrossRef]

17. Ng, A.H.M.; Chang, H.C.; Ge, L.; Rizos, C.; Omura, M. Assessment of radar interferometry performance for ground subsidence monitoring due to underground mining. *Earth Planet. Space* **2009**, *61*, 733–745. [CrossRef]

18. Nikolaos, L.; Prashanth, R.; Kosmas, M.; Ouarda, P.; Taha, B.M.J. Ground subsidence monitoring with SAR interferometry techniques in the Rural Area of Al Wagan, UAE. *Remote Sens. Environ.* **2018**, *216*, 276–288.

19. Berardino, P.; Fornaro, G.; Lanari, R.; Sansosti, E. A new algorithm for surface deformation monitoring based on small baseline differential SAR interferograms. *Inst. Electr. Electron. Eng. Trans. Geosci. Remote Sens.* **2002**, *40*, 2375–2383. [CrossRef]

20. Choi, J.K.; Won, J.S.; Lee, S.; Kim, S.W.; Kim, K.D.; Jung, H.S. Integration of a subsidence model and SAR interferometry for a coal mine subsidence hazard map in Taebaek, Korea. *Int. J. Remote Sens.* **2011**, *32*, 8161–8181. [CrossRef]

21. Dong, S.; Samsonov, S.; Yin, H.; Yao, S.; Xu, C. Spatio-temporal analysis of ground subsidence due to under-ground coal mining in Huainan coalfield, China. *Environ. Earth Sci.* **2015**, *73*, 5523–5534. [CrossRef]

22. Plank, S. Rapid damage assessment by means of multi-temporal SAR-a comprehensive review and outlook to Sentinel-1. *Remote Sens.* **2014**, *6*, 4870–4906. [CrossRef]

23. Du, Z.; Ge, L.; Ng, A.H.-M.; Li, X. Investigation on mining subsidence over Appin–West Cliff Colliery using time-series SAR interferometry. *Int. J. Remote Sens.* **2017**, *39*, 1528–1547. [CrossRef]

24. Zheng, M.; Deng, K.; Fan, H.; Du, S.; Zou, H. Monitoring and analysis of mining 3D time-series deformation based on multi-track SAR data. *Int. J. Remote Sens.* **2018**, *40*, 1409–1425. [CrossRef]

25. Xing, X.; Chang, H.-C.; Chen, L.; Zhang, J.; Yuan, Z.; Shi, Z. Radar interferometry time series to investigate deformation of soft clay subgrade settlement—A case study of Lungui Highway, China. *Remote Sens.* **2019**, *11*, 429. [CrossRef]

26. Lanari, R.; Mora, O.; Manunta, M.; Mallorquí, J.J.; Berardino, P.; Sansosti, E. A small-baseline approach for investigating deformations on full-resolution differential SAR interferograms. *Inst. Electr. Electron. Eng. Trans. Geosci. Remote Sens.* **2004**, *42*, 1377–1386. [CrossRef]

27. Casu, F.; Manzo, M.; Lanari, R. A quantitative assessment of the SBAS algorithm performance for surface deformation retrieval from DInSAR data. *Remote Sens. Environ.* **2006**, *102*, 195–210. [CrossRef]

28. Ferretti, A.; Prati, C.; Rocca, F. Nonlinear subsidence rate estimation using permanent scatters in differential SAR interferometry. *Inst. Electr. Electron. Eng. Trans. Geosci. Remote Sens.* **2000**, *38*, 2202–2212.

29. Ferretti, A.; Prati, C.; Rocca, F. Permanent scatterers in SAR interferometry. *Inst. Electr. Electron. Eng. Trans. Geosci. Remote Sens.* **2001**, *39*, 8–19. [CrossRef]

30. Gama, F.F.; Cantone, A.; Mura, J.C.; Pasquali, P.; Paradella, W.R.; dos Santos, A.R.; Silva, G.G. Monitoring subsidence of open pit iron mines at Carajás Province based on SBAS interferometric technique using TerraSAR-X data. *Remote Sens. Appl. Soc. Environ.* **2017**, *8*, 199–211. [CrossRef]

31. Hunan Institute of Geology. *Mine Geological Environment Impact Assessment Report of Lixian Mirabilite Mine*; Xinli Chemical co., Ltd.: Changsha, China, 2007.

32. Liang, W.; Zhao, Y. Analysis on mechanism of salt deposit solution mining. *J. Taiyuan Univ. Technol.* **2002**, *33*, 234–237.

33. Goldstein, R.M.; Werner, C.L. Radar interferogram filtering for geophysical applications. *Geophys. Res. Letters* **1998**, *25*, 4035–4038. [CrossRef]

34. Costantini, M.; Rosen, P.A. A Generalized Phase Unwrapping Approach for Sparse Data. In Proceedings of the IEEE International Geoscience & Remote Sensing Symposium, Hamburg, Germany, 28 June–2 July 1999.

35. Dai, K.; Li, Z.; Tomás, R.; Yu, B.; Wang, X.; Cheng, H.; Chen, J.; Stockamp, J. Monitoring activity at the Daguangbao Mega-landslide (China) using Sentinel-1 TOPS time series interferometry. *Remote Sens. Environ.* **2016**, *186*, 501–513. [CrossRef]

36. Du, Y.; Zhang, L.; Feng, G.; Lu, Z.; Sun, Q. On the accuracy of topographic residuals retrieved by MTInSAR. *Inst. Electr. Electron. Trans. Geosci. Remote Sens.* **2017**, *55*, 1053–1065. [CrossRef]

37. Ma, Z.; Nie, C.; Wang, R. *Well Mineral. Salt Geology and Mining Process*; Sichuan Salt Society, National Well Mineral Salt Science and Technology Information Station: Chengdu, China, 1992.

38. Fomelis, M.; Papageorgiou, E.; Stamatopoulos, C. Episodic ground deformation signals in Thessaly Plain (Greece) revealed by data mining of SAR interferometry time series. *Int. J. Remote Sens.* **2016**, *37*, 3696–3711. [CrossRef]

39. Ran, C. Temperature Drop and Heat Dissipation Calculation of Hot Water Heating Pipeline. In Proceedings of the 2000 Annual Meeting of National HVAC Refrigeration, Nanning, China, 31 October–2 November 2000; pp. 114–117.

40. Qiu, Z. Mechanism analysis of surface collapse in the area of solution salt mining. *J. Saf. Sci. Technol.* **2011**, *7*, 27–31.

Article

Phase Difference Measurement of Under-Sampled Sinusoidal Signals for InSAR System Phase Error Calibration

Zhihui Yuan [1,2,3]**, Yice Gu** [1,2]**, Xuemin Xing** [1,4,]*** and Lifu Chen** [1,2]

[1]   Laboratory of Radar Remote Sensing Applications, Changsha University of Science and Technology, Changsha 410114, China; yuanzhihui@csust.edu.cn (Z.Y.); guyice@stu.csust.edu.cn (Y.G.); lifu_chen@csust.edu.cn (L.C.)
[2]   School of Electrical and Information Engineering, Changsha University of Science and Technology, Changsha 410114, China
[3]   Roy M. Huffington Department of Earth Sciences, Southern Methodist University, Dallas, TX 75275, USA
[4]   School of Traffic and Transportation Engineering, Changsha University of Science and Technology, Changsha 410114, China
*   Correspondence: xuemin.xing@csust.edu.cn; Tel.: +86-18874761257

Received: 5 September 2019; Accepted: 1 December 2019; Published: 3 December 2019

Abstract: Phase difference measurement of sinusoidal signals can be used for phase error calibration of the spaceborne single-pass interferometric synthetic aperture radar (InSAR) system. However, there are currently very few papers devoted to the discussion of phase difference measurement of high-frequency internal calibration signals of the InSAR system, especially the discussion of sampling frequency selection and the corresponding measuring method when the high-frequency signals are sampled under the under-sampling condition. To solve this problem, a phase difference measurement method for high-frequency sinusoidal signals is proposed, and the corresponding sampling frequency selection criteria under the under-sampling condition is determined. First, according to the selection criteria, the appropriate under-sampling frequency was chosen to sample the two sinusoidal signals with the same frequency. Then, the sampled signals were filtered by limited recursive average filtering (LRAF) and coherently accumulated in the cycle of the baseband signal. Third, the filtered and accumulated signals were used to calculate the phase difference of the two sinusoidal signals using the discrete Fourier transform (DFT), digital correlation (DC), and Hilbert transform (HT)-based methods. Lastly, the measurement accuracy of the three methods were compared respectively by different simulation experiments. Theoretical analysis and experiments verified the effectiveness of the proposed method for the phase error calibration of the InSAR system.

Keywords: interferometric synthetic aperture radar (InSAR); phase error calibration; phase difference measurement; under-sampling; coherent accumulation

1. Introduction

Phase difference measurement of sinusoidal signals [1–9] is one of the most important research topics in applications such as phase error calibration of the spaceborne single-pass interferometric synthetic aperture radar (InSAR) system [10–13], power system monitoring [14], radio frequency communication [15], and laser ranging [16]. For the spaceborne single-pass InSAR system, a possible interferometric phase error can arise from relative phase differences between the two receiver channels, because the two signal receivers are not identical mechanically or thermally, and the signal path length from receiving antenna to electronics is vastly different because of the 60 m baseline [12]. Therefore, an internal calibration signal with common reference is distributed to the antennas over an optical fiber cable to the deployed antenna [10–13], and the phase difference of the internal calibration signals (usually sinusoidal signals) received separately from the primary and secondary antennas needs to be measured. More than that, the frequency of the calibration signal is generally high. For example, the frequency of the calibration signal of the InSAR system on the Shuttle Radar Topography Mission (SRTM) is as high as 263 MHz [10]. Due to the limitation of the A/D converter itself, the sampling frequency cannot be made too high, so the signal can only be sampled by under-sampling [17].

Regarding the phase difference measurement of sinusoidal signals, many different methods have been proposed, including discrete Fourier transform (DFT) [18,19], digital correlation (DC) [20], Hilbert transform (HT) [21], least squares (LS) [22], independent component analysis (ICA) [23], and zero cross detection (ZCD) [24] based methods. In Reference [18], considering the negative frequency contribution, a new DFT-based algorithm for phase difference measurement of extreme frequency signal is proposed. The phase difference calculation formula under different windows is deduced in detail. Compared with the traditional DFT-based phase difference measurement algorithm, the new algorithm has stronger spectral leakage suppression capability and higher precision. In Reference [19], considering the spectral superposition of real signals, a new modulation and DFT-based estimation method is proposed which obtains the phase difference by combining the estimated signal frequency and four DFT samples of the modulated signal. However, the above DFT-based phase difference measurement methods have a drawback in that a complete sampling cycle is required for calculation. In Reference [20], an all-digital phase measurement method based on cross-correlation analysis is proposed, and the measurement errors caused by sampling quantization, intrinsic white noise, and non-whole-cycle sampling are analyzed. This method is named the digital correlation (DC)-based method in this paper. In Reference [21], a phase difference estimation method based on data expansion and HT is proposed. This method obtains the phase difference estimation by data expansion, HT, cross-correlation, autocorrelation, and weighted phase averaging which can suppress the end effect of the HT effectively. In Reference [22], a new algorithm for phase difference measurement of sinusoidal signals based on LS is proposed. The algorithm uses digitized samples of the input signal and can determine the amplitude and phase of the two signals simultaneously. Compared with the DFT-based method, this algorithm not only has the advantages of good filtering characteristics and high precision, but also filters out high-frequency components, direct current components, and white noise and can adjust the length of the data window according to the requirements of accuracy and calculation speed. In Reference [23], a robust phase difference measurement method is proposed which uses ICA to separate sinusoidal signals and noise and has strong robustness and accuracy. The ZCD-based method proposed in Reference [24] has a relatively simple principle and is relatively easy to implement in hardware and software, but it is susceptible to interference from noise and harmonics and has poor real-time performance.

However, there are currently very few papers devoted to the discussion of phase difference measurement of high-frequency internal calibration signals of the InSAR system, especially the discussion of sampling frequency selection when the high-frequency signals are sampled under the under-sampling condition. Under such conditions, the initial phases of the sampled signal and the original high-frequency internal calibration signal will be the same, opposite or irrelevant which is different from the general situation. Therefore, the selection of the sampling frequency becomes very important.

In response to the problems mentioned above, the phase difference measurement of high-frequency sinusoidal signals is discussed in this paper, and the corresponding sampling frequency selection criteria under the under-sampling condition is also determined. According to the previous analysis, the DFT-based method is the classical frequency domain measurement method which can be realized by fast Fourier transform (FFT) and can effectively suppress the influence of random noise and harmonics. The DC-based method is the classical time domain measurement method which has a strong ability to suppress random noise; the HT-based method can make real-time measurement of phase difference, and, with the progress of the computer and signal processing technology, the method will continue to overcome the difficulty in instrument design and improve the measurement accuracy. In view of the advantages and representativeness of these three methods, we chose to apply them to the phase difference measurement of high-frequency signals in the phase error calibration of the InSAR system and analyzed and compared them. The specific application process was as follows: Firstly, according to the selection criteria, the appropriate under-sampling frequency was chosen to sample the two sinusoidal signals with the same frequency. Then, the sampled signals were filtered by the limited recursive average filtering (LRAF) and coherently accumulated in the cycle of the baseband signal. Thirdly, the filtered and accumulated sampled signals were used to calculate the phase difference of the two sinusoidal signals by using the DFT-, DC-, and HT-based methods. Lastly, the measurement accuracy of the three methods were compared, respectively, by the different simulation experiments. The experimental results showed that the proposed method in this paper is suitable for the phase difference measurement of the high-frequency internal calibration signals in the InSAR system and can improve the accuracy of the phase difference measurement results.

2. Selection of Sampling Frequency

In this section, the selection criteria of the sampling frequency for the sinusoidal signal under the under-sampling condition is deduced by mathematical formulas and diagrams.

Considering a sinusoidal signal $s(t)$ and its mathematical expression:

$$s(t) = A\cos(2\pi f t + \varphi) \tag{1}$$

where A is the unknown amplitude, f the frequency, t the time, and φ the unknown initial phase ($-\pi < \varphi \leq \pi$). Assuming that the sinusoidal signal is sampled with the frequency f_s, it can be known from the Nyquist sampling theorem that f_s must be greater than or equal to $2f$ to accurately recover the original signal. Especially when it is necessary to measure the phase difference between two sinusoidal signals, f_s must be much larger than $2f$. However, when the signal frequency itself is very high, as the signal frequency increases, the sampling frequency will also become higher and higher. When the sampling frequency is high to a certain extent, it will be difficult to achieve under the existing equipment and technical conditions, which makes it difficult to sample the high frequency signal. Therefore, it is necessary to reduce the sampling frequency according to the band-pass sampling theorem [25], that is, to use the under-sampling method to sample the signal. Next, we will discuss the selection of the sampling frequency and its value range.

The spectrum of the signal $s(t)$ is shown in Figure 1a, where ω means the angular frequency, f is the frequency of the signal, the vertical upward arrow represents the amplitude spectrum, and the solid black dot represents the phase spectrum. Figure 1b is the spectrum of the sampled signal $s_s(t)$. The spectral expression of the sampled signal, $s_s(t)$, is as follows:

$$
\begin{aligned}
S_s(\omega) &= \left(\pi e^{-j\phi}\delta(\omega + 2\pi f) + \pi e^{j\phi}\delta(\omega - 2\pi f)\right) * \\
&\qquad f_s \sum_{n=-\infty}^{+\infty} \delta(\omega - n \cdot 2\pi f_s) \\
&= \pi f_s e^{-j\phi} \sum_{n=-\infty}^{+\infty} \delta(\omega + 2\pi f - n \cdot 2\pi f_s) + \\
&\qquad \pi f_s e^{j\phi} \sum_{n=-\infty}^{+\infty} \delta(\omega - 2\pi f - n \cdot 2\pi f_s)
\end{aligned} \tag{2}
$$

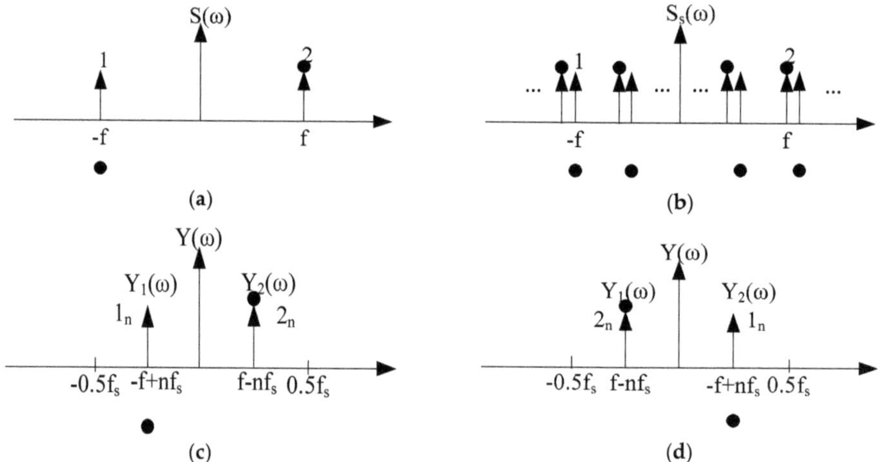

Figure 1. Signal spectrum schematic: (**a**) Original signal spectrum; (**b**) signal spectrum after sampling; (**c**) signal spectrum after low-pass filtering (case 1); (**d**) signal spectrum after low-pass filtering (case 2).

Obviously, in order to avoid spectral aliasing of the sampled signal, the following condition must be met between the sampling frequency, f_s, and the signal frequency, f:

$$-f + nf_s \neq f, n = 1, 2, 3, \cdots \tag{3}$$

That is:

$$f_s \neq \frac{2f}{n} = \frac{f}{n/2}, n = 1, 2, 3, \cdots \tag{4}$$

After passing through a filter with a gain of $1/f_s$ and a passband range of $0 \sim 0.5f_s$, the rest is the spectrum of the baseband signal. At this time, there may be two cases, as shown in Figure 1c,d, where the part marked with "1_n" is the result of shifting the spectrum of the original signal to the right by n times, and the part marked with "2_n" is the result of shifting the spectrum of the original signal by n times.

(1) In the case shown in Figure 1c, the condition as follows must be met:

$$0 < f - nf_s < 0.5f_s, n = 1, 2, 3, \cdots \tag{5}$$

That is:

$$\frac{f}{n+0.5} < f_s < \frac{f}{n}, n = 1, 2, 3, \cdots \tag{6}$$

The resulting baseband signal spectrum at this time is:

$$Y(\omega) = \pi e^{-j\phi}\delta(\omega + 2\pi f - 2\pi n f_s) + \pi e^{j\phi}\delta(\omega - 2\pi f + 2\pi n f_s) \tag{7}$$

The reconstructed baseband signal after inverse Fourier transform is:

$$y(t) = \cos(2\pi(f - nf_s)t + \varphi) = \cos(2\pi f_0 t + \varphi_0) \tag{8}$$

where f_0 is the frequency of $y(t)$ and φ_0 is the initial phase of $y(t)$. Then, as can be seen from Equation (8):

$$\begin{cases} f_0 = f - nf_s \\ \varphi_0 = \varphi \end{cases} \tag{9}$$

That is to say, the initial phase of the baseband signal, $y(t)$, is the same as the initial phase of the signal $s(t)$.

(2) In the case shown in Figure 1d, the condition as follows must be met:

$$0 < -f + nf_s < 0.5f_s, n = 1, 2, 3, \cdots \tag{10}$$

That is:

$$\frac{f}{n} < f_s < \frac{f}{n - 0.5}, n = 1, 2, 3, \cdots \tag{11}$$

The resulting baseband signal spectrum at this time is:

$$Y(\omega) = \pi e^{-j\phi}\delta(\omega - 2\pi f + 2\pi nf_s) + \pi e^{j\phi}\delta(\omega + 2\pi f - 2\pi nf_s) \tag{12}$$

The reconstructed baseband signal after inverse Fourier transform is:

$$y(t) = \cos(-2\pi(f - nf_s)t - \varphi) = \cos(2\pi f_0 t + \varphi) \tag{13}$$

As can be seen from Equation (13):

$$\begin{cases} f_0 = -f + nf_s \\ \varphi_0 = -\varphi \end{cases} \tag{14}$$

That is to say, the initial phase of the baseband signal, $y(t)$, is opposite to the initial phase of the signal, $s(t)$.

From the above analysis, the following conclusions can be drawn: high-frequency sinusoidal signals can be reconstructed based on the frequency and initial phase of the low frequency baseband signal, and the phase difference of the two sinusoidal signals with the same frequency can be measured by selecting the sampling frequency that satisfies the conditions of Equations (6) or (11).

3. Signal Processing Based on Limited Recursive Average Filtering and Coherent Accumulation

In this section, the signal processing process based on limited recursive average filtering (LRAF) and coherent accumulation (CA) under under-sampling conditions is discussed. For a detection system, the preprocessing of the collected signals is an essential part in the whole measurement process. If we want to measure the phase difference, the collected signals should be preprocessed to eliminate the effects of the noise to some extent. In order to minimize the influence of the noise on the phase difference measurement, the preprocessing step used in this paper is divided into two parts: LRAF and CA.

3.1. Signal Sampling

For the case where the frequency of the calibration signal in the InSAR system is high, under-sampling should be selected to sample the signal according to the band-pass sampling theorem [25]. Therefore, the two sinusoidal signals with the same frequency can be sampled by selecting the appropriate sampling frequency according to the selection criteria described in Section 2. Here, we assume that the sampling frequency satisfies the condition in Equation (6), the total length of the sampled signal is N points, the number of sampling points in the baseband signal's period is N_0, and the relationship between N and N_0 is $N = m \cdot N_0$ (m is a positive integer). Then, the two sampled signals are:

$$\begin{aligned} \hat{s}_1(kT) &= A_1 \cos(2\pi(f_0 + nf_s)kT + \varphi_1) + n_1(kT) \\ &= A_1 \cos(2\pi f_0 kT + 2\pi nk + \varphi_1) + n_1(kT) \\ &= A_1 \cos(2\pi f_0 kT + \varphi_1) + n_1(kT), \ k = 0, 1, 2, \cdots, N \end{aligned} \tag{15}$$

$$\begin{aligned}
\hat{s}_2(kT) &= A_2 \cos(2\pi(f_0 + nf_s)kT + \varphi_2) + + n_2(kT) \\
&= A_2 \cos(2\pi f_0 kT + 2\pi nk + \varphi_2) + n_2(kT) \\
&= A_2 \cos(2\pi f_0 kT + \varphi_2) + n_2(kT), \; k = 0, 1, 2, \cdots, N
\end{aligned} \tag{16}$$

where T is the sampling period ($T = 1/f_s$), $n_1(kT)$ and $n_2(kT)$ are the noises of the two receiving channels, and the physical meaning of other parameters are shown in the explanation part of Equation (1) in Section 1.

3.2. Limited Recursive Average Filtering

There are many ways to remove signal noise, including the seasonal model method, autoregressive summation moving average model method, limited recursive average filtering method, etc. In this paper, the LRAF method was used to deal with high-frequency interference. In this method, N_w sampling points continuously obtained from each receiving channel were treated as a queue; then, the abnormal sampling points with clearly distorted amplitudes were deleted according to the preset threshold, and then the remaining sampling points in the queue were arithmetically averaged. The calculated arithmetic average value was taken as the new sample value of the sampling point at the center of the queue, so that the filtering function was implemented. The process was done point by point. When a new sampling point was obtained, it was placed at the end of the queue, and the sampling point at the beginning of the original queue (first in first out, FIFO) was discarded, and then the same operation as before was performed.

The specific steps for performing the LRAF process on $s_1(kT)$ and $s_2(kT)$ are as follows:

(1) Observing the characteristics of the sampling signals from the two receiving channels, determining the maximum allowable amplitude difference among adjacent sampling points, respectively, recorded as the threshold values A_{th1} and A_{th2};

(2) The length a of the queue, N_w, is determined based on the total number of samples in a baseband signal period;

(3) From the first sampling point, the limited average filtering is performed point by point. The queue corresponding to the ith sampling point is $[i - N_w/2, \cdots, i, \cdots, i + N_w/2]$, the abnormal sampling points whose amplitudes are clearly distorted are deleted according to A_{th1} and A_{th2}, then the remaining sampling points in the queue are arithmetically averaged, and then the calculated arithmetic average value is taken as the new sample value of the ith sampling point.

3.3. Coherent Accumulation

Coherent accumulation refers to the addition or accumulation of the signal-to-noise ratio equal to the signal-to-noise ratio of a single pulse multiplied by the pulse number of the pulse train. In this paper, a pulse was equivalent to a signal with a baseband period length. Theoretically, CA improves the signal-to-noise ratio by a factor of N (N is the number of accumulated pulses). By coherently accumulating the filtered signal with the period $T_0(T_0 = N_0/f_s)$ of the baseband signal, $y(t)$, more Gaussian noise can be further filtered out, i.e.,:

$$\begin{aligned}
\hat{s}_{1a}(kT) &= A_1 \cos(2\pi f_0 kT + \varphi_1) + A_1 \cos(2\pi f_0(k + N_0)T + \varphi_1) \\
&\quad + \cdots + A_1 \cos(2\pi f_0(k + (m-1)N_0)T + \varphi_1) + n_{1a}(kT), \\
k &= 0, 1, 2, \cdots, N_0 - 1
\end{aligned} \tag{17}$$

$$\begin{aligned}
\hat{s}_{2a}(kT) &= A_2 \cos(2\pi f_0 kT + \varphi_2) + A_2 \cos(2\pi f_0(k + N_0)T + \varphi_2) \\
&\quad + \cdots + A_2 \cos(2\pi f_0(k + (m-1)N_0)T + \varphi_2) + n_{2a}(kT), \\
k &= 0, 1, 2, \cdots, N_0 - 1
\end{aligned} \tag{18}$$

Most of the noise interference was already filtered out at this time, so the filtered signals, $\hat{s}_{1a}(kT)$ and $\hat{s}_{2a}(kT)$, can be directly used for the next processing step: phase difference measurement.

4. Phase Difference Measurement

At present, the measurement methods used to estimate the phase difference between two sinusoidal signals can be divided into two categories. The first category is the model-based parametric measurement algorithm, such as the LS, HT, and correlation analysis methods. The second is the model-based non-parametric measurement algorithm, such as the DFT method. In this paper, the DFT, DC, and HT methods were used to measure the phase difference of the signals that were processed by LRAF and CA, and the performance of these methods are compared and analyzed in Section 5. Below we introduce the three methods separately.

4.1. DFT-Based Method

Among the many phase difference measurement methods, the DFT-based method is widely used because of its physical meaning, simple implementation, high measurement accuracy, and fast response speed. This method can transform the signal from the time space to frequency domain and can effectively suppress the influence of random noise and harmonics. The DFT operations are performed on the accumulated signals $s_{1a}(kT)$ and $s_{2a}(kT)$ separately, so that the initial phases φ_1 and φ_2 of the two sinusoidal signals can be obtained by:

$$
\begin{aligned}
\varphi_1 &= \angle\left\{\mathrm{DFT}(\hat{s}_{1a}(nT))\big|_{k=1}\right\} \\
&= \angle\left\{\left(\sum_{n=0}^{N_0-1} \hat{s}_{1a}(nT)e^{-j\frac{2\pi}{N_0}nk}\right)\Bigg|_{k=1}\right\}
\end{aligned}
\tag{19}
$$

$$
\begin{aligned}
\varphi_2 &= \angle\left\{\mathrm{DFT}(\hat{s}_{2a}(nT))\big|_{k=1}\right\} \\
&= \angle\left\{\left(\sum_{n=0}^{N_0-1} \hat{s}_{2a}(nT)e^{-j\frac{2\pi}{N_0}nk}\right)\Bigg|_{k=1}\right\}
\end{aligned}
\tag{20}
$$

Then, the phase difference between the two sinusoidal signals is obtained based on the initial phase of the two sinusoidal signals:

$$
\varphi = \varphi_2 - \varphi_1
\tag{21}
$$

4.2. DC-Based Method

The DC is a digitized version of the correlation analysis method. In the DC-based method, because the correlation between the noise signal and the effective signal is very small, the method has a good noise suppression ability. Using correlation analysis to calculate the phase difference is considered to be one of the optimal phase difference calculation methods which has the advantages of fast calculation speed, strong anti-noise interference ability, and high accuracy. In this method, the phase difference is obtained by sampling the two noised sinusoidal signals in a full cycle and then performing cross-correlation operations on them. The analytical expression for the cross-correlation operation of the two signals is as follows:

$$
R_{xy}(\tau) = \frac{1}{T_0}\int_0^{T_0} \hat{s}_{1a}(t)\hat{s}_{2a}(t+\tau)dt
\tag{22}
$$

where $R_{xy}(\tau)$ is the correlation coefficient of the two signals $\hat{s}_{1a}(t)$ and $\hat{s}_{2a}(t)$, τ is the time delay between the two signals, T_0 is the period of the baseband signal $y(t)$. Ideally, the signal and noise are not related to each other, and the noises of the two receiving channels are also uncorrelated. Therefore, when $\tau = 0$, the correlation coefficient $R_{xy}(\tau)$ will reach the maximum value, and its expression can be simplified as:

$$
R_{xy}(0) = \frac{A_1 A_2}{2}\cos(\varphi_2 - \varphi_1)
\tag{23}
$$

Thus, the phase difference between the two sinusoidal signals is:

$$\varphi = \varphi_2 - \varphi_1 = \arccos\left(\frac{2R_{xy}(0)}{A_1 A_2}\right) \tag{24}$$

4.3. HT-Based Method

The HT-based method can make real-time measurement of the phase difference and improve the measurement accuracy. The HT technology was successfully applied to the instantaneous frequency measurement of signals very early, but its application to phase difference measurement is rarely seen. The phase difference measurement method based on HT can make real-time measurements of phase difference, and with the progress of computer and signal processing technology, the method will continue to overcome the difficulty in instrument design and improve the measurement accuracy. Therefore, it is more suitable for intelligent detection equipment and other modern detection equipment.

Suppose that the HT of $s_{1a}(kT)$ and $s_{2a}(kT)$ are $y_1(t)$ and $y_2(t)$, respectively, and let:

$$z_1(t) = s_{1a}(kT) \times y_2(t) \tag{25}$$

$$z_2(t) = s_{2a}(kT) \times y_1(t) \tag{26}$$

$$z = z_1(t) - z_2(t) \tag{27}$$

$$r_1(t) = s_{1a}(kT) \times s_{2a}(kT) \tag{28}$$

$$r_2(t) = y_1(t) \times y_2(t) \tag{29}$$

$$r = r_1(t) + r_2(t) \tag{30}$$

At last, the phase difference between the two sinusoidal signals can be obtained by:

$$\varphi = \varphi_2 - \varphi_1 = \text{arctg}\frac{z}{r} \tag{31}$$

5. Experiments and Results

In order to verify the effectiveness of the method proposed in this paper, some experiments were carried out using simulated data. The parameters used in the experiments are shown in Table 1.

Table 1. Parameters used in the experiments.

Parameters	Value Size
signal-to-noise ratio (SNR)	2 dB
signal frequency (f)	200 MHz
sampling frequency (f_s)	33 MHz
total length (N)	10,240
number of points in one baseband signal period (N_0)	1024
amplitude of signal 1 (A_1)	0.25
amplitude of signal 2 (A_2)	0.2
initial phase of signal 1 (φ_1)	30°
initial phase of signal 2 (φ_2)	45°

One of the two simulated sinusoidal signals with noise is shown in Figure 2a. Figure 2b shows the zoomed-in view of one cycle of Figure 2a. Figure 2c is one cycle of the signal filtered by LRAF, and Figure 2d is one cycle of the signal filtered by CA. Comparing Figure 2c,d with Figure 2b, respectively, it can be seen that both the LRAF and CA have obvious filtering effects, because the noise is greatly weakened, but the effect of CA is better than the LRAF.

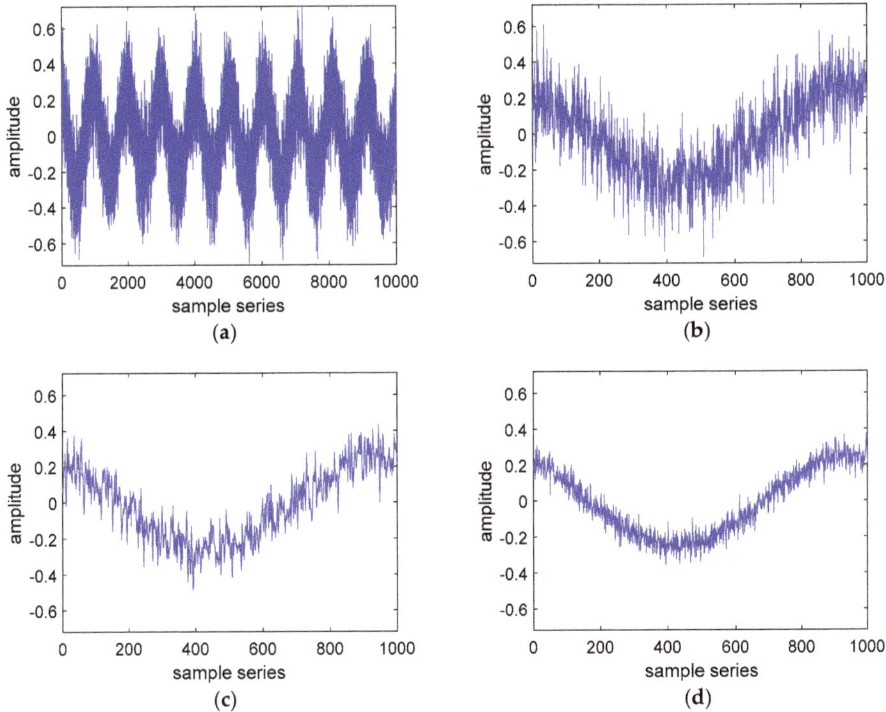

Figure 2. Comparison of the signals before and after limited recursive average filtering and coherent accumulation: (**a**) simulated sinusoidal signal with noise; (**b**) zoomed-in view of one cycle of (**a**); (**c**) one cycle of the filtered signal by limited recursive average filtering (LRAF); (**d**) one cycle of the filtered signal by coherent accumulation (CA).

Ten thousand phase difference measurement simulation experiments were carried out, and the phase difference measurement errors by the DFT, DC, and HT-based methods before and after the LRAF and CA are shown in Figure 3. Figure 3a shows the measurement error of the conventional DFT-based method, Figure 3b shows the measurement error of the DFT-based method after performing the LRAF, and Figure 3c shows the measurement error of the DFT-based method after performing the CA. Figure 3d shows the measurement error of the DC-based method, Figure 3e shows the measurement error of the DC-based method after performing the LRAF, and Figure 3f shows the measurement error of the DC-based method after performing the CA. Figure 3g shows the measurement error of the HT-based method, Figure 3h shows the measurement error of the HT-based method after performing the LRAF, and Figure 3i shows the measurement error of the HT-based method after performing the CA. It can be seen from Figure 3a–c that the preprocessing of the received signal had the most obvious effect on the DFT-based method for the measurement accuracy improvement, and the coherent accumulation had a significant effect which reduced the error by five times, but the LRAF had no effect at all. However, the contribution of these two filtering strategies to the DC- and HT-based methods was not as obvious as the DFT-based method. From Figure 3d–i, we know that the phase difference measurement accuracy of the DC- and HT-based methods had only a certain degree of improvement after the LRAF and CA completed, and the degree of improvement for the two methods was similar.

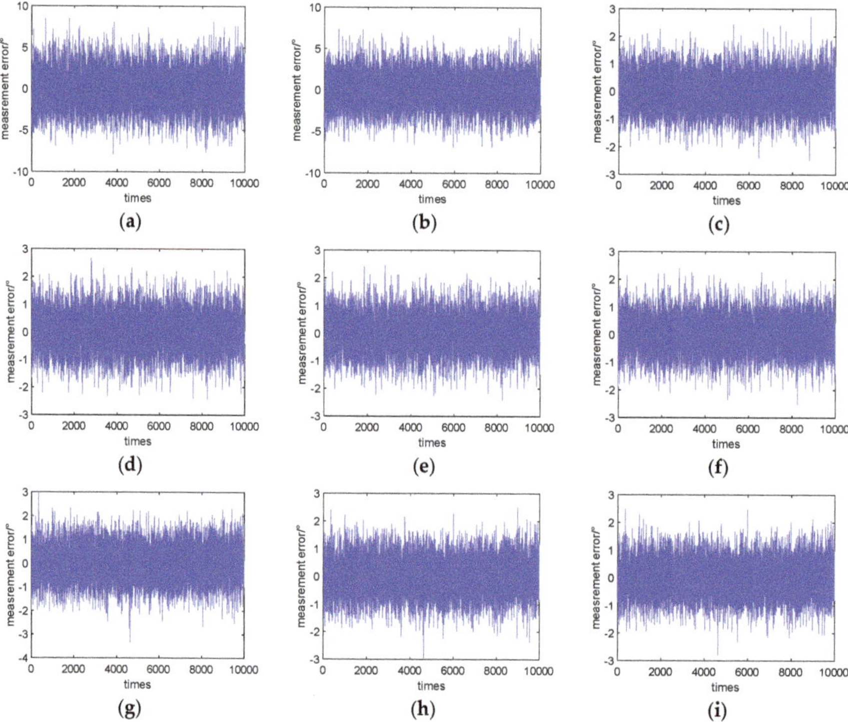

Figure 3. Phase difference measurement error by discrete Fourier transform (DFT)-, digital correlation (DC)-, Hilbert transform (HT) based methods before and after the limited recursive average filtering and coherent accumulation: (**a**) measurement error of the traditional DFT method; (**b**) measurement error of the DFT method after performing the limited recursive average filtering; (**c**) measurement error of the DFT method after performing coherent accumulation; (**d**) measurement error of the traditional DC method; (**e**) measurement error of the DC method after performing the limited recursive average filtering; (**f**) measurement error of the DC method after performing coherent accumulation; (**g**) measurement error of the HT method; (**h**) measurement error of the HT method after performing the limited recursive average filtering; (**i**) measurement error of the HT method after performing the coherent accumulation.

Figure 4 shows the effect of the preprocessing on the performance of the DFT-, DC-, and HT-based phase difference measurement methods under different SNRs. In this experiment, the total number of accumulation cycles was 10, and the SNR varied from 1 dB to 50 dB. Figure 4a,b shows the mean and standard deviation of the measurement error of the phase difference which is measured by the DFT-based method after adding different preprocessing steps, respectively. It can be seen from the two figures that, when the SNR varies from 1 dB to 50 dB, the mean and standard deviation of the measurement error gradually decreased and approached zero at last. However, the measurement accuracy was not improved after the two received signals were filtered by the LRAF, but it was greatly improved after the two received signals were filtered by the CA. More than that, the measurement error of the phase difference was almost negligible when the SNR was greater than 12 dB. Therefore, we can conclude that the CA is very helpful for the performance improvement of the DFT-based phase difference measurement method if the SNR of the signal is poor, while LRAF does not make much sense. Figure 4c,d shows the mean and standard deviation of the measurement error of the phase difference which is measured by the DC-based method after adding different preprocessing steps, respectively. Figure 4e,f shows the mean and standard deviation of the measurement error of the phase difference which is measured by the HT-based method after adding different preprocessing steps, respectively. From Figure 4c–f, we know that the phase difference measurement accuracy of the DC- and HT-based methods is better than the DFT-based method, but it has only a certain degree of improvement after the LRAF and CA are completed, and the degree of improvement for the two methods is similar. Similar to the DFT-based method, the measurement error of the phase difference is almost negligible when the SNR is greater than 12 dB. Therefore, we can conclude that LRAF and CA do not contribute much to the performance improvement of the CA- and HT-based phase difference measurement methods. In general, when the signal-to-noise ratio of the signal is greater than 12 dB, the phase difference measurement can be directly performed using the DFT-, DC-, and HT-based methods.

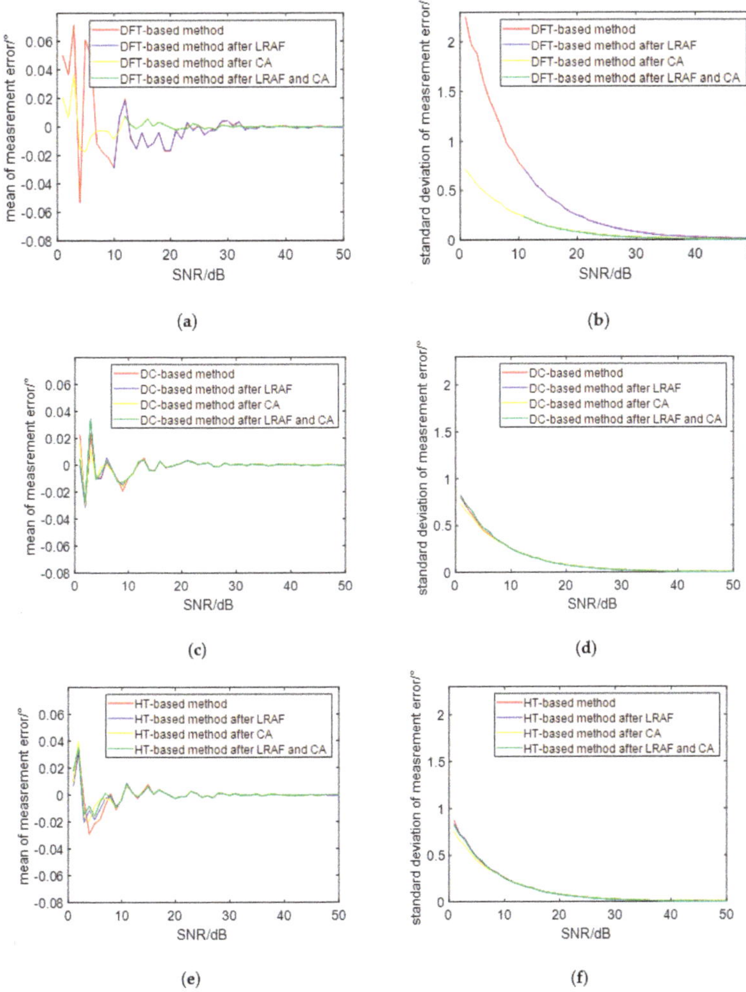

Figure 4. The effect of the preprocessing on the performance of the DFT-, DC-, and HT-based phase difference measurement methods with different SNR: (**a**) mean of the measurement error of the DFT-based method; (**b**) standard deviation of the measurement error of the DFT-based method; (**c**) mean of the measurement error of the DC-based method; (**d**) standard deviation of the measurement error of the DC-based method; (**e**) mean of the measurement error of the HT-based method; (**f**) standard deviation of the measurement error of the HT-based method.

Figure 5a,b show the mean and standard deviation of the phase difference measurement error with a SNR of 2 dB and an accumulative cycle number from 1 to 100, respectively. As can be seen from Figure 5a,b, the mean and standard deviation of the phase error also become smaller and smaller as the accumulative cycle number increases, and even negligible when the accumulative cycle number is greater than 20.

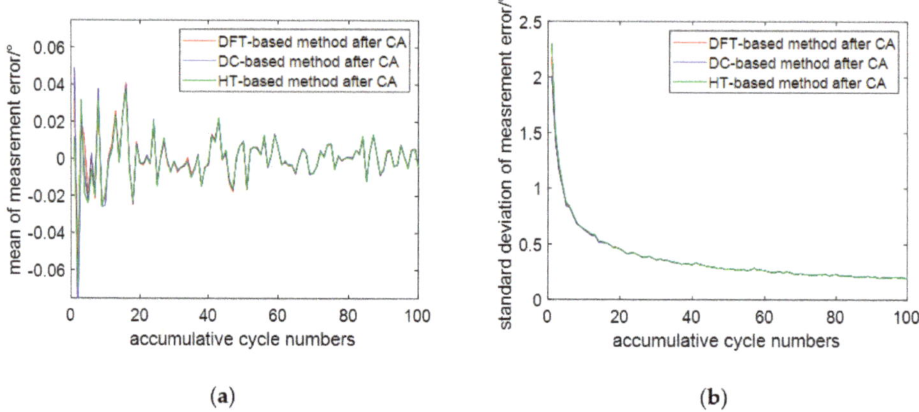

Figure 5. Effects of the different accumulation cycles on each method: (**a**) mean of measurement error; (**b**) standard deviation of measurement error.

Table 2 shows the mean and standard deviation of the measurement error by different phase difference measurement methods with a SNR of 2 dB and an accumulative cycle number of 10. As can be seen from the table, the measurement accuracy was improved after LRAF and CA compared with the direct measurement of the phase difference of the original sinusoidal signal. However, it can also be seen that LRAF had no effect on the DFT-based method but had an effect on the other two phase difference measurement methods; CA can greatly help improve the accuracy of various phase difference measurement methods and has the most obvious effect on DFT method. However, it can be seen that LRAF had no effect on the DFT-based method but had an effect on the other two methods; CA is helpful for improving the measurement accuracy of various phase difference measurement methods and had the most obvious effect on the DFT-based method.

Table 2. The mean and standard deviation of the measurement error by different phase difference measurement methods.

Measurement Methods	Measurement Error	Original Signal	LRAF Only	CA Only	LRAF and CA
DC-based method	Mean (°)	−0.0281	−0.0276	−0.0235	−0.0217
	Standard deviation (°)	0.6852	0.6391	0.6348	0.6292
DFT-based method	Mean (°)	−0.0644	−0.0601	−0.0365	−0.0305
	Standard deviation (°)	1.9611	1.9578	0.6257	0.6252
HT-based method	Mean (°)	−0.0451	−0.0426	−0.0357	−0.0361
	Standard deviation (°)	0.7114	0.6447	0.6374	0.6278

6. Discussion

According to the experimental results in Section 5, both LRAF and CA can effectively filter out noise, but the effect of CA is much better than LRAF. We think that this is mainly because CA makes use of the consistency of the waveform of each period of the sinusoidal signal, but LRAF only uses the method of finding the local average of the adjacent sampling points, and the filtering effect is limited.

Secondly, both LRAF and CA can help the DFT-, DC-, and HT-based phase difference measurement methods improve their measurement accuracy, but they are not very helpful for the DC- and HT-based methods. The main reason may be that the DC- and HT-based phase difference measurement methods themselves have a strong ability to suppress random noise.

Third, when the SNR is small, both LRAF and CA have obvious filtering effects on the signal, but when the SNR is large, the preprocessing has no effect on the measurement accuracy. That is because LRAF and CA only play the role of filtering or suppressing noise; the noise in the signal is relatively small when the SNR is relatively large, so there is no noise that can be filtered even with LRAF and CA.

Fourth, the number of CA cycles has a great influence on the phase difference measurement results. The higher the number of cycles, the more obvious the filtering effect and the higher the accuracy of the corresponding phase difference measurement. This is in line with the law: the larger the number of samples, the more accurate the measurement results.

In addition, it is worth mentioning that the effects of LRAF and CA were only verified on the DFT-, DC-, and HT-based phase difference measurement methods in this paper, so further work can be done in the future to verify them on other phase difference measurement methods, such as the least squares (LS) method, independent component analysis (ICA) method, and zero cross-detection (ZCD) method.

7. Conclusions

In order to solve the phase difference measurement problem of the high-frequency internal calibration signal of the InSAR system, a phase difference measurement method based on LRAF and CA under under-sampling conditions was proposed in this paper, and the sampling frequency selection criteria under the under-sampling condition were determined. Experimental results confirmed the validity of the method. Through theoretical analysis and experiments, the conclusions obtained in this paper are as follows:

(1)   The sampling frequency used to under-sample high-frequency sinusoidal signals should meet the conditions in Equations (6) or (11).

(2)   Both LRAF and CA can effectively filter out noise, but the effect of CA is much better than LRAF.

(3)   Both LRAF and CA can help the DFT-, DC-, and HT-based phase difference measurement methods improve their measurement accuracy, but they are not very helpful for the DC- and HT-based methods.

(4)   When the SNR is small (<12 dB under the simulation condition of this paper), both LRAF and CA have obvious filtering effects on the signal, but when the SNR is large, the preprocessing has no effect on the measurement accuracy.

(5)   The number of CA cycles has a great influence on the phase difference measurement results. The higher the number of cycles, the more obvious the filtering effect and the higher the accuracy of the corresponding phase difference measurement.

In summary, the phase difference measurement method proposed in this paper is suitable for the phase difference measurement of the high-frequency internal calibration signal of the InSAR system for phase error calibration. This method can effectively filter out noise in the sinusoidal signal, improve the phase difference measurement accuracy of the sinusoidal signal, and greatly reduce the phase error. The simulation experiments in Section 5 demonstrate the effectiveness of the proposed method.

Author Contributions: Conceptualization, Z.Y. and X.X.; methodology, Z.Y. and X.X.; software, Y.G.; validation, Y.G. and Z.Y.; formal analysis, Y.G. and Z.Y.; investigation, Y.G. and Z.Y.; resources, Z.Y.; data curation, Y.G. and Z.Y.; writing—original draft preparation, Y.G. and Z.Y.; writing—review and editing, Z.Y.; visualization, Y.G. and Z.Y.; supervision, L.C.; project administration, L.C., Z.Y. and X.X.; funding acquisition, L.C., Z.Y. and X.X.

Funding: This research was funded in part by the National Natural Science Foundation of China, grant numbers 61701047 and 41701536; the Hunan Provincial Natural Science Foundation of China, grant numbers 2019JJ50639 and 2017JJ3322; the Scientific Research Fund of Hunan Provincial Education Department, grant numbers 18A148 and 16B004; and in part by the China Scholarship Council, grant number 201800800006.

Conflicts of Interest: The authors declare no conflict of interest.

References

1. Bertotti, F.L.; Hara, M.S.; Abatti, P.J. A simple method to measure phase difference between sinusoidal signals. *Rev. Sci. Instrum.* **2010**, *81*, 115106. [CrossRef] [PubMed]
2. So, H.C.; Zhou, Z. Two accurate phase-difference estimators for dual-channel sine-wave model. *EURASIP J. Adv. Signal Process.* **2013**, *2013*, 122. [CrossRef]
3. Vucijak, N.M.; Saranovac, L.V. A Simple Algorithm for the Estimation of Phase Difference Between Two Sinusoidal Voltages. *IEEE Trans. Instrum. Meas.* **2010**, *59*, 3152–3158. [CrossRef]
4. Yang, J.R. Measurement of Amplitude and Phase Differences Between Two RF Signals by Using Signal Power Detection. *IEEE Microw. Wirel. Compon. Lett.* **2014**, *24*, 206–208. [CrossRef]
5. Ignatjev, V.; Stankevich, D. A Fast Estimation Method for the Phase Difference Between Two Quasi-harmonic Signals for Real-Time Systems. *Circuits Syst. Signal Process.* **2017**, *36*, 3854–3863. [CrossRef]
6. Bai, L.; Su, X.; Zhou, W.; Ou, X. On precise phase difference measurement approach using border stability of detection resolution. *Rev. Sci. Instrum.* **2015**, *86*, 015106. [PubMed]
7. Chen, N.; Fan, S.C.; Zheng, D.Z. A phase difference measurement method based on strong tracking filter for Coriolis mass flowmeter. *Rev. Sci. Instrum.* **2019**, *90*, 075003. [CrossRef] [PubMed]
8. Zhang, M.; Wang, H.; Qin, H.B.; Zhao, W.; Liu, Y. Phase Difference Measurement Method Based on Progressive Phase Shift. *Electronics* **2018**, *7*, 86. [CrossRef]
9. Choi, U.G.; Kim, H.Y.; Han, S.T.; Yang, J.R. Measurement Method of Amplitude Ratios and Phase Differences Based on Power Detection Among Multiple Ports. *IEEE Trans. Instrum. Meas.* **2019**, *68*, 4615–4617. [CrossRef]
10. Werner, M.; Häusler, M. X-SAR/SRTM instrument phase error calibration. In Proceedings of the IEEE 2001 International Geoscience and Remote Sensing Symposium, Sydney, Australia, 9–13 July 2001.
11. McWatters, D.A.; Lutes, G.; Caro, E.R.; Tu, M. Optical calibration phase locked loop for the Shuttle Radar Topography Mission. *IEEE Trans. Instrum. Meas.* **2001**, *50*, 40–46. [CrossRef]
12. Farr, T.G.; Rosen, P.A.; Caro, E.; Crippen, R.; Duren, R.; Hensley, S.; Kobrick, M.; Paller, M.; Rodriguez, E.; Roth, L.; et al. The Shuttle Radar Topography Mission. *Rev. Geophys.* **2007**, *45*, 1–33. [CrossRef]
13. Wang, Y.; Liang, X.; Wu, Y. A comparison of internal calibration schemes for spaceborne single-pass InSAR applications. In Proceedings of the IEEE International Geoscience and Remote Sensing Symposium, Barcelona, Spain, 23–28 July 2007; pp. 1573–1792.
14. Ree, J.D.L.; Centeno, V.; Thorp, J.S.; Phadke, A.G. Synchronized Phasor Measurement Applications in Power Systems. *IEEE Trans. Smart Grid* **2010**, *1*, 20–27.
15. Wang, Z.; Mao, L.; Liu, R. High-Accuracy Amplitude and Phase Measurements for Low-Level RF Systems. *IEEE Trans. Instrum. Meas.* **2012**, *61*, 912–921. [CrossRef]
16. Yoon, H.; Park, K. Development of a laser range finder using the phase difference method. In Proceedings of the Optomechatronic Sensors & Instrumentation, International Society for Optics and Photonics, Sapporo, Japan, 5–7 December 2005; pp. 230–237.
17. David, S.M.; Francisco, M.M.; Ernesto, M.G.; José, L.G. SNR Degradation in Undersampled Phase Measurement Systems. *Sensors* **2016**, *16*, 1772.
18. Shen, T.; Tu, Y.; Li, M.; Zhang, H. A new phase difference measurement algorithm for extreme frequency signals based on discrete time Fourier transform with negative frequency contribution. *Rev. Sci. Instrum.* **2015**, *86*, 015104. [CrossRef] [PubMed]
19. Wang, K.; Tu, Y.; Shen, Y.; Xiao, W.; Des, M. A modulation based phase difference estimator for real sinusoids to compensate for incoherent sampling. *Rev. Sci. Instrum.* **2018**, *89*, 085120. [CrossRef] [PubMed]
20. Liang, Y.R.; Duan, H.Z.; Yeh, H.C.; Luo, J. Fundamental limits on the digital phase measurement method based on cross-correlation analysis. *Rev. Sci. Instrum.* **2012**, *83*, 095110. [CrossRef] [PubMed]
21. Shen, Y.L.; Tu, Y.Q.; Chen, L.J.; Shen, T.A. Phase difference estimation method based on data extension and Hilbert transform. *Meas. Sci. Technol.* **2015**, *26*, 095003. [CrossRef]
22. Micheletti, R. Phase angle measurement between two sinusoidal signals. *IEEE Trans. Instrum. Meas.* **1991**, *40*, 40–42. [CrossRef]
23. Gong, G.; Lu, H.; Chen, G.; Jin, M.; Chen, X. Phase Difference Measurement Method for Sine Signals Based on Fast ICA. In Proceedings of the 2013 Fourth Global Congress on Intelligent Systems, Hong Kong, China, 3–4 December 2013; pp. 249–253.

24. Lin, D.Y.; Lu, J.F.; Jia, R.C.; Yang, L. Research on the Technology of Phase Difference Measurement Based on FPGA. In Proceedings of the 2014 Fourth International Conference on Instrumentation and Measurement, Computer, Communication and Control, Harbin, China, 18–20 September 2014; pp. 731–734.
25. Oppenheim, A.V.; Schafer, R.W. *Discrete-Time Signal Processing*, 3rd ed.; Prentice Hall: Upper Saddle River, NJ, USA, 2009; pp. 792–889.

Article

Permafrost Deformation Monitoring Along the Qinghai-Tibet Plateau Engineering Corridor Using InSAR Observations with Multi-Sensor SAR Datasets from 1997–2018

Zhengjia Zhang [1,2], Mengmeng Wang [1], Zhijie Wu [3] and Xiuguo Liu [1,*]

[1] School of Geography and Information Engineering, China University of Geosciences, 388 Lumo Road, Wuhan 430074, China; zhangzj@cug.edu.cn (Z.Z.); wangmm@cug.edu.cn (M.W.)
[2] Artificial Intelligence School, Wuchang University of Technology, No. 16 Jiangxia Avenue, Wuhan 430223, China
[3] College of Resources Engineering, Longyan University, Longyan 264012, China; wuzhijiefj@163.com
*   Correspondence: liuxg318@163.com

Received: 18 October 2019; Accepted: 28 November 2019; Published: 2 December 2019

Abstract: As the highest elevation permafrost region in the world, the Qinghai-Tibet Plateau (QTP) permafrost is quickly degrading due to global warming, climate change and human activities. The Qinghai-Tibet Engineering Corridor (QTEC), located in the QTP tundra, is of growing interest due to the increased infrastructure development in the remote QTP area. The ground, including the embankment of permafrost engineering, is prone to instability, primarily due to the seasonal freezing and thawing cycles and increase in human activities. In this study, we used ERS-1 (1997–1999), ENVISAT (2004–2010) and Sentinel-1A (2015–2018) images to assess the ground deformation along QTEC using time-series InSAR. We present a piecewise deformation model including periodic deformation related to seasonal components and interannual linear subsidence trends was presented. Analysis of the ERS-1 result show ground deformation along QTEC ranged from −5 to +5 mm/year during the 1997–1999 observation period. For the ENVISAT and Sentinel-1A results, the estimated deformation rate ranged from −20 to +10 mm/year. Throughout the whole observation period, most of the QTEC appeared to be stable. Significant ground deformation was detected in three sections of the corridor in the Sentinel-1A results. An analysis of the distribution of the thaw slumping region in the Tuotuohe area reveals that ground deformation was associated with the development of thaw slumps in one of the three sections. This research indicates that the InSAR technique could be crucial for monitoring the ground deformation along QTEC.

Keywords: InSAR; Qinghai-Tibet Engineering Corridor; deformation; permafrost

---

1. Introduction

Permafrost, defined as soil or rock ground that remains frozen (ground temperature below 0 °C) for two or more consecutive years [1,2], has the potential to affect the global climate [3,4], carbon balance [5], and water-heat balance [6]. The Qinghai-Tibet Plateau (QTP) has the largest extent of permafrost outside the polar regions, with 50% of the QTP's area underlain by permafrost. With the implementation of western development strategy and the One Belt and One Road strategy, several key engineering projects have been conducted on this fragile and harsh environmental plateau, such as the Qinghai-Tibet Railway (QTR) [7,8], the Qinghai-Tibet Highway (QTH) [9], oil pipelines [10] and electric transmission lines [11]. Along the QTR from the Chumaerhe to Fenghuo Mountain is the significant section of Qinghai-Tibet Engineering corridor (QTEC) [12,13]. In recent years, with the global warming, the increase of human activities, and the operation of permafrost engineering, the permafrost has become seriously degraded, intensifying the permafrost engineering instability, land desertification and soil moisture loss [14]. Therefore, long-term permafrost measurement along the QTEC is of great importance for permafrost environment protection, climate change and cold-region hazard prevention.

Traditional geodetic measurement methods such as levelling and the global position system (GPS) surveys, can achieve high-precision monitoring. However, these point-based geodetic measured methods are limited to discrete points on fixed routes and are time consuming. Compared with those methods, the satellite remote sensing provides a valuable tool for observing large and hard-to-access areas with high spatial and temporal resolution [15]. Synthetic aperture radar interferometry (InSAR) is a promising technique that can be used to monitor slow ground deformation with millimeter accuracy by analyzing the phase information from two SAR images [16]. Due to the advantages of large coverage, high resolution and measurement accuracy, InSAR has been used to measure surface deformation over larger areas induced by earthquake [17], volcanoes [18], and land subsidence [19,20]. It has also been adopted to determine the ground deformation in permafrost regions [15,21,22].

To mitigate the intrinsic limitations of the traditional differential InSAR (DInSAR) (spatial-temporal decorrelations and atmospheric delay) [23], time-series InSAR techniques such as persistent scatterer interferometry (PSI) [24,25], the small baseline subset interferometry (SBAS) [26,27], multi-temporal InSAR (MTInSAR) [28], have been proposed by analyzing the time series interferometric phase on stable objects, such as buildings, rocks and roads.

Due to the merits of time-series InSAR, many studies have used it to retrieve surface deformation information related to permafrost thawing and freezing in QTP [29–39] (Table 1) and other permafrost regions [21,22,40].

Table 1. Permafrost deformation studies in the Qinghai-Tibet Plateau (QTP) using synthetic aperture radar interferometry (InSAR) technologies.

Study Areas	InSAR Method	SAR Dataset	Observation Period	Deformation Rate (mm/year)	References
Beiluhe	PSI	ENVISAT	August 2003–May 2007	−20 to 3	[29]
Beiluhe	IPTA and SBAS	ALOS-1 and ENVISAT	November 2004–December 2010	−20 to 20	[30]
Beiluhe	SBAS	ALOS-1	June 2007–December 2010	−20 to 20	[31]
Beiluhe	SBAS	ENVISAT	April 2003–July 2010	−16 to 2	[32]
Tanggula	PSI	ENVISAT	February 2007–September 2009	−10 to 0	[33]
Yangbajing	MTInSAR	TerraSAR-X	December 2011–November 2012	−30 to 10	[34]
Yangbajing	SBAS	ENVISAT	May 2007–September 2010	−50 to 10	[35]
Wudaoliang	SBAS	ALOS-1	May 2007–March 2009	−2 to 0	[36]
Beiluhe	MTInSAR	TerraSAR-X	July 2014–March 2017	−20 to 0	[37]
Wudaoliang-Fenghuo Mountain	MTInSAR	Sentinle-1A	November 2017–December 2018	—	[38]
Northwestern Tibet	NSBAS (new small baseline subset)	ENVISAT	2003–2011	−4 to 4	[39]

The studies listed in Table 1 preliminarily explored the deformation of the permafrost region using time-series InSAR. Unfortunately, the most of the above-mentioned literatures have only focused on permafrost deformation monitoring in QTP over a short period of time such as from 2004 to 2009 with ENVISAT images or from 2007–2010 with ALOS-1 images, or from 2014 to 2016 with TerraSAR-X images. Long-term Permafrost thaw deformation on the QTP and the relationship between permafrost deformation and QTP engineering are still poorly quantified and understood. It is necessary to focus on the latest development and the temporal evolution of ground deformation of the permafrost region in QTP. With the launch of new SAR satellites such as Sentinel-1A/B [41], more SAR images with short repeat cycles (six days) can be obtained, which are suitable for determining the ground deformation in permafrost regions. Daout et al., developed a method to enhance InSAR performances for such difficult terrain conditions and construct an 8 year timeline of the surface deformation over a 60,000 km^2 area [39]. Rouyet et al. used the InSAR to investigate the seasonal ground deformation in and around Adventdalen with TerraSAR-X StripMap Mode (2009–2017) and Sentinel-1 Interferometric Wide Swath Mode (2015–2017) SAR images [15]. Combining the archived SAR images, long-term ground deformation in the permafrost region can be determined.

The objectives of this paper were to retrieve the surface deformation along QTEC from the Wudaoliang to the Tuotuohe section over a 20-years period using time-series InSAR technique and to analyze temporal evolution of the QTEC deformation. More than 90 SAR images, including ERS-1, ENVISAT, and Sentinel-1A, were collected to jointly retrieve the feature of ground deformation from 1997 to 2018. A hybrid time-series methodology taking advantage of the merits of PSI and SBAS was used to identify more measurement points [42]. Moreover, a piecewise deformation model combining a seasonal deformation term related to active layer thawing and freezing and linear subsidence component related to permafrost thawing is introduced. The spatiotemporal feature of the ground deformation along the QTEC and its relationships with permafrost engineering and permafrost distribution were analyzed.

2. Study Area and Datasets

2.1. Study Area

The permafrost region along the Wudaoliang-Tuotuohe section of QTEC was chosen as the study area. The area is in the in the Hoh Xil mountain area between the Kunlun Mountain and Tanggula Mountain ranges and is the source area of the Yangtze river, which is in the northern part of the QTP [12]. The QTR is a high-elevation railway connecting Xining to Lhasa, with the length of 1956 km. About 550 km length of the QTR is laid on the discontinuous permafrost [7]. The QTR from Wudaoliang to Tuotuohe section began to construct in 2001 and completed in 2006. Figure 1 provide a topographic map of the study area, with an average elevation of more than 4500 m above the sea level. Underground ice developed extensively in this region [5]. Several thermokarst lakes have developed, such as Zuonai Lake, Kusai Lake and Salt Lake (Figure 1). The active layer thickness (ALT) varies from 0.8 to 4 m with a mean of about 2 m [43]. The typical ground features in our study area can be classified into six landcover types: alpine meadow, alpine desert, Thermokarst Lake, QTR, QTH, and electric transmission power line (Figure 2). This area is dominated by sub frigid semi-arid climate with the mean temperature of about −3.8 °C [44]. The annual mean precipitation varies from 50 mm to 400 mm, concentrated in the summer season [36]. The amplitude of Sentinel-1A, shuttle radar topography mission (SRTM) digital elevation map (DEM) data [45] and the slope of the QTEC are show in the bottom of Figure 1. In the QTEC, several permafrost engineering structures have been constructed that have considerably influenced the stability of the permafrost. In the SAR images, the QTR is a bright line as shown in Figure 1. The other engineering structures cannot be easily observed in medium resolution SAR images.

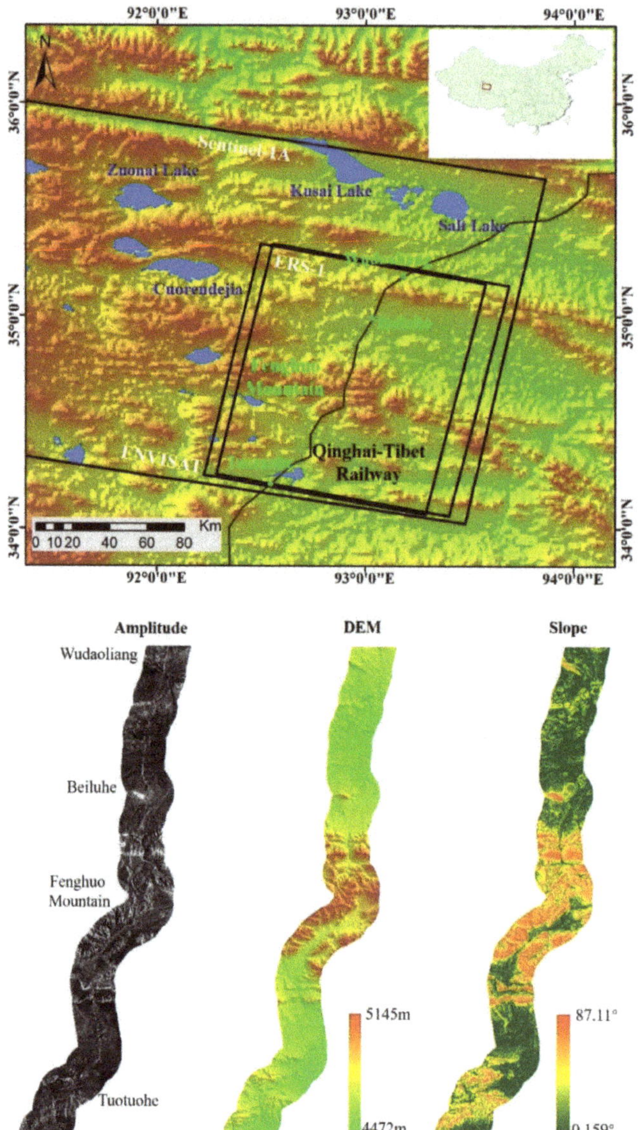

Figure 1. Top: Coverage of radar data stacks (black squares) on the shuttle radar topography mission (SRTM) digital elevation map (DEM) over the study area. The black lines show the Qinghai-Tibet Railway (QTR). The green points represent the railway station in the study area. The blue polygons are the large lakes. **Bottom**: the amplitude of Sentinel-1A, DEM, and slope of the selected Qinghai-Tibet Engineering corridor (QETC) section from Wudaoliang to Tuotuohe.

Within the QTEC coverage, several key developmental projects have been constructed, such as QTR, QTH, and electric transmission power line (Figure 2d–f). Due to the constructions of those permafrost engineering structures, the original hydrothermal balance of permafrost has been destroyed and the permafrost has begun to degrade. Studies have showed that the ground deformation rate of the permafrost along QTR can reach −10 mm/year in some sections [46]. The study area is an overlap of the available SAR images. About 110 km of the QTEC region from Wudaoliang to Tuotuohe was selected as the study object. The daily air temperature in Wudaoliang weather station from 1997 to 2018 was collected. Figure 3 shows the daily air temperature in our study area from 1997 to 2018.

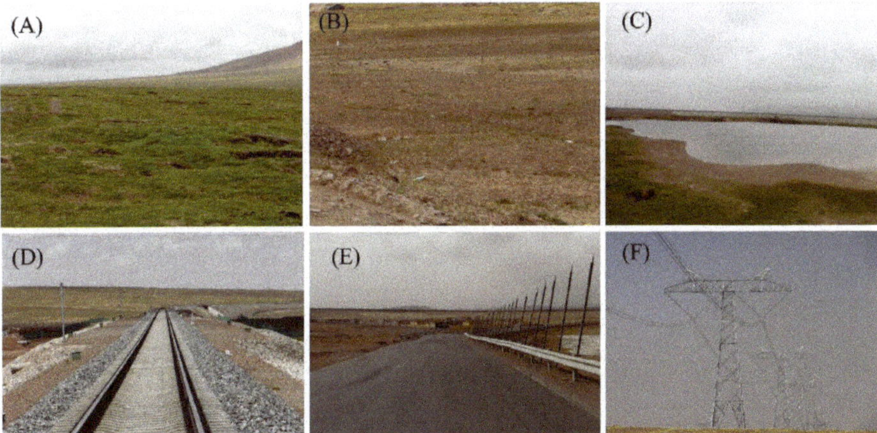

Figure 2. Field photos of the study area in August 2014. (**A**) alpine meadow, (**B**) alpine desert, (**C**) thermokarst lake, (**D**) QTR, (**E**) Qinghai-Tibet Highway (QTH), and (**F**) electric transmission power line.

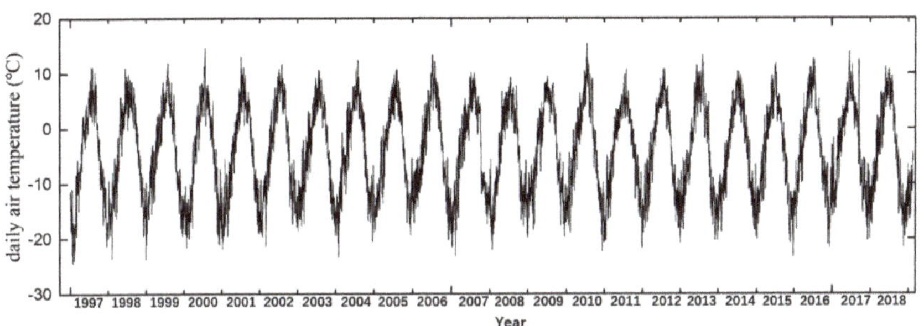

Figure 3. Daily air temperature in Wudaoliang from 1997 to 2018.

2.2. Datasets

To reveal the ground deformation in the study area in the selected 20-year period, SAR images from three different satellites were collected. There are ERS-1 SAR images acquired from October 1997 to December 1999; ENVISAT SAR images from November 2004 to July 2010, and Sentinel-1A SAR images from April 2015 to December 2018. The coverages of the above SAR stacks are shown in Figure 1. The amplitude of the Sentinel-1A along QTEC is show in the bottom of Figure 1. The QTR and QTH can been easily observed due to their strong back scattering. The acquisition parameters of the three SAR images are listed in Table 2. Unfortunately, time gaps, where no SAR images are acquired, exist 2002–2004 and 2009–2015. SRTM DEM data with a spatial resolution of 30 m were adopted to remove the topographic phase.

Table 2. SAR image numbers and parameters used in this study.

Sensors	Start and End Date	Acquisitions (n)	Incidence Angle (°)	Polarization	Pixel Spacing/ Range (m)	Pixel Spacing/ Azimuth (m)
ERS-1	1997-04-24 to 1999-12-30	9	19.3~26.5	VV	7.9	3.9
ENVISAT	2004-11-18 to 2010-07-15	39	18.6~26.2	VV	7.8	4
Sentinel-1A	2015-04-13 to 2018-12-17	40	30.7~37.6	VV	5	20

3. Methodology

3.1. InSAR Processing

Studies have demonstrated that the main challenges and limitations of the InSAR technique in detecting the ground deformation in the permafrost region are the serious temporal decorrelation and non-linear deformation trends caused by the seasonal thaw-freeze process of active layer [31,37,46,47]. It is difficult to obtain sufficiently stable measurement points due to the above limitations. In this study, the small baseline strategy was applied to suppress the temporal decorrelation.

Firstly, all the SAR images were co-registered. Then, a multi-temporal InSAR data processing strategy was used to retrieve ground deformation. Considering the different attribute of SAR stacks with different wave lengths, different small baseline strategies were adopted for those SAR stacks [48]. Through previous studies, the temporal decorrelation is serious in permafrost region, so the temporal baseline (350 days) is no longer than one year. Consideration the orbit accuracy of different sensors and the time sampling of SAR images, the normal baseline threshold values are 800 m, 500 m, and 200 m for ERS, ENVISAT and Sentinel-1A, respectively. For ERS-1 and ENVISAT, the multi-looking with 5×1 looks in the azimuth and range direction was performed, respectively. For Sentinel-1A, the multi-looking with 1×4 looks in the azimuth and range direction were performed. After all the interferograms have been generated, each of the interferograms were checked, and the interferograms with serious temporal decorrelations were rejected for deformation retrieval. Finally, we obtained a total number of 17 ERS-1 interferograms (normal baseline < 800 m and temporal baseline < 350 days), 105 ENVISAT ASAR interferograms (normal baseline < 500 m and temporal baseline < 350 days), and 131 Sentinel-1A interferograms (normal baseline < 200 m and temporal baseline < 350 days). Figure 4 shows the spatial and temporal baseline configuration of the three SAR stacks. The differential interferometric phase is generated by removing the topographic phase from the interferograms using the SRTM DEM data. To suppress the noise in the interferograms, the Goldstein filtering method was applied [49].

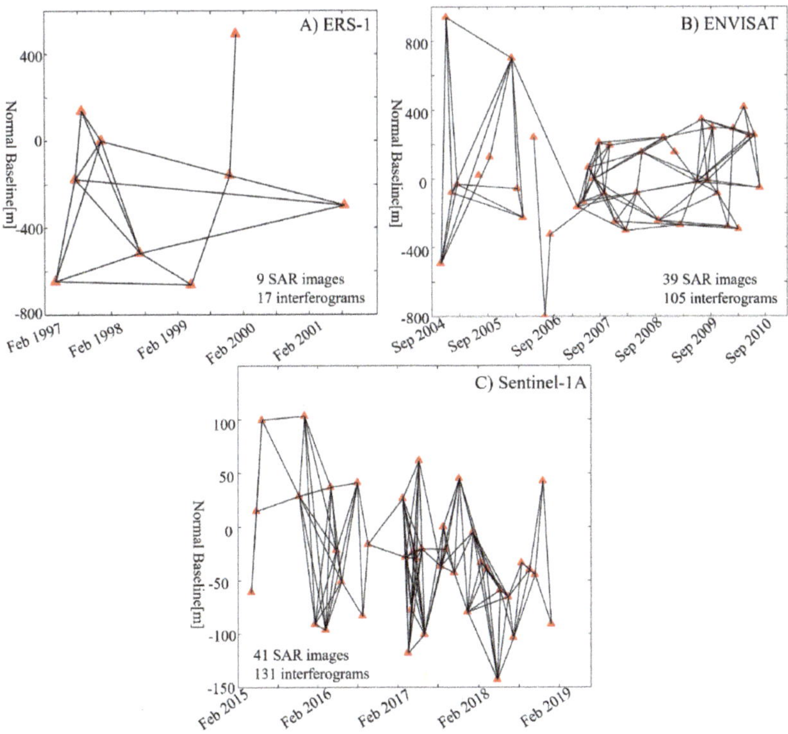

Figure 4. Generated interferometric pairs of (**A**) ERS-1, (**B**) ENVISAT, and (**C**) Sentinel-1A. All the lines represent the interferograms used to monitor the time-series ground deformation. All the points represent the SAR images.

3.2. Seasonal and Long-Term Deformation Model

The thaw-freeze process of the active layer is complex and correlated with many factors, such as vegetation, snow, soil moisture, soil properties and temperature [7]. In permafrost regions, the seasonal deformation component is larger than the annual deformation. Therefore, using an appropriate seasonal phase model to monitor the thawing-freezing process of the permafrost is essential. Mathematical models, such as the sinusoidal model [33,36,50] and cubic term model [31] have been proposed to retrieved the seasonal deformation of permafrost. However, the seasonal deformation term is much complicated and is closely related to Environmental and climatic factors, such as temperature, soil moisture. These environmental and climatic factors should be considered. Liu et al. [51] introduced a seasonal model based on the Stefan model in the Alaska permafrost region, which describes the relationship between the thaw depth and the square root of the accumulated degree days of thawing (ADDT). The Stefan equation is widely used to estimate the thaw depth. This deformation model is based on the cumulative temperature and is reasonable, which have been successfully applied in QTP regions [37,38]. In this study, we adopted a deformation model combining a linear subsidence term for the long-term permafrost thaw subsidence and seasonal deformation term for the seasonal thawing and freezing of the active layer.

The deformation model is defined as follows:

$$d_s = R{\cdot}t + A_t{\cdot}\sqrt{\mathrm{ADDT}(t_1)} - A_f{\cdot}\sqrt{\mathrm{ADDF}(t_2)} + \varepsilon \tag{1}$$

where, R is the long-term deformation rate, A_t and A_f are the thawing and freezing deformation coefficients, respectively; and ADDT and ADDF are the accumulated degree days of thawing and freezing, respectively. ADDT reaches its the maximum at the end of the thawing season. The daily ADDT and ADDF were calculated based on the air temperature measured at the Wudaoliang Meteorological station. Due to the sporadic acquisitions of ERS-1 images, a deformation model without a seasonal term was used for ERS-1 datasets.

3.3. Calculation of ADDT and ADDF

The thawing and freezing onsets of the active layer are fixed as 1 May and 15 September, respectively [21,37]. However, the freezing and thawing onsets change every year in the QTP. Error would occur if we assume that the thawing and freezing onsets were the same in every year. In QTP, the length of freezing season is longer than the of thawing season. Generally, a uniform thawing and freezing onsets of the active layer are chosen based on temperature observation data. In this study, we first used the following model to monitor temperature:

$$T(t) = a_0 + a_1 \cos(t{\cdot}w) + a_2 \sin(t{\cdot}w) \tag{2}$$

where, $T(t)$ is the temperature on day t. a_0, a_1, a_{03}, *and* w are the parameters. For each year, we used this model to monitor the annual temperature and identify the thawing and freezing onsets in each year.

Figure 5 shows the time-series temperature of each year from 1997 to 2018. Most of the 20 years of temperature data were modeled accuracy with a coefficient of determination (R^2) > 0.9 and root mean square error (RMSE) < 2.7. Figure 5 shows that the onsets of thawing and freezing changed every year. Through the monitoring results, we identified the onsets of thawing and freezing and relatively accurately calculate the ADDT and ADDF each year.

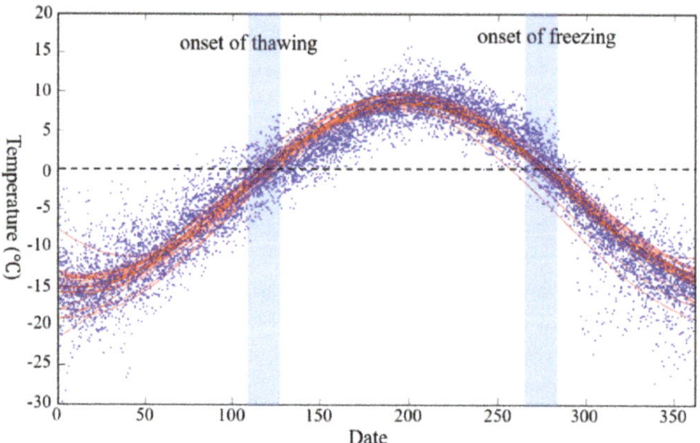

Figure 5. Seasonal pattern temperature of each year from 1997 to 2018 and the monitored models each year.

3.4. Time-Series InSAR Method

3.4.1. Coherence Point (CP) Selection

CPs are those points with high coherence and stable amplitude value during the whole observation period. In permafrost area, the ground feature includes four types: permafrost engineering, Thermokarst Lake, alpine meadow, alpine desert (Figure 2). In order to exclude the water bodies, vegetated areas and other decorrelated areas from the CPs, the thresholds of coherence and the dispersion of amplitude are both used to identify the CPs. In this paper, the coherence threshold value is 0.65 and the dispersion amplitude threshold value is 0.25.

3.4.2. Topographic and Orbit Error Removal

The atmospheric delay is influential in high latitude mountainous regions. Our study area has an average elevation of over 4400 m with some mountains. In the mountainous areas, the stratified troposphere can produce serious atmospheric delays in the interferograms. Obvious residual orbital phase was in some interferograms. In this paper, to remove those phase ramps, we applied a phase ramps correction model combining a biquadratic model for orbital phase ramps and a linear model for elevation dependents errors [52]:

$$\varphi(x,y)_{ramp} = a_0 + a_1 \cdot x + a_2 \cdot x^2 + a_3 \cdot x \cdot y + a_4 \cdot y + a_5 \cdot y^2 + a_6 \cdot h + \varepsilon(x,y) \qquad (3)$$

where, $\varphi(x,y)_{ramp}$ is the modeled phase ramps, $\varepsilon(x,y)$ is the random phase error, a_i represents the estimated parameters. The interferograms with obvious phase ramps were corrected using this model. After that correction process, we assumed that most of the topography related phase errors (DEM error and atmospheric delay) have been removed.

3.4.3. Atmospheric Phase Screen (APS) Removal

The residual phases for each interferogram were calculated by subtracting the estimated LP deformation and topographic error phase from the differential interferograms and unwrapped by the sparse Minimum Cost Flow (MCF) method [53]. The atmospheric phase was considered to consist of two components: topography related and non-topographic related [54]. The two components were estimated separately. The topography related component can be estimated by the M-estimated. The non-topographic related atmospheric phase component is highly correlated in space but poorly in time, which can be estimated from the resultant phase based on the low pass filtering operation in spatial domain and high pass filtering operation in the temporal domain. After removing the APS from each interferogram and applying additional least-square estimation, we obtained the time-series deformation map.

3.4.4. Parameter Estimation

After identification of the CPs, all the CPs were connected to further remove the effects of the atmospheric delay. The differences of those differential interferometric phase between the neighboring CPs in the ith interferograms can be written as:

$$\Delta\varphi^i{}_{model} = \Delta\varphi_{def,i}(\Delta R, \Delta A) + \Delta\varphi_{topo,i}(\Delta\tau) + \Delta\varepsilon_i \qquad (4)$$

where, $\Delta\varphi^i{}_{model}$ is the model phase difference of the neighboring two CPs in the ith interferograms. ΔR and ΔA are the differential rate of linear deformation and seasonal deformation (Equation (1)), respectively; $\Delta\varphi_{topo,i}$ is the residual topographic phase due to the DEM error ($\Delta\tau$); $\Delta\varepsilon_i$ denotes the phase noise.

The identified CPs were firstly connected based on the Delaunay triangulation network. Then, the differential phase of all the edges were calculated, which is beneficial to further remove atmospheric and orbital errors. The parameters $\Delta R, \Delta A$, and $\Delta\tau$ were optimally estimated for all of the edges using the periodogram approach [24]. After the differential parameters of all the edges had been estimated, a quality test was performed to reject links with temporal coherence lower than the threshold. In our experiment, the temporal coherence the threshold value is 0.7. Moreover, the edges with the length larger than 3 km were also rejected to mitigate the spatially-correlated phase errors, such as atmospheric delay. Then, a reference point was selected and we used the least-squares estimation to derive the parameters (R, A, and τ) of each point. We applied the temporal coherence as a weighting function during the inversion process. The estimated R and A are along the slant light of sight (LOS) direction. We assumed that the detected deformation is in vertical direction, and the LOS estimated deformation was converted to the vertical direction by dividing the cosine of the average incidence angles. The specific procedures of the approach are illustrated in Figure 6.

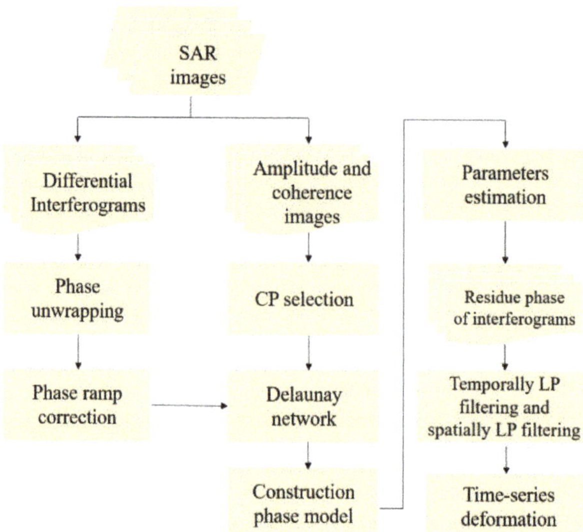

Figure 6. The flowchart of the time-series InSAR approach.

4. Results and Analysis

4.1. InSAR Results

Using the time-series InSAR method described above, the estimated average ground deformation rate along QTEC from the Wudaoliang to Tuotuohe sections using three C-band SAR stacks from 1997 to 2018 have been generated, including the deformation rate from 1997 to 1999 calculated with ERS-1 data (see Figure 7a), the deformation rate from 2004 to 2010 calculated using ENVISAT data (see Figure 7b), and the deformation rate from 2015 to 2018 calculated using Sentinel-1A data (see Figure 7c). The reference point (red star, Figure 7) was selected at the railway bridge. Negative deformation velocity represents an increasing distance with time away from the radar satellite; and positive deformation velocity indicates a decreasing distance towards the radar satellite. About over 100 km of the QTEC have been monitored. 40,760 CPs were detected for the ERS-1 along QTEC, and 125,522 CPs were detected for the ENVISAT, 217,096 CPs were selected for the Sentinel-1A. Figure 7d–f depict the estimated DEM errors of ERS-1, ENVISAT, and Sentinel-1A, respectively. The estimated DEM error ranged from −20 m to 10 m in most of the study area, which is consistent with the relative accuracy of the SRTM DEM. Most of the CPs are corresponded to QTR and QTH embankments, rocky mountains, and other artificial engineering structures. Before the 1999, the QTR and QTH were not completely constructed and few CPs were detected for ERS-1 data. For the Sentinel-1A, more SAR images are collected per year and more interferograms with less baseline were generated, so more CPs were detected.

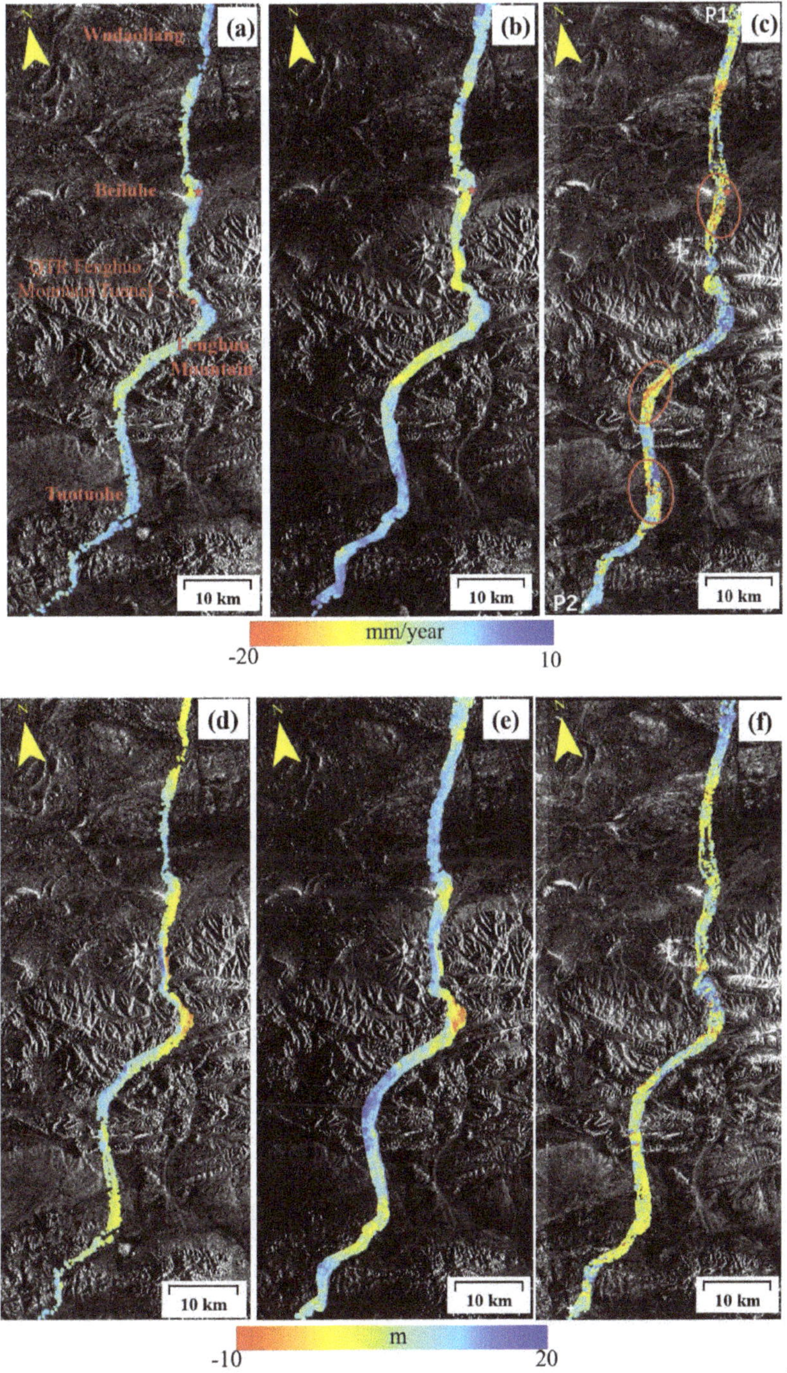

Figure 7. Estimated average ground deformation rate along the QTEC in (**a**) 1997–1999, (**b**) 2004–2010, and (**c**) 2015–2018 derived from the ERS-1, ENVISAT and Sentinel-1A data, respectively. The red star is the reference point. (**d**–**f**) are the corresponding estimated DEM error.

The ground deformation rate along the QTEC ranges from −10 to +10 mm/year during the 1997–1999 observation period derived from ERS-1 data. For the ENVISAT and Sentinel-1A experiments, the estimated deformation velocity was primarily in the range of −20 to +10 mm/year. The spatial distribution of the deformation before 2004 was quite different from those after 2004, and the deformation rate of the ERS-1 was inaccurate due to the few SAR datasets and heterogeneous spatial-temporal baseline.

In this study, we choose the QTR as an example to analyze the deformation of permafrost engineering. Figure 8 shows the deformation rate profile of QTR (from points P1 to P1 in Figure 7c) from 1997 to 2018. Through the above result, we found that before the opening of the QTR in 2006, the ground deformation along was relatively minimal. After the opening of the QTR, the overall mean deformation rate at the beginning and the end of QTR was within 10 mm/year. Four regions with obvious ground deformation in recent year have been detected. Regions A (Beiluhe) and B (south of Fenghuo Mountain) showed an obvious subsidence area, with the largest deformation rate being 15 mm/year. From 2015 to 2018, two more QTR section with ground deformation, Region C (Tuotuohe) and D, were detected, with the maximum ground displacement velocity over 17 mm/year. In some sections of the QTR, some cracks were found on the embankment shoulder and slopes through our field investigation. Long term monitoring is necessary in those areas. The surface subsidence along the embankment of QTR was primarily in the range of −20 mm/year to 5 mm/year. Human activities, such as embankment construction and railway operation, disrupt the original hydrothermal balance of the active layer, contributing to the obvious ground settlements [31].

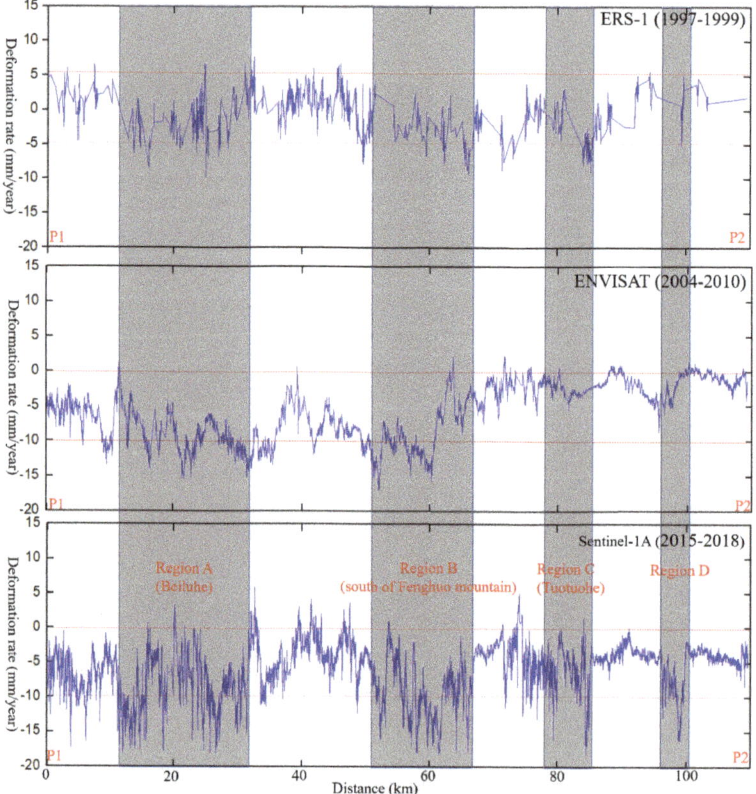

Figure 8. Deformation rate profile along the QTR, from point P1 to point P2 in Figure 7c.

4.2. Regional Analysis

Obvious deformation along QTEC was detected in three areas as enclosed by the red dashed ellipses in Figure 7 corresponding to the regions prone to subsidence based on previous investigations [37,46], i.e., Beiluhe, Fenghuo Mountain, and Tuotuohe areas. To analyze the deformation pattern along the QTEC, the detected obvious ground deformation regions in Beiluhe (Figures 9 and 10), Fenghuo Mountain (Figure 11), and Tuotuohe (Figure 12) are analyzed in detailed. A closer analysis of those three areas is provided below.

4.2.1. Beiluhe

The Beiluhe basin region is in the tundra of Hoh Xil and is underlain by cold permafrost. The terrain is relatively flat and most of the slope is less than 40°. The soil moisture content in the surface is high in the summer season and can reach 0.3 [55]. The vegetation coverage ranges from 0.3 to 0.9, which would contribute to serious temporal decorrelation. The Beiluhe permafrost region has been undergoing serious ground deformation in recent decades [5,7].

Figure 9 shows the mean LOS displacement rate in the Beiluhe permafrost region. Most of the selected points were located on the embankment of QTR and QTH. Fewer CPs are located on the alpine meadow areas due to serious temporal decorrelation. The primary displacement rate was in the range of −6 to 5 mm/year during 1997–1999 from ERS-1 dataset. The ENVISAT and Sentinel-1A results showed obvious ground deformation trend, with the larger deformation rate of −10 mm/year and −15mm/year respectively. Most of the deformation points are in the south of the region, which is consistent with the finding reported in previous studies [37,43].

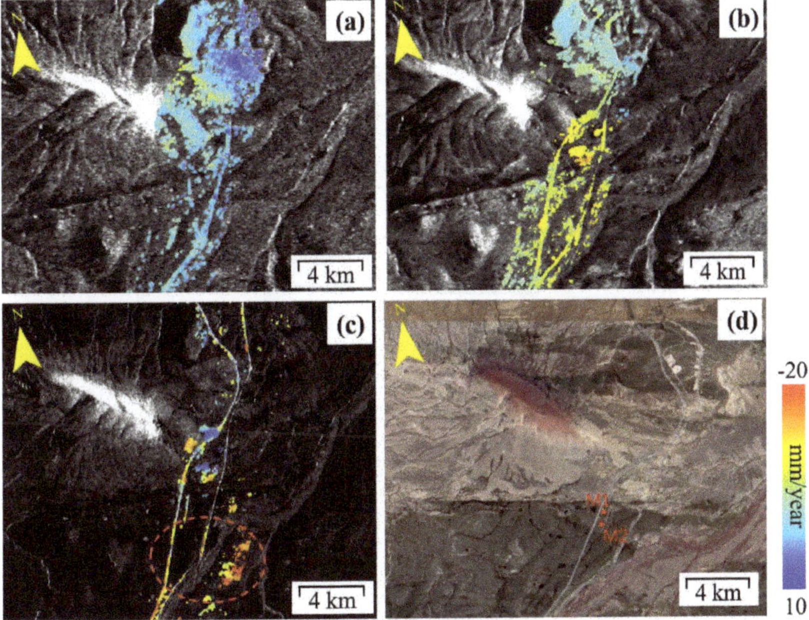

Figure 9. Permafrost ground deformation rate at the Beiluhe region from InSAR in 1997–2018. (**a**) ERS-1 1997–1999, (**b**) ENVISAT 2004–2010, (**c**) Sentinel-1A 2015–2018, and (**d**) the corresponding Google map.

During the field investigations in 2014 and 2015, some surface cracks or fissures of about 20 cm along the QTEC and alpine meadow regions were found in the Beiluhe regions, as shown in Figure 10a–c. The long-term active layer thawing-freezing effect caused long cracks in the alpine meadow areas.

Figure 10. Field photos with surface cracks and fissures in the Beiluhe region (March 2015). (**a**) Photo taken at the point M1 in Figure 9d. (**b**) and (**c**) Photos taken at the point M2 in Figure 9d.

4.2.2. Southern of Fenghuo Mountain

The Fenghuo Mountain, with an average elevation of more than 5000 m, is to the southeast of Hoh Xil, 380 km away from the city of Golmud. The 1.33 km long QTR Fenghuo Mountain tunnel was successfully traversed on 19 October, 2002. Figure 7 shows that the ground along the QTR Fenghuo Mountain tunnel was stable from 2004 to 2018 and no obvious deformation trend has detected. In the south of Fenghuo Mountain, visible ground displacement was found per the InSAR results. Figure 11 shows the mean LOS displacement rate at the south of the Fenghuo Mountain region. The ERS-1 result in Figure 11a shows that the ground is stable and the displacement rate is mostly less than −5 mm/year. During 2004–2010, the InSAR results showed obvious ground deformation in the north. During 2015–2018, the surface deformation was more severe, and obvious deformations have occurred throughout the region. The largest was −20 mm/year during 2015–2018. Most of the QTR embankment showed a minor deformation rate.

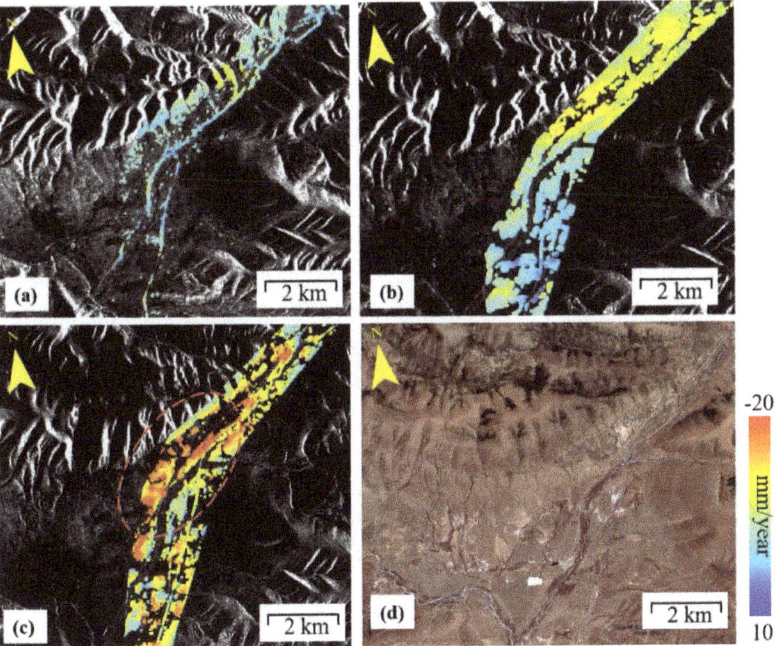

Figure 11. Permafrost ground deformation rate in the south of Fenghuo Mountain from InSAR during 1997–2018. (**a**) ERS-1 1997–1999, (**b**) ENVISAT 2004–2010, (**c**) Sentinel-1A 2015–2018, and (**d**) the corresponding Google map.

4.2.3. Tuotuohe

The average elevation of Tuotuohe region is about 4780 m. The ALT ranges from 1 to 4 m. Figure 12 shows the average deformation rate of Tuotuohe region during 1997–2018 from the ERS-1, ENVISAT and Sentinel-1A datasets. In the ERS-1 and ENVISAT deformation results (Figure 12a,b respectively), no obvious deformation area was found. From 2015 to 2018, serious deformation was found in this area, marked by red dotted ellipses. The largest deformation rate was over −20 mm/year per the Sentinel-1A results. Subsiding regions were found around the embankment of the QTR, which will be analyzed in the following section.

Figure 13 shows the time series displacement of the three selected points in this region. Because the number of the ERS was small, the timeseries displacements were analyzed for ENVISAT and sentinel-1A. The long-term subsidence was probably caused by melting of ground ice near the permafrost table [21]. The seasonal trend was remarkable, reflecting the effects of the thawing and freezing of the active layer. Points A, B and C exhibited the accumulative deformation less than 40 mm from 1997 to 2010. For the Sentinel-1A results, the time series displacement of the three points showed a similar seasonal trend, with the deformation rates of −8.5 mm/year, −20.1 mm/year and −11.9 mm/year, respectively. The accumulative displacement of Point B was 120 mm from 2015 to 2018. With time, deformations in parts of the Tuotuohe regions intensified. An increasing deformation trend was found in this region.

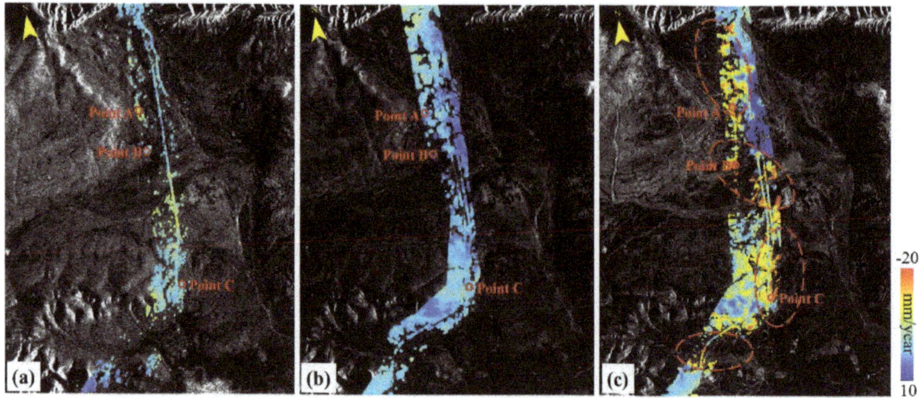

Figure 12. Permafrost ground deformation rate in the Tuotuohe region from InSAR during 1997–2018. (**a**) ERS-1 1997–1999, (**b**) ENVISAT 2004–2010, (**c**) Sentinel-1A 2015–2018.

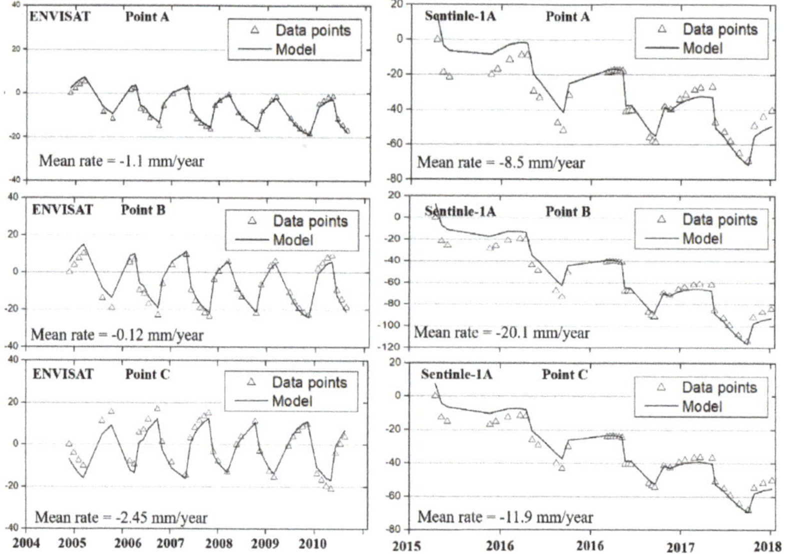

Figure 13. Time-series deformation of Point A, B, and C from 2004–2010 (ENVISAT) and 2015–2018 (Sentinel-1A). The hollow triangle indicates the time-series displacement using the InSAR method, and the black polylines denote the modeled deformation.

4.3. Deformation Analysis

4.3.1. Deformation and Permafrost Thermal Regimes

In the last 2010s, the permafrost in QTP underwent serious degradation due to global warming. During the period from 1961 to 2007, the observed air temperatures over the QTP showed a rising trend, with a mean increasing rate of 0.037 °C/year [56]. Against the background of global warming, the air temperatures over the QTP continued to rise. The ground deformation was a manifestation of the degeneration of the permafrost. The mean annual ground temperature (MAGT) is often used for permafrost thermal regime mapping on a large scale. The MAGT is correlated with the elevation, local slope, soil properties, vegetation, location, and other factors [43]. Lu et al. [57] proposed a relationship model between MAGT and the elevation, latitude and slope aspects from 29 boreholes along the QTEC from Beiluhe to Fenghuo Mountain. The multi-correlation coefficient is significant with a value of 0.936. The study area in Lu et al. [57] is the same as our study site and the model is easy to application. So, the model is used to monitor the MAGT and evaluate the stability of the permafrost in our study site.

The modeled MAGT of the study site is shown in Figure 14. The modeled MAGTs were the lowest for the Fenghuo Mountain areas with the temperature of less than −2.0 °C and the highest for the river valley areas Tuotuohe with the temperature above 0 °C. For the Beiluhe basin areas, the relatively warm MAGTs ranged from −2.0 to 0 °C. The modeled MAGTs are consistent with the latest researches on MAGTs in QTP [58,59]. Comparing Figures 7 and 14, we found that the subsiding regions are consistent with the ground with high MAGT value; the Tuotuohe and Beiluhe regions have experienced undergone serious ground deformation in recent years. High MAGTs would contribute to the acceleration of permafrost thawing and then increase the settlement of the ground.

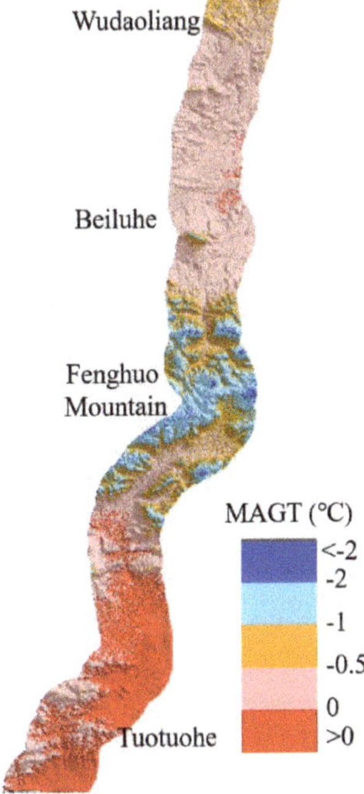

Figure 14. Map of the modeled mean annual ground temperature (MAGT) along the QTEC from Wudaoliang to Tuotuohe.

4.3.2. Deformation and Thaw Slumping

Thermokarst lakes have been developing along the QTEC as a result of increased human activity and ongoing climate warming [60]. The thermokarst lakes and thaw slumping have been observed more frequently in permafrost areas, such as the Beiluhe region and Fenghuo Mountain [61]. Thaw slumping has occurred near the embankments of QTR and QTH. In the regions with obvious ground deformation in our study area, some thaw slumps have been observed in the Tuotuohe region through the time series SAR amplitude maps.

Figure 15 shows the time-series amplitude maps of the Tuotuohe area from 2007 to 2018, the same area as that shown in Figure 12. At least three areas, marked as R1, R2 and R3, underwent thaw slumping throughout the whole observation period. By comparing Figures 12 and 15, we found that from 2007 to 2018, the areas experiencing thaw slumping in the three regions have increased by 0.435, 0.679, 0.317 km² respectively (Table 3). The distributions of thaw slumps areas are consistent with the ground deformation. The formation of thaw slumps may be initiated by several processes that expose ice-rich permafrost sloping terrain, which contributes to serious ground deformation [59]. The observed increase in areas of thermokarst lakes or thaw slumping regions indirectly validates our retrieved ground deformation result.

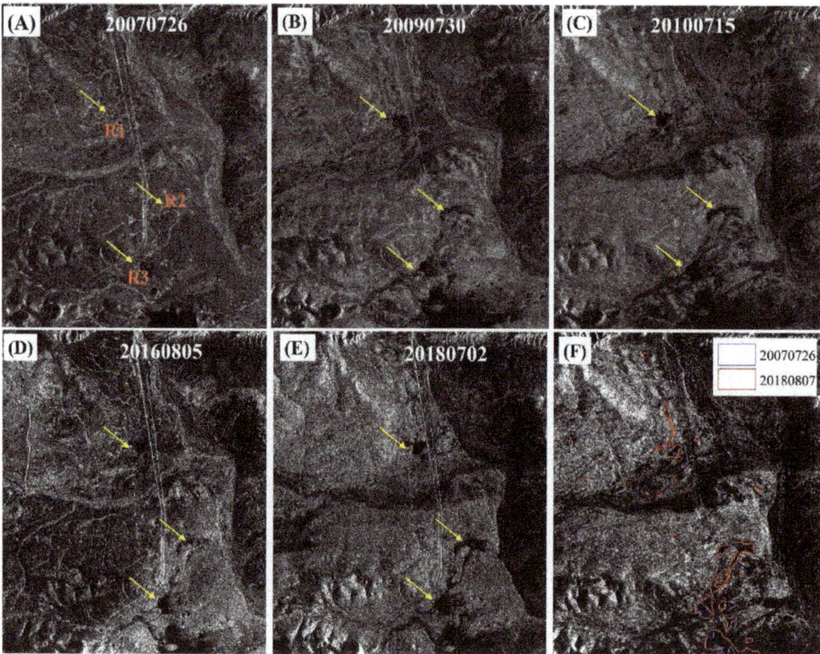

Figure 15. The time-series amplitude maps of Tuotuohe area, the same location as that shown in Figure 9: (**a**) 26 July 2007, (**b**) 30 July 2009, (**c**) 15 July 2010, (**d**) 5 August 2016, (**e**) 2 July 2018, and (**f**) water regions between 26 July 2007 and 7 August 2018. The blue polygon indicates 26 July 2007 and the red polygon indicates 7 August 2018. The yellow arrows indicate the water regions: R1, R2, and R3.

Table 3. Areas of regions R1, R2 and R3 between 2007-07-26 and 2018-08-07.

Region	Areas (km^2)		
	26 July 2007	**7 August 2018**	**Change**
R1	0.023	0.458	0.435
R2	0.068	0.747	0.679
R3	0.244	0.561	0.317

5. Discussion

We think that most of the embankments and foundations of the permafrost along QTEC are stable, but some sections are still experiencing obvious deformation. Based on the 20 years of InSAR observations, at least three regions have been identified as undergoing serious ground deformation, consistent with the previous studies in the QTP [30,31]. The ground deformation tends to expand. The embankments of QTR and QTH around Fenghuo Mountains should be reinforced as should points A, B and C near the Tuotuohe regions.

To evaluate the estimated results, the levelling measurement data should be collected. Because it is difficult for us to collect the levelling data in QTEC region, the estimated results could not be directly validated. However, several pieces of ground deformations evidences have been found in our field investigations that indirectly verifies the results. In the Beiluhe sections, visible fissures have been found in the QTR subgrade and alpine meadow region (Figure 10). We also compared our results with the previous studies in the QTP permafrost area (Table 1). In the Beiluhe area, several studies have been conducted on the deformation of permafrost using InSAR. Chen et al. [31] retrieved the ground deformation along the QTR in the Beiluhe area using C- and L-band small SAR interferometry. The estimated surface deformation rate along embankment ranges from −20 to +20 mm/year. Li et al., [32] monitored the surface deformation in the Beiluhe area using InSAR with ENVISAT images. The deformation velocity near the QTR embankment was larger than −10 mm/year. Similarly, our previous studies in the Beiluhe regions with TerraSAR-X ST mode images showed the similar deformation trends, with the deformation rate ranging from −20 to 0 mm/year [37]. Our retrieved ground deformation rate is consistent with those studies. The small differences between our findings and those reported by the previous studies are due to the following aspects: (1) different band SAR images and the InSAR processing method were used, which contributed to this difference, and (2) the observation periods were difference. Despite these case studies being conducted at different time periods, the gradual subsidence trends were all in the order of centimeters per year, similar to our reported subsidence trends. Most of the previous studies used the SAR images acquired before 2010. In this study, the latest ground deformation along QTEC were obtained.

There are three limitations in this study. Firstly, due to the complexity of the permafrost thawing and freezing process, monitoring the ground deformation using a physical equation was challenging. Linear [29,46], cubic polynomial [31], seasonal [33,36,49], and equation with climatic factors [35] and temperature [21,37,51] phase models have been used. These models have been applied successfully in some permafrost regions. Many other factors, such as vegetation coverage, soil moisture, snow cover, and solar radiation, should be considered in the future when monitoring the permafrost deformation.

Another limitation of the InSAR applications on permafrost regions is the temporal decorrelations [31,46]. The permafrost surface experiences dynamic environmental conditions and severe climate change from summer to winter season, which result in the dramatic temporal variations in the ground surface. Many studies used the SAR datasets acquired in the winter season [21,51] or use the L-band SAR images [62] to suppress the temporal decorrelations. Some methods and advanced methods have been proposed to solve this difficulty. Daout et al., 2017 used a PCA approach to help for the unwrapping in the decorrelated permafrost environment [39]. With the launches of satellites with long-wavelength SAR sensors such as ALOS-2, and the shortening of the satellite revisit cycle, and the development of advanced algorithm, InSAR technology (distributed scatterer interferometry, DSI) [63], the temporal decorrelation will be largely suppressed.

Last, comparing the estimated deformation rate and DEM error term, we found that they are the trade-offs for the ERS-1 images. We think at least two factors contribute to this. Firstly, a covariance exists between the temporal and perpendicular baseline, especially for ERS-1 data. The smaller the spatial perpendicular baseline, the higher the quality of the interferograms. The smaller the temporal baseline, the higher the quality of the interferograms. However, in the permafrost areas, the quality of the interferograms would be better between two images acquired in the same season with large temporal baseline and some interferograms with small temporal baseline are rejected due to serious temporal decorrelation. Secondly, in the permafrost region, the deformation may be correlated with the topography. Most of the subsiding areas are the plane regions (Beiluhe and Tuotuohe). In the mountainous areas, the deformation rate is small and stable. More SAR images with short revisit cycle are needed in the future research.

6. Conclusions

In this paper, we presented an application using the time-series InSAR technique with multisensory SAR datasets to monitor the permafrost ground deformation along the QTEC from 1997 to 2018. A deformation model combining a linear subsidence term and seasonal deformation term was adopted in the time-series InSAR method to exploit the permafrost ground deformation. Three deformation rate maps along a 100 km stretch of the QTEC were generated from 9 ERS-1, 39 ENVISAT, and 41 Sentinel-1A images. The three independent InSAR measurement results showed a consistent deformation trend and most of the ground surface along the QTEC was stable with the deformation rate ranging from −10 to 10 mm/year. The conclusions are summarized as follows:

(1) Before the operation of the QTR, the QTEC from Wudaoliang to Tuotuohe was in stable with a deformation velocity of less than −5 mm/year from ERS-1 images. The embankment of the engineering structure was considered stable. The thawing and freezing of the active layer were the main deformation driving-forces. After the QTR started operation and the human activities increased, some sections of the QTEC were underwent obvious deformation, and the deformation has increased more recently.

(2) From 2015 to 2018, obvious deformation was found in three areas: Beiluhe, southern of Fenghuo Mountain, and Tuotuohe, with the large deformation rates of over −20 mm/year. Real-time deformation monitoring must be conducted in these sections. The subsiding areas are consistent with the permafrost areas with large MAGTs.

(3) This work demonstrated the potential of the time-series InSAR for the surveillance of the state of QTEC on a large scale. Interferometric decorrelation is still one of the problems for time-series InSAR monitoring of the ground deformation in permafrost region. With the proposed innovative methods and newly-launched SAR systems with shorter revisit cycles (Sentinel-1A/1B and TerraSAR-L), higher temporal sampling allows us to better characterize the ground deformation related to the process of permafrost thawing and freezing.

In future work we will focus on investigating the three-dimension deformation in permafrost regions using multiple satellites SAR images, and retrieving the geophysical parameters of permafrost such as the active layer thickness, on a larger scale.

Author Contributions: All four authors contributed to this work. M.W. collected the SAR data and provided valuable suggestions for the revision. Z.W. provided the validation and analysis. X.L. implemented the methodology, results interpretation. Z.Z. designed the research program and supervised the research and finished the manuscript.

Funding: Please add: This work was supported in part by the funded the National Natural Science Foundation of China under Grant (41801348, 61801443, 41771467), the Open Research Fund of Key Laboratory of Digital Earth Science, Institute of Remote Sensing and Digital Earth, Chinese Academy of Sciences (No. 2017LDE006) and the State Key Laboratory of Resources and Environmental Information System.

Acknowledgments: The authors would like to thank European Space Agency (ESA) for providing free and open Sentinel-1A data, ERS-1 and Envisat Historical archived data.

Conflicts of Interest: The authors declare no conflict of interest.

References

1. Short, N.; Brisco, B.; Couture, N.; Pollard, W.; Murnaghan, K.; Budkewitsch, P. A comparison of Terrasar-x, Radarsat-2 and ALOS-PALSAR interferometry for monitoring permafrost environments, case study from Herschel Island, Canada. *Remote Sens. Environ.* **2011**, *115*, 3491–3506. [CrossRef]

2. Widhalm, B.; Bartsch, A.; Leibman, M.; Khomutov, A. Active-layer thickness estimation from x-band SAR backscatter intensity. *Cryosphere* **2017**, *11*, 483–496. [CrossRef]

3. Liu, X.; Chen, B. Climatic warming in the Tibetan Plateau during recent decades. *Int. J. Climatol.* **2000**, *20*, 1729–1742. [CrossRef]

4. Osterkamp, T.E. Characteristics of the recent warming of permafrost in Alaska. *J. Geophys. Res. Earth Surf.* **2007**, *112*. [CrossRef]

5. Wu, Q.; Zhang, T.; Liu, Y. Permafrost temperatures and thickness on the Qinghai-Tibet Plateau. *Global Planet. Chang.* **2010**, *72*, 32–38. [CrossRef]

6. Niu, F.; Lin, Z.; Liu, H.; Lu, J. Characteristics of thermokarst lakes and their influence on permafrost in Qinghai–Tibet Plateau. *Geomorphology* **2011**, *132*, 222–233. [CrossRef]

7. Cheng, G. A roadbed cooling approach for the construction of Qinghai–Tibet Railway. *Cold Reg. Sci. Technol.* **2005**, *42*, 169–176. [CrossRef]

8. Ma, W.; Mu, Y.; Wu, Q.; Sun, Z.; Liu, Y. Characteristics and mechanisms of embankment deformation along the Qinghai–Tibet railway in permafrost regions. *Cold Reg. Sci. Technol.* **2011**, *67*, 178–186. [CrossRef]

9. Wu, Q.; Shi, B.; Fang, H.Y. Engineering geological characteristics and processes of permafrost along the Qinghai–Xizang (Tibet) Highway. *Eng. Geol.* **2003**, *68*, 387–396. [CrossRef]

10. He, R.; Jin, H. Permafrost and cold-region environmental problems of the oil product pipeline from Golmud to Lhasa on the Qinghai–Tibet Plateau and their mitigation. *Cold Reg. Sci. Technol.* **2010**, *64*, 279–288. [CrossRef]

11. Yu, Q.; Zhang, Z.; Wang, G.; Lei, G.; Wang, X.; Wang, P.; Bao, Z. Analysis of tower foundation stability along the qinghai–tibet power transmission line and impact of the route on the permafrost. *Cold Reg. Sci. Technol.* **2016**, *121*, 205–213. [CrossRef]

12. Jin, H.; Wei, Z.; Wang, S.; Yu, Q.; Lü, L.; Wu, Q.; Ji, Y. Assessment of frozen-ground conditions for engineering geology along the Qinghai–Tibet highway and railway, China. *Eng. Geol.* **2008**, *101*, 96–109. [CrossRef]

13. Cheng, G.; Wu, T. Responses of permafrost to climate change and their environmental significance, qinghai-tibet plateau. *J. Geophys. Res.* **2007**, *112*. [CrossRef]

14. Pastick, N.J.; Jorgenson, M.T.; Wylie, B.K.; Nield, S.J.; Johnson, K.D.; Finley, A.O. Distribution of near-surface permafrost in Alaska: Estimates of present and future conditions. *Remote Sens. Environ.* **2015**, *168*, 301–315. [CrossRef]

15. Rouyet, L.; Lauknes, T.R.; Christiansen, H.H.; Strand, S.M.; Larsen, Y. Seasonal dynamics of a permafrost landscape, Adventdalen, Svalbard, investigated by InSAR. *Remote Sens. Environ.* **2019**, *231*, 111236. [CrossRef]

16. Amelung, F.; Galloway, D.L.; Bell, J.W.; Zebker, H.A.; Laczniak, R.J. Sensing the ups and downs of lavages: InSAR reveals structural control of land subsidence and aquifer-system deformation. *Geology* **1999**, *27*, 483. [CrossRef]

17. Massonnet, D.; Briole, P.; Arnaud, A. Deflation of Mount Etna monitored by spaceborne radar interferometry. *Nature* **1995**, *375*, 567–570. [CrossRef]

18. Massonnet, D.; Rossi, M.; Carmona, C.; Adragna, F.; Peltzer, G.; Feigl, K.; Rabaute, T. The displacement field of the Landers earthquake mapped by radar interferometry. *Nature* **1993**, *364*, 138–142. [CrossRef]

19. Liu, Y.; Zhao, C.; Zhang, Q.; Yang, C.; Zhang, J. Land subsidence in Taiyuan, China, monitored by InSAR technique with multisensor SAR datasets from 1992 to 2015. *IEEE J. Sel. Top. Appl. Earth Obs. Remote Sens.* **2018**, *11*, 1509–1519. [CrossRef]

20. Ma, P.; Wang, W.; Zhang, B.; Wang, J.; Shi, G.; Huang, G.; Chen, F.; Jiang, L.; Lin, H. sensing large- and small-scale ground subsidence: A case study of the Guangdong–Hong Kong–Macao Greater Bay Area of China. *Remote Sens. Environ.* **2019**, *232*, 111282. [CrossRef]

21. Liu, L.; Schaefer, K.; Zhang, T.; Wahr, J. Estimating 1992–2000 average active layer thickness on the Alaskan North Slope from remotely sensed surface subsidence. *J. Geophys. Res. Earth Surf.* **2012**, *117*. [CrossRef]

22. Short, N.; Leblanc, A.M.; Sladen, W.; Oldenborger, G.; Mathon-Dufour, V.; Brisco, B. Radarsat-2 D-InNSAR for ground displacement in permafrost terrain, validation from Iqaluit airport, Baffin Island, Canada. *Remote Sens. Environ.* **2014**, *141*, 40–51. [CrossRef]

23. Zebker, H.A.; Villasenor, J. Decorrelation in interferometric radar echoes. *IEEE Trans. Geosci. Remote Sens.* **1992**, *30*, 950–959. [CrossRef]

24. Ferretti, A.; Prati, C.; Rocca, F. Nonlinear subsidence rate estimation using permanent scatterers in differential SAR interferometry. *IEEE Trans. Geosci. Remote Sens.* **2002**, *38*, 2202–2212. [CrossRef]

25. Hooper, A.; Zebker, H.; Segall, P.; Kampes, B. A new method for measuring deformation on volcanoes and other natural terrains using INSAR persistent scatterers. *Geophys. Rese. Lett.* **2004**, *31*. [CrossRef]

26. Berardino, P.; Fornaro, G.; Lanari, R.; Sansosti, E. A new algorithm for surface deformation monitoring based on small baseline differential SAR interferograms. *IEEE Trans. Geosci. Remote Sens.* **2002**, *40*, 2375–2383. [CrossRef]

27. Mora, O.; Mallorqui, J.J.; Broquetas, A. Linear and nonlinear terrain deformation maps from a reduced set of interferometric SAR images. *IEEE Trans. Geosci. Remote Sens.* **2003**, *41*, 2243–2253. [CrossRef]

28. Perissin, D.; Wang, T. Repeat-pass SAR interferometry with partially coherent targets. *IEEE Trans. Geosci. Remote Sens.* **2012**, *50*, 271–280. [CrossRef]

29. Xie, C.; Li, Z.; Xu, J.; Li, X. Analysis of deformation over permafrost regions of Qinghai-Tibet plateau based on permanent scatterers. *Int. J. Remote Sens.* **2010**, *31*, 1995–2008. [CrossRef]

30. Chen, F.; Lin, H.; Li, Z.; Chen, Q.; Zhou, J. Interaction between permafrost and infrastructure along the Qinghai–Tibet Railway detected via jointly analysis of C-and L-band small baseline SAR interferometry. *Remote Sens. Environ.* **2012**, *123*, 532–540. [CrossRef]

31. Chen, F.; Lin, H.; Zhou, W.; Hong, T.; Wang, G. Surface deformation detected by ALOS PALSAR small baseline SAR interferometry over permafrost environment of Beiluhe section, Tibet Plateau, China. *Remote Sens. Environ.* **2013**, *138*, 10–18. [CrossRef]

32. Li, Z.; Tang, P.; Zhou, J.; Tian, B.; Chen, Q.; Fu, S. Permafrost environment monitoring on the Qinghai-Tibet Plateau using time series ASAR images. *Int. J. Digit. Earth* **2015**, *8*, 840–860. [CrossRef]

33. Chang, L.; Hanssen, R.F. Detection of permafrost sensitivity of the Qinghai–Tibet railway using satellite radar interferometry. *Int. J. Remote Sens.* **2015**, *36*, 691–700. [CrossRef]

34. Li, Y.; Zhang, J.; Li, Z.; Luo, Y.; Jiang, W.; Tian, Y. Measurement of subsidence in the Yangbajing geothermal fields, Tibet, from TerraSAR-X InSAR time series analysis. *Int. J. Digit. Earth* **2016**, *9*, 697–709. [CrossRef]

35. Zhao, R.; Li, Z.W.; Feng, G.C.; Wang, Q.J.; Hu, J. Monitoring surface deformation over permafrost with an improved SBAS-InSAR algorithm: With emphasis on climatic factors modeling. *Remote Sens. Environ.* **2016**, *184*, 276–287. [CrossRef]

36. Jia, Y.; Kim, J.W.; Shum, C.; Lu, Z.; Ding, X.; Zhang, L.; Erkan, K.; Kuo, C.-Y.; Shang, K.; Tseng, K.-H.; et al. Characterization of active layer thickening rate over the northern Qinghai-Tibetan plateau permafrost region using ALOS interferometric synthetic aperture radar data, 2007–2009. *Remote Sens.* **2017**, *9*, 84. [CrossRef]

37. Wang, C.; Zhang, Z.; Zhang, H.; Zhang, B.; Tang, Y.; Wu, Q. Active Layer Thickness Retrieval of Qinghai–Tibet Permafrost Using the TerraSAR-X InSAR Technique. *IEEE J. Sel. Top. Appl. Earth Obs. Remote Sens.* **2018**, *11*, 4403–4413. [CrossRef]

38. Zhang, X.; Zhang, H.; Wang, C.; Zhang, B.; Wu, F.; Wang, J.; Zhang, Z. Time-Series InSAR Monitoring of Permafrost Freeze-Thaw Seasonal Displacement over Qinghai–Tibetan Plateau Using Sentinel-1 Data. *Remote Sens.* **2019**, *11*, 1000. [CrossRef]

39. Daout, S.; Doin, M.P.; Peltzer, G.; Socquet, A.; Lasserre, C. Large-scale InSAR monitoring of permafrost freeze-thaw cycles on the Tibetan Plateau. *Geophys. Rese. Lett.* **2017**, *44*, 901–909. [CrossRef]

40. Bartsch, A.; Leibman, M.; Strozzi, T.; Khomutov, A.; Widhalm, B.; Babkina, E.; Mullanurov, D.; Ermokhina, K.; Kroisleitner, C.; Bergstedt, H. Seasonal Progression of Ground Displacement Identified with Satellite Radar Interferometry and the Impact of Unusually Warm Conditions on Permafrost at the Yamal Peninsula in 2016. *Remote Sens.* **2019**, *11*, 1865. [CrossRef]

41. Dai, K.; Li, Z.; Tomás, R.; Liu, G.; Yu, B.; Wang, X.; Cheng, H.; Chen, J.; Stockamp, J. Monitoring activity at the Daguangbao mega-landslide (China) using Sentinel-1 TOPS time series interferometry. *Remote Sens. Environ.* **2016**, *186*, 501–513. [CrossRef]

42. Hooper, A. A multi-temporal InSAR method incorporating both persistent scatterer and small baseline approaches. *Geophys. Rese. Lett.* **2008**, *35*. [CrossRef]

43. Wu, Q.; Zhang, T.; Liu, Y. Thermal state of the active layer and permafrost along the Qinghai-Xizang (Tibet) Railway from 2006 to 2010. *Cryosphere* **2012**, *6*, 607–612. [CrossRef]

44. Wang, M.; He, G.; Zhang, Z.; Wang, G.; Zhang, Z.; Cao, X.; Wu, Z.; Liu, X. Comparison of spatial interpolation and regression analysis models for an estimation of monthly near surface air temperature in China. *Remote Sens.* **2017**, *9*, 1278. [CrossRef]

45. Jarvis, A.; Reuter, H.I.; Nelson, A.; Guevara, E. Hole-Filled Seamless SRTM Data V4. International Centre for Tropical Agriculture (CIAT). 2008. Available online: http://srtm.csi.cgiar.org (accessed on 30 October 2018).

46. Wang, C.; Zhang, Z.; Zhang, H.; Wu, Q.; Zhang, B.; Tang, Y. Seasonal deformation features on Qinghai-Tibet railway observed using time-series InSAR technique with high-resolution TerraSAR-X images. *Remote Sens. Lett.* **2017**, *8*, 1–10. [CrossRef]

47. Zhang, Z.; Wang, C.; Zhang, H.; Tang, Y.; Liu, X. Analysis of permafrost region coherence variation in the Qinghai–Tibet Plateau with a high-resolution TerraSAR-X image. *Remote Sens.* **2018**, *10*, 298. [CrossRef]

48. Lanari, R.; Mora, O.; Manunta, M.; Mallorquí, J.J.; Berardino, P.; Sansosti, E. A small-baseline approach for investigating deformations on full-resolution differential SAR interferograms. *IEEE Trans. Geosci. Remote Sens.* **2004**, *42*, 1377–1386. [CrossRef]

49. Goldstein, R.M.; Werner, C.L. Radar interferogram filtering for geophysical applications. *Geophys. Res. Lett.* **1998**, *25*, 4035–4038. [CrossRef]

50. Li, Z.; Zhao, R.; Hu, J.; Wen, L.; Feng, G.; Zhang, Z.; Wang, Q. InSAR analysis of surface deformation over permafrost to estimate active layer thickness based on one-dimensional heat transfer model of soils. *Sci. Rep.* **2015**, *5*, 15542. [CrossRef]

51. Liu, L.; Schaefer, K.; Gusmeroli, A.; Grosse, G.; Jones, B.M.; Zhang, T.; Parsekian, A.D.; Zebker, H.A. Seasonal thaw settlement at drained thermokarst lake basins, Arctic Alaska. *Cryosphere* **2014**, *8*, 815–826. [CrossRef]

52. Sun, Q.; Zhang, L.; Ding, X.L.; Hu, J.; Li, Z.W.; Zhu, J.J. Slope deformation prior to Zhouqu, China landslide from InSAR time series analysis. *Remote Sens. Environ.* **2015**, *156*, 45–57. [CrossRef]

53. Costantini, M. A novel phase unwrapping method based on network programming. *IEEE Trans. Geosci. Remote Sens.* **1998**, *36*, 813–821. [CrossRef]

54. Ge, L.; Ng, A.H.M.; Li, X.; Abidin, H.Z.; Gumilar, I. Land subsidence characteristics of Bandung Basin as revealed by ENVISAT ASAR and ALOS PALSAR interferometry. *Remote Sens. Environ.* **2014**, *154*, 46–60. [CrossRef]

55. Wang, C.; Zhang, Z.; Paloscia, S.; Zhang, H.; Wu, F.; Wu, Q. Permafrost Soil Moisture Monitoring Using Multi-Temporal TerraSAR-X Data in Beiluhe of Northern Tibet, China. *Remote Sens.* **2018**, *10*, 1577. [CrossRef]

56. Li, X.; Cheng, G.; Jin, H.; Kang, E.; Che, T.; Jin, R.; Wu, L.; Nan, Z.; Wang, J.; Shen, Y. Cryospheric change in China. *Global Planet. Chang.* **2008**, *62*, 210–218. [CrossRef]

57. Lu, J.H.; Niu, F.J.; Lin, Z.J.; Liu, H.; Luo, J. Permafrost modeling and mapping along the Qinghai–Tibet engineering corridor considering slope-aspect. *Geogr. Geoinf. Sci.* **2012**, *28*, 63–67. (In Chinese)

58. Yin, G.; Zheng, H.; Niu, F.; Luo, J.; Lin, Z.; Liu, M. Numerical Mapping and Modeling Permafrost Thermal Dynamics across the Qinghai-Tibet Engineering Corridor, China Integrated with Remote Sensing. *Remote Sens.* **2018**, *10*, 2069. [CrossRef]

59. Niu, F.; Yin, G.; Luo, J.; Lin, Z.; Liu, M. Permafrost distribution along the Qinghai-Tibet Engineering Corridor, China using high-resolution statistical mapping and modeling integrated with remote sensing and GIS. *Remote Sens.* **2018**, *10*, 215. [CrossRef]

60. Niu, F.; Lin, Z.; Lu, J.; Luo, J.; Wang, H. Assessment of terrain susceptibility to thermokarst lake development along the Qinghai–Tibet engineering corridor, China. *Environ. Earth Sci.* **2015**, *73*, 5631–5642. [CrossRef]

61. Luo, J.; Niu, F.; Lin, Z.; Liu, M.; Yin, G. Recent acceleration of thaw slumping in permafrost terrain of Qinghai-Tibet Plateau: An example from the Beiluhe Region. *Geomorphology* **2019**, *341*, 79–85. [CrossRef]

62. Dini, B.; Daout, S.; Manconi, A.; Loew, S. Classification of slope processes based on multitemporal DInSAR analyses in the Himalaya of NW Bhutan. *Remote Sens. Environ.* **2019**, *233*, 111408. [CrossRef]

63. Ferretti, A.; Fumagalli, A.; Novali, F.; Prati, C.; Rocca, F.; Rucci, A. A new algorithm for processing interferometric data-stacks: SqueeSAR. *IEEE Trans. Geosci. Remote Sens.* **2011**, *49*, 3460–3470. [CrossRef]

Article

ScanSAR Interferometry of the Gaofen-3 Satellite with Unsynchronized Repeat-Pass Images

Zaoyu Sun *, Anxi Yu, Zhen Dong and Hui Luo

College of Electronic Science and Technology, National University of Defense Technology, No. 109 Deya Road, Changsha 410073, China; yu_anxi@nudt.edu.cn (A.Y.); dongzhen@nudt.edu.cn (Z.D.); luohui@nudt.edu.cn (H.L.)
* Correspondence: sunzaoyu@nudt.edu.cn or sun_zyu@163.com; Tel.: +86-1346-761-1715

Received: 9 August 2019; Accepted: 23 October 2019; Published: 28 October 2019

Abstract: Gaofen-3 is a Chinese remote sensing satellite with multiple working modes, among which the scanning synthetic aperture radar (ScanSAR) mode is used for wide-swath imaging. synthetic aperture radar (SAR) interferometry in the ScanSAR mode provides the most rapid way to obtain a global digital elevation model (DEM), which can also be realized by Gaofen-3. Gaofen-3 ScanSAR interferometry works in the repeat-pass mode, and image pair non-synchronizations can influence its performance. Non-synchronizations can include differences of burst central times, satellite velocities, and burst durations. Therefore, it is necessary to analyze their influences and improve the interferometric coherence. Meanwhile, interferometric phase compensation and rapid DEM geolocation also need to be considered in interferometric processing. In this paper, interferometric coherence was analyzed in detail, followed by an iterative filtering method, which helped to improve the interferometric performance. Further, a phase compensation method for Gaofen-3 was proposed to compensate for the phase error caused by the unsynchronized azimuth time offset of image pair, and a closed-form solution of DEM geolocation with ground control point (GCP) information was derived. Application of our methods to a pair of Gaofen-3 interferometric images showed that these methods were able to process the images with good accuracy and efficiency. Notably, these analysis and processing methods can also be applied to other SAR satellites in the ScanSAR mode to obtain DEMs with high quality.

Keywords: Gaofen-3 satellite; ScanSAR; interferometry; interferometric coherence; phase compensation; DEM geolocation

---

1. Introduction

Launched on 10 August 2016, Gaofen-3 is a Chinese high-resolution remote-sensing satellite with a C-band multi-polarization synthetic aperture radar (SAR) payload [1]. Since then, it has been widely used in ocean surveillance, land management, ship detection, disaster reduction, and so on [2–8]. It can also be used with the SAR interferometry technique to extract a digital elevation model (DEM) of the Earth. SAR interferometry utilizes image phases, which contain topographic information, to obtain three-dimensional coordinates of the Earth's surface. Because of its outstanding performance, it has become an important DEM mapping technique.

Gaofen-3 works in a sun-synchronous orbit, and its altitude is about 755 km. The revisiting period of Gaofen-3 is 29 days. Gaofen-3 can work in many working modes with different resolutions and swath characteristics, such as stripmap mode, spotlight mode, and scanning synthetic aperture radar (ScanSAR) mode. In the spotlight mode, the resolution is 1 m and the swath is 10 km × 100 km. In the ultra-fine stripmap mode, the resolution is 3 m and the swath is 30 km. In the standard stripmap mode, the resolution is 25 m and the swath is 130 km. In the narrow ScanSAR mode, the resolution is 50 m and the swath is 300 km. In the wide ScanSAR mode, the resolution is 100 m and the swath is 500 km.

Among these modes, the ScanSAR mode is important as it can achieve wide-swath SAR images. SAR interferometry in ScanSAR mode can be used for wide-area topographic mapping because of this capability. This technique is worthy of in-depth research as a rapid global DEM-mapping method. In SAR interferometry, at least two images are needed, and this paper only considered two. The two SAR images used for Gaofen-3 interferometry are achieved in a repeat-pass mode.

For spaceborne remote sensing toward the Earth, the ScanSAR mode was first used in Spaceborne Imaging Radar-C (SIR-C) to acquire several experimental data. The SIR-C system was installed on a space shuttle and the mission was carried out in 1994 [9]. The Canadian satellite RADARSAT launched in 1995 was the first spaceborne SAR system with an operational ScanSAR mode [9]. Subsequently, SAR interferometry in ScanSAR mode has been deeply studied and widely used. The concept of ScanSAR interferometry was proposed in 1995 by Guarnieri [10]. He detailed ScanSAR interferometry and verified the interferometric method using simulated ERS-1 SAR data [11]. Bamler presented a ScanSAR interferogram using real RADARSAT data for the first time in 1999 [12], and in 2002, a complete description of RADARSAT ScanSAR interferometry was published [13]. In 2000, the Shuttle Radar Topography mission (SRTM) was carried out to map the world's landmass. This project demonstrated the rapid mapping ability of ScanSAR interferometry, which was able to map the landmass of the Earth in 10 days [14]. SAR interferometry in ScanSAR mode has also been used in other satellites, such as ENVISAT [15,16], ALOS [17], ALOS-2 [18], and TerraSAR-X [19]. The Gaofen-3 satellite can also work in ScanSAR mode, and it is necessary to study its interferometry. In the above studies, the master and slave images used the same observing parameters. However, in Gaofen-3 ScanSAR interferometry, the images are unsynchronized and may have different pulse repetition frequencies (PRFs), velocities, and burst durations. These differences, together with the burst central time difference, influence the interferometric coherence. It is necessary to analyze these influences and present a corresponding filtering method to improve the interferometric coherence. Between the master and slave images, the unsynchronized azimuth time offset causes a phase error when there is no phase adjustment during imaging; thus, interferometric phase compensation is needed. This compensation is a problem that has not yet been studied. From the compensated interferometric phase, we can determine the DEM. In DEM geolocation integrated with the absolute phase calculation and calibration, the most efficient method is to determine a closed-form solution. It is necessary to derive a closed-form solution for DEM geolocation combined with absolute phase calculation and phase error compensation.

This paper discusses several questions in Gaofen-3 ScanSAR interferometry, and is divided into seven sections. Section 2 analyzes the interferometric performance of Gaofen-3 in ScanSAR mode. Section 3 presents the iterative filtering method to improve interferometric performance. Section 4 proposes a compensation for the interferometric phase in Gaofen-3 ScanSAR interferometry. Section 5 derives a closed-form solution of DEM geolocation with ground control point (GCP) information. Processing experiments with Gaofen-3 interferometric images in ScanSAR mode were made to verify the analyses and methods in Section 6. Conclusions are drawn in Section 7.

2. Interferometric Model and Performance of Gaofen-3

The ScanSAR mode is a SAR mode with a width swath. By beam scanning, ScanSAR can observe several sub-swathes simultaneously. These sub-swathes are located at different points along the range direction. Together, they can cover a whole wide swath. Because only a single beam is used in ScanSAR, the observing time must be separated and allocated to different sub-swathes. Thus, for a single sub-swath, the observing signals are in the burst mode. During bursts, signal pulses are transmitted to the sub-swath, but no signal pulses are used between bursts. In burst mode, the azimuth resolution is decreased. ScanSAR can cover a width swath but with low resolution. Thus, ScanSAR is suitable for rapid mapping, but not suitable for subtle measurement. ScanSAR interferometry is based on the ScanSAR mode, so it has similar characteristics.

The ScanSAR interferometry of Gaofen-3 works in a repeat-pass mode. The two images achieved from the two passes may be unsynchronized. In this paper, we used a pair of Gaofen-3 interferometric images over Kunlun Mountain. These images were achieved in ScanSAR mode for wide-swath remote sensing. The main parameters are listed as Table 1.

Table 1. Main parameters of the Gaofen-3 interferometric images. PRF: pulse repetition frequency.

Parameters	Master Image	Slave Image
Central frequency (GHz)	5.4	5.4
Center look angle (°)	34.7	34.7
PRF (Hz)	1185.637085	1190.421753
Satellite velocity (km/s)	7.5674	7.5679
Band width (MHz)	30	30
Pulse width (μs)	45	45
Pulse number	100	100

In the parameters, the PRFs and velocities are different. Burst durations are decided by PRFs and pulse numbers, so they were also different. Because there was no burst synchronization between the two images, a burst central time difference also existed. These unsynchronized characteristics influence interferometric performance.

From the ScanSAR principle, the ScanSAR mode observes the Earth's surface only in bursts. It is different from the normal stripmap mode, which uses continuous observation. Thus, the interferometric performance of the ScanSAR mode needs to consider these burst characteristics in the signal model. The burst characteristics also include the above-mentioned unsynchronizations. This paper analyzed the interferometric performance of the ScanSAR signal model [20]:

$$s(t, \tau) = \sum_{n=0}^{N} rect\left(\frac{t-T_c-nT_d}{T_b}\right) \cdot \iint_D V_d(r_p, t_p) W(t - t_p - t_x) \exp\left[-j\frac{4\pi}{\lambda} R(r_p, t - t_p)\right]$$
$$\cdot a\left[\tau - \frac{2R(r_p, t-t_p)}{c}\right] \cdot \exp\left\{j\pi k\left[\tau - \frac{2R(r_p, t-t_p)}{c}\right]^2\right\} dr_p dt_p \tag{1}$$

where

$$R(r_p, t - t_p) = \sqrt{r_p^2 + \left[v_r(t - t_p)\right]^2} \tag{2}$$

where t is the slow time, τ is the fast time, V_d is the scattering coefficient, $W(\cdot)$ stands for azimuth envelope, $a(\cdot)$ stands for pulse envelope, $rect(\cdot)$ stands for rectangular function, r_p is the vertical distance from the orbit to target p, t_p is the moment when the vertical sight line passing target p, t_x is the time offset caused by squint, v_r is the equivalent velocity, T_c is the burst central time, T_d is the burst cycle time, T_b is the burst duration, and k is the pulse modulation rate.

The ScanSAR signal can be processed by the extended chirp scaling (ECS) algorithm [20–22]. In this algorithm, the signal is first translated into the range-Doppler domain and processed along the range dimension, and then processed by azimuth scaling and focusing. The range processing includes chirp scaling, bulk range cell migration correction (RCMC), range compressing, and second-range compressing. After range processing, we retrieve the processed signal in the range-Doppler domain. Considering a single burst, the processed signal can be expressed as follows [20–22]:

$$S(f_t, \tau) = \iint_D V_d(r_p, t_p) A \exp(-j2\pi f_t t_p) \text{sinc}\left[kT_r\left(\tau - \frac{2r_p}{cD(f_{t_{ref}})}\right)\right]$$
$$\cdot W_a(f_t - f_{t_x}) W_b(f_t - f_{t_c}) \exp\left(-j\frac{4\pi r_p f_0 D(f_t)}{c}\right) dr_p dt_p \tag{3}$$

where

$$\text{sinc}(x) = \sin(\pi x)/(\pi x) \tag{4}$$

$$D(f_t) = \sqrt{1 - \frac{c^2 f_t^2}{4v_r^2 f_0^2}}, \quad W_a(f_t) = W\left(\frac{-cr_p f_t}{2f_0 v_r^2 D(f_t)}\right), \quad W_b(f_t) = rect\left(\frac{-cr_p f_t}{2f_0 v_r^2 D(f_t)T_b}\right) \tag{5}$$

$$f_{t_x} = -\frac{2v_r^2 f_0 t_x}{c\sqrt{r_p^2 + v_r^2 t_x^2}}, \quad f_{t_c} = -\frac{2v_r^2 f_0(T_c - t_p)}{c\sqrt{r_p^2 + v_r^2(T_c - t_p)^2}}, \quad f_{t_{ref}} = f_{t_x} \tag{6}$$

where f_t is the azimuth frequency, A is a constant coefficient, T_r is the pulse duration of the transmitted signal, and f_0 is the central frequency of the chirp signal.

By azimuth scaling processing, the second phase term of the range processed signal can be transformed as follows:

$$\exp\left(-j\frac{4\pi r_p f_0 D(f_t)}{c}\right) \xrightarrow{\text{azimuth scaling}} \exp\left(-j\frac{4\pi r_p f_0}{c}\right)\exp\left(j\frac{\pi r_{p_{ref}} c}{2v_r^2 f_0}f_t^2\right) \tag{7}$$

where $r_{p_{ref}}$ is the referenced range distance.

In order to focus the signal along the azimuth dimension, the signal can be processed by the spectral analysis (SPECAN) algorithm [20,23]. According to the algorithm, the signal needs to be transformed into the time domain. In this domain, the signal can be dechirped by multiplying $\exp(j\pi k_a t^2)$. The dechirped signal is expressed as [20,23]:

$$\begin{aligned}
S(t,\tau) = \iint_D V_d(r_p, t_p) A \operatorname{sinc}\left[kT_r\left(\tau - \frac{2r_p}{cD(f_{t_{ref}})}\right)\right] W\left(t - t_p + \frac{f_{t_x}}{k_a}\right) \\
\cdot rect\left[\frac{1}{T_b}\left(t - t_p + \frac{f_{t_c}}{k_a}\right)\right] \exp\left(-j\frac{4\pi r_p f_0}{c}\right) \exp\left(-j\pi k_a t_p^2 + j2\pi k_a t_p t\right) dr_p dt_p
\end{aligned} \tag{8}$$

where

$$k_a = \frac{2v_r^2 f_0}{cr_{p_{ref}}} \tag{9}$$

The signal can then be transformed by Fourier-transform (FT) along the azimuth dimension. The transformed signal is [20,23]:

$$\begin{aligned}
S(t',\tau) = \iint_D V_d(r_p, t_p) A' \operatorname{sinc}\left[kT_r\left(\tau - \frac{2r_p}{cD(f_{t_{ref}})}\right)\right] W\left(T_c - t_p - t_x\right) \exp\left(-j\frac{4\pi r_p f_0}{c}\right) \\
\cdot \exp\left(-j\pi k_a t_p^2\right) \exp\left[j2\pi k_a(t_p - t')\left(t_p - \frac{f_{t_c}}{k_a}\right)\right] \operatorname{sinc}\left[k_a T_b(t' - t_p)\right] dr_p dt_p
\end{aligned} \tag{10}$$

By multiplying $\exp(j\pi k_a t'^2)$, the azimuth phase of the signal can be compensated. This is the last step of the SPECAN algorithm. At this stage, we can acquire the focused image, the expression of which can be approximated by the following equation [20,23]:

$$\begin{aligned}
S(t',\tau) = \iint_D V_d(r_p, t_p) A' \operatorname{sinc}\left[kT_r\left(\tau - \frac{2r_p}{cD(f_{t_{ref}})}\right)\right] W\left(T_c - t_p - t_x\right) \exp\left(-j\frac{4\pi r_p f_0}{c}\right) \\
\cdot \exp\left[j\pi k_a(t' - t_p)^2\right] \exp\left[-j2\pi k_a(t' - t_p)(T_c - t_p)\right] \operatorname{sinc}\left[k_a T_b(t' - t_p)\right] dr_p dt_p
\end{aligned} \tag{11}$$

In ScanSAR interferometry, the interferometric image pair can also be expressed as Equation (11) with slow time t_i, fast time τ_i, target time t_{pi}, target range r_{pi}, burst central time T_{ci}, Doppler modulation rate k_{ai}, burst duration T_{bi} and equivalent velocity v_{ri} instead of t', τ, t_p, r_p, T_c, k_a, T_b, and v_r, where the subscript "$*_i$" means the image index. "$i = 1$" indicates the master image and "$i = 2$" means the slave image.

After image co-registration, the slave image can be expressed as:

$$S_{2p}(t_1, \tau_1) = \iint_D V_d(r_{p1}, t_{p1})A'\,\text{sinc}\left[kT_r\left(\tau_1 - \frac{2r_{p1}}{cD(f_{t_{ref}})}\right)\right]\text{sinc}\left[\frac{k_{a2}v_{r1}}{v_{r2}}T_{b2}(t_1 - t_{p1})\right]W(T_{c2} - t_{p2} - t_x)$$

$$\cdot \exp\left(-j\frac{4\pi f_0 r_{p2}}{c}\right)\exp\left[j\pi\frac{k_{a2}v_{r1}^2}{v_{r2}^2}(t_1 - t_{p1})^2\right]\exp\left[-j2\pi\frac{k_{a2}v_{r1}^2}{v_{r2}^2}(t_1 - t_{p1})\left(\frac{v_{r2}T_{c2}}{v_{r1}} - t_{p1}\right)\right]dr_{p1}dt_{p1} \tag{12}$$

Then we substitute Equations (11) and (12) into the expression of interferometric coherence [24,25]:

$$\gamma(t_1, \tau_1) = \frac{< S_1(t_1, \tau_1) \cdot S_{2p}^*(t_1, \tau_1) >}{\sqrt{< S_1(t_1, \tau_1) \cdot S_1^*(t_1, \tau_1) >< S_{2p}(t_1, \tau_1) \cdot S_{2p}^*(t_1, \tau_1) >}} \tag{13}$$

We can get:

$$\gamma(t_1, \tau_1) = \gamma_a(t_1, \tau_1) \cdot \gamma_b(t_1, \tau_1) \tag{14}$$

$$\gamma_a(t_1, \tau_1) = \left\{\int_T \exp\left[-j2\pi k_{a1}(t_1 - t_{p1})\left(T_{c1} - \frac{v_{r2}T_{c2}}{v_{r1}}\right)\right]\text{sinc}\left[k_{a1}T_{b1}(t_1 - t_{p1})\right]\right.$$

$$\left.\cdot \text{sinc}\left[k_{a1}\frac{v_{r2}T_{b2}}{v_{r1}}(t_1 - t_{p1})\right]dt_{p1}\right\} \bigg/ \sqrt{\int_T \text{sinc}^2\left[k_{a1}T_{b1}(t_1 - t_{p1})\right]dt_{p1}}$$

$$\bigg/ \sqrt{\int_T \text{sinc}^2\left[k_{a1}\frac{v_{r2}T_{b2}}{v_{r1}}(t_1 - t_{p1})\right]dt_{p1}} \tag{15}$$

$$\gamma_b(t_1, \tau_1) = \left\{\int_R \text{sinc}^2\left[kT_r\left(\tau_1 - \frac{2r_{p1}}{cD(f_{t_{ref}})}\right)\right]\exp\left(-j\frac{4\pi f_0(r_{p1} - r_{p2})}{c}\right)dr_{p1}\right\}$$

$$\bigg/ \left\{\int_R \text{sinc}^2\left[kT_r\left(\tau_1 - \frac{2r_{p1}}{cD(f_{t_{ref}})}\right)\right]dr_{p1}\right\} \tag{16}$$

where γ_b is the coherence caused by the baseline and γ_a is the coherence caused by burst central time difference. γ_b is the same as the corresponding coherence in stripmap mode, and γ_a can be simplified as follows:

$$\gamma_a(t_1, \tau_1) = \begin{cases} \min(T_{b1}, T'_{b2})/\sqrt{T_{b1}T'_{b2}}, & |T_{c1} - T'_{c2}| \leq \frac{|T_{b1}-T'_{b2}|}{2} \\ \frac{1}{\sqrt{T_{b1}T'_{b2}}}\left[\min(T_{b1}, T'_{b2}) - \left(|T_{c1} - T'_{c2}| - \frac{|T_{b1}-T'_{b2}|}{2}\right)\right], & \frac{T_{b1}+T'_{b2}}{2} > |T_{c1} - T'_{c2}| > \frac{|T_{b1}-T'_{b2}|}{2} \\ 0, & |T_{c1} - T'_{c2}| \geq \frac{T_{b1}+T'_{b2}}{2} \end{cases} \tag{17}$$

where $T'_{c2} = v_{r2}T_{c2}/v_{r1}$ and $T'_{b2} = v_{r2}T_{b2}/v_{r1}$.

From Equation (17), we can see that the interferometric coherence is influenced by the burst central time difference $|\Delta T| = |T_{c1} - T'_{c2}|$. The difference needs to be kept low relative to the burst duration. The interferometric coherence is also influenced by the burst duration difference as a secondary factor. The velocity difference and PRF difference relate to the burst central time difference and burst duration difference.

3. Increasing the Interferometric Coherence by Iterative Filtering

From the analysis of ScanSAR interferometry above, when the burst central time difference is non-negligible, the reduction of coherence should be considered. In this situation, the interferometric coherence can be increased by signal filtering.

In this method, the focused images should be transformed to the signal forms expressed in Equation (8). After that, the echoes from each target in the corresponding signal possess the same azimuth range, which facilitates the application of the filtering method. This filter can be expressed as:

$$f(t) = rect\left[\frac{1}{\min(T_{b1}, T'_{b2}) - |\Delta T| + |T_{b1} - T'_{b2}|/2}\left(t - \frac{T_{c1} + T'_{c2}}{2} \pm \frac{|T_{b1} - T'_{b2}|}{4}\right)\right] \tag{18}$$

where the sign '±' is determined by the property of $T_{c1} - T'_{c2}$ and $T_{b1} - T'_{b2}$ to be positive or negative. When using this filter, the azimuth time array of the slave image should be calibrated as the time array of the master image.

Multiplying the transformed signals by this filter, the burst central times of the master and slave images become equal, leading to an increased coefficient γ_a.

In most cases, we do not actually know the burst central time difference, and so the filter is not precise. In this situation, iterative searches are required to find an accurate filter. The steps are shown in the following diagram (Figure 1).

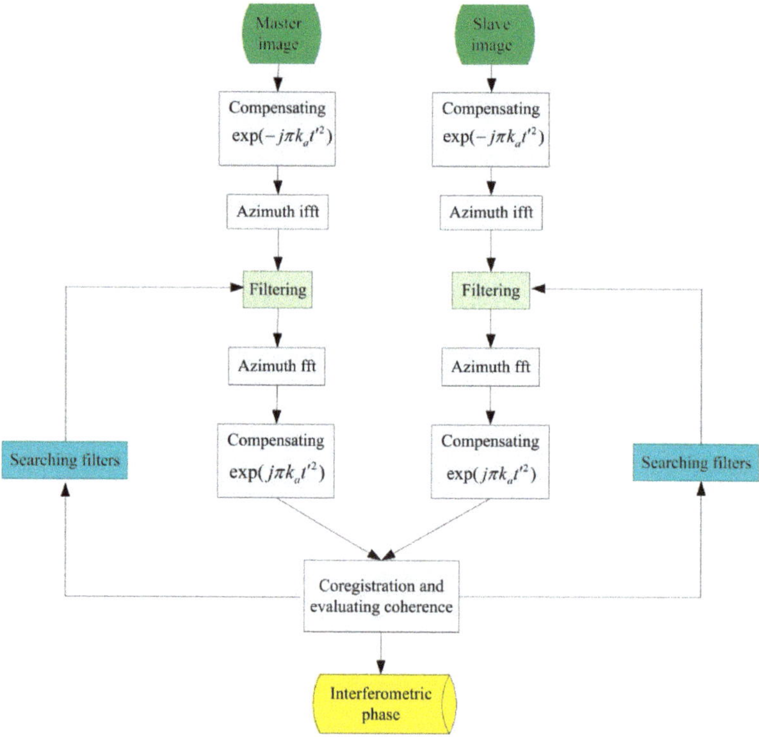

Figure 1. The iterative filtering method diagram.

In these steps, the master and slave images are first inversely processed to the signals in Equation (8). The signals can then be filtered to remove the signal parts irrelevant to interferometry. The selection of filters depends on the coherence value. We should choose the filter with the best coherence. After signal filtering, we can continue interferometric processing. The interferometric phase can then be obtained with better coherence.

During the ScanSAR signal processing, some other windows can also increase interferometric coherence (such as the Hanning window). Thus, in the iterative filtering method, a combined filter $f_c(t) = f(t) \cdot hanning(t)$ can be used to get better interferometric performance. In the combined filter, $f(t)$ is the above-mentioned rectangular window, and $hanning(t)$ is a Hanning window.

4. Phase Compensation of Gaofen-3 Interferometry

It was stated in Section 2 that a compensation phase $\exp(j\pi k_a t'^2)$ is required in the standard steps of ScanSAR imaging for interferometry. However, this step is not carried out because the Gaofen-3 ScanSAR images are used mainly with their amplitude information [26]. Thus, the phase $\exp(j\pi k_a t'^2)$

is not important in this situation. However, these images can still be used for interferometry if further corresponding processing is done. In this section, we analyzed the influence of this characteristic and designed a phase compensation method for the images.

In practical images, there is an unsynchronized azimuth time offset Δt between the master and slave images. After range processing and azimuth scaling for a ScanSAR echo, considering Δt, the signal in the time domain can be expressed as follows:

$$
\begin{aligned}
S(t,\tau) = {} & \iint_D V_d(r_p,t_p) A\mathrm{sinc}\left[kT_r\left(\tau - \frac{2r_p}{cD(f_{t_{ref}})}\right)\right] W\left(t - t_p - \Delta t + \frac{f_{tx}}{k_a}\right) \\
& \cdot rect\left[\frac{1}{T_b}\left(t - t_p - \Delta t + \frac{f_{tc}}{k_a}\right)\right] \exp\left(-j\frac{4\pi r_p f_0}{c}\right) \exp\left[-j\pi k_a(t - t_p - \Delta t)^2\right] dr_p dt_p
\end{aligned}
\tag{19}
$$

Multiplied by $\exp(j\pi k_a t^2)$ and transformed by FT, the focused image is:

$$
\begin{aligned}
S(t',\tau) = {} & \iint_D V_d(r_p,t_p) A'\mathrm{sinc}\left[kT_r\left(\tau - \frac{2r_p}{cD(f_{t_{ref}})}\right)\right] W\left(T_c - t_p - t_x\right) \exp\left(-j\pi k_a(t_p + \Delta t)^2\right) \\
& \cdot \exp\left(-j\frac{4\pi r_p f_0}{c}\right) \exp\left[j2\pi k_a(t_p + \Delta t - t')(t_p + \Delta t - \frac{f_{tc}}{k_a})\right] \mathrm{sinc}\left[k_a T_b(t' - t_p - \Delta t)\right] dr_p dt_p
\end{aligned}
\tag{20}
$$

If the image is compensated by a multiplying factor $\exp(j\pi k_a t'^2)$, we can describe the image as:

$$
\begin{aligned}
S(t',\tau) = {} & \iint_D V_d(r_p,t_p) A'\mathrm{sinc}\left[kT_r\left(\tau - \frac{2r_p}{cD(f_{t_{ref}})}\right)\right] W\left(T_c - t_p - t_x\right) \exp\left(-j\frac{4\pi r_p f_0}{c}\right) \\
& \exp\left[j\pi k_a(t' - t_p - \Delta t)^2\right] \exp\left[-j2\pi k_a(t' - t_p - \Delta t)(-\frac{f_{tc}}{k_a})\right] \mathrm{sinc}\left[k_a T_b(t' - t_p - \Delta t)\right] dr_p dt_p
\end{aligned}
\tag{21}
$$

From this equation, we can see that the azimuth time offset Δt can be handled by azimuth shifting, and the interferometric phase will not be influenced.

However, if the phase term is not compensated, the interferometric image $S_1(t_1,\tau_1) \cdot S_{2p}^*(t_1,\tau_1)$ will have an uncompensated phase term:

$$
P(t_1) = \exp\left(j2\pi k_{a1} t_1 \Delta t - j\pi k_{a1} \Delta t^2\right) = \exp(j2\pi k_{a1} t_1 \Delta t) \cdot \exp\left(-j\pi k_{a1} \Delta t^2\right)
\tag{22}
$$

This phase term is useless and will influence the interferometric phase. It can be divided into two terms: $\exp(-j\pi k_{a1}\Delta t^2)$ is a constant term, and only the linear phase term $\exp(j2\pi k_{a1} t_1 \Delta t)$ is needed for compensation.

However, this compensation is not sufficient, because the velocities and PRFs are different in Gaofen-3 interferometric images. In this situation, Δt is a variant along the azimuth direction. Without considering high orders, variant Δt can be approximated as $\Delta t = \Delta t_0 + k_t t_1$. The main term of $P(t_1)$ then becomes $\exp(j2\pi k_{a1} t_1 \Delta t_0) \cdot \exp(j2\pi k_{a1} k_t t_1^2)$. In the main term, $\exp(j2\pi k_{a1} t_1 \Delta t_0)$ is compensated in the above-mentioned step as a linear phase term, so the second-order sub-term $\exp(j2\pi k_{a1} k_t t_1^2)$ should be compensated along the azimuth direction sequentially.

5. DEM Geolocation of Gaofen-3 Interferometry

The above sections discussed the coherence of Gaofen-3 ScanSAR interferometry, and proposed several methods to solve unsynchronized problems. Together with these discussions, the interferometric processing steps can be expressed as Figure 2.

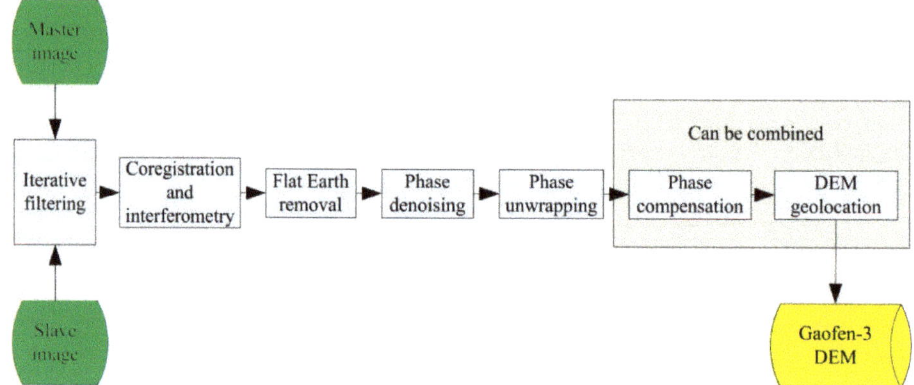

Figure 2. The interfeometric processing diagram. DEM: digital elevation model.

In the processing, the interferometric images are iteratively filtered to increase their coherence. After co-registration and interferometry, an interferometric phase image can then be achieved. The phase image should be processed by flat Earth removal, phase denoising, and phase unwrapping in sequence. After phase compensation and DEM geolocation, a Gaofen-3 DEM can then be retrieved.

DEM geolocation is the last step of interferometric processing. Using the compensated unwrapped phase together with the system geometric parameters and the payload parameters, the DEM of the Earth's surface can be extracted. This processing is based on three equations [27,28]:

$$\mathbf{v} \cdot (\mathbf{T} - \mathbf{S}) = \lambda f_{dc} r_s / 2 \tag{23}$$

$$|\mathbf{T} - \mathbf{S}| = r_s \tag{24}$$

$$|\mathbf{T} - \mathbf{S}_b| = r_s + \lambda \phi / 4\pi \tag{25}$$

where $\mathbf{v} = (v_x, v_y, v_z)$ is the velocity of the satellite, $\mathbf{T} = (T_x, T_y, T_z)$ is the position of the target, $\mathbf{S} = (S_x, S_y, S_z)$ is the position of the satellite in the first pass, and $\mathbf{S}_b = (S_{bx}, S_{by}, S_{bz})$ is the position of the satellite in the second pass. These four vectors are defined in Earth-centered fixed coordinates. λ is the wave length, f_{dc} is the Doppler central frequency, r_s is the range distance, and ϕ is the interferometric phase.

By solving Equations (23)–(25), the target coordinates $\mathbf{T}$ can be obtained. One of the calculation methods able to solve the equations involves using the Newton iteration method, but this method remains time intensive. In a Gaofen-3 interferometric situation, another calculation method requires a closed-form solution to be acquired [28,29]. Because this kind of method does not use iteration, its calculation efficiency is better. According to the Gaofen-3 parameter settings, we can describe the closed-form solution as follows:

$$\mathbf{T} = (c_{1x} T_z + c_{0x}, c_{1y} T_z + c_{0y}, T_z), \quad T_z = (-c_b \pm \sqrt{c_b^2 - 4 c_a c_b}) / (2 c_a) \tag{26}$$

where the sign "$\pm$" is determined by the satellite's looking direction. In Equation (26), the parameters can be expressed as [28]:

$$c_a = c_{1x}^2 + c_{1y}^2 + 1, \quad c_b = 2 c_{1x} c_{0x} + 2 c_{1y} c_{0y} - 2 S_x c_{1x} - 2 S_y c_{1y} - 2 S_z,$$
$$c_c = c_{0x}^2 + c_{0y}^2 - r_s^2 - 2 S_x c_{0x} - 2 S_y c_{0y} + \mathbf{S} \cdot \mathbf{S} \tag{27}$$

The parameters c_{0i} and c_{1i} ($i = x, y$) in the above equations are expressed as:

$$c_{0i} = m_{0i}r_1 + m_{1i}r_2, \quad c_{1i} = -m_{0i}v_z - m_{1i}(S_z - S_{bz}) \tag{28}$$

where

$$\mathbf{M} = \begin{bmatrix} m_{0x} & m_{1x} \\ m_{0y} & m_{1y} \end{bmatrix} = \begin{bmatrix} v_x & v_y \\ S_x - S_{bx} & S_y - S_{by} \end{bmatrix}^{-1} \tag{29}$$

$$r_1 = \lambda f_{dc} r_s / 2 + \mathbf{v} \cdot \mathbf{S}, \quad r_2 = \{ \mathbf{S} \cdot \mathbf{S} - \mathbf{S}_b \cdot \mathbf{S}_b + [\lambda \phi / (4\pi)]^2 + \lambda \phi r_s / (2\pi) \} / 2 \tag{30}$$

During processing, the absolute interferometric phase ϕ is required. However, from the compensated unwrapped phase, only the relative phase can be achieved. A system phase ϕ_0 should be compensated to the relative phase. We use GCPs to determine the phase ϕ_0. The point heights can be derived from known DEM data, such as SRTM DEM. The height of a GCP can be expressed as:

$$|\mathbf{T}| = h \tag{31}$$

Combining and solving Equations (23), (24), and (31), we can find coordinates $\mathbf{T}$ of a GCP. The closed-form solution of the equations is the same as Equation (26), except that some parameters should be replaced:

$$\mathbf{M} = \begin{bmatrix} m_{0x} & m_{1x} \\ m_{0y} & m_{1y} \end{bmatrix} = \begin{bmatrix} v_x & v_y \\ S_x & S_y \end{bmatrix}^{-1} \tag{32}$$

$$r_2 = (\mathbf{S} \cdot \mathbf{S} + h^2 - r_s^2) / 2 \tag{33}$$

Substituting the GCP coordinates into Equation (25), we can find the absolute interferometric phase ϕ of a GCP. Subtracting the relative interferometric phase from ϕ, phase ϕ_0 can be obtained. Phases ϕ_0 from multiple GCPs can be then averaged. We can then acquire the absolute interferometric phase image of all the points by compensating the average phase ϕ_0.

In the above method, system errors are not considered. In presence of some system errors, the system phase ϕ_0 varies along the range and azimuth directions, and it can be expressed as $\phi_e(t_1, r_1)$. From the system phases of GCPs, some system errors can be estimated and then compensated for, including the azimuth phase error discussed in Section 4. Thus, the phase compensation discussed in Section 4 can be combined with the DEM geolocation processing.

The system phase $\phi_e(t_1, r_1)$ can be expressed as:

$$\begin{aligned} \phi_e(t_1, r_1) &= \phi_0 + k_{ae1}t_1 + k_{ae2}t_1^2 + k_{re1}r_1 + k_{c1}t_1r_1 + k_{c2}t_1^2 r_1 \\ k_{ae1} &= 2\pi k_{a1}\Delta t_0 + k_{ab1}, \quad k_{ae2} = 2\pi k_{a1}k_t + k_{ab2} \end{aligned} \tag{34}$$

where $k_{re1}, k_{ab1}, k_{ab2}, k_{c1}$, and k_{c2} are phase error coefficients caused by baseline error.

If we obtain the system phase values of multiple GCPs, we can estimate the compensation coefficients using the least square method:

$$\mathbf{K} = \left(\mathbf{P}' \mathbf{P} \right)^{-1} \mathbf{P}' \mathbf{\Phi}_e \tag{35}$$

$$\mathbf{K} = (\phi_0, k_{ae1}, k_{ae2}, k_{re1}, k_{c1}, k_{c2})', \quad \mathbf{\Phi}_e = (\phi_{e1}, \cdots, \phi_{ei}, \cdots, \phi_{eN})' \tag{36}$$

$$\mathbf{P} = \begin{pmatrix} 1 & t_{11} & t_{11}^2 & r_{11} & t_{11}r_{11} & t_{11}^2 r_{11} \\ \cdots & \cdots & \cdots & \cdots & \cdots & \cdots \\ 1 & t_{1i} & t_{1i}^2 & r_{1i} & t_{1i}r_{1i} & t_{1i}^2 r_{1i} \\ \cdots & \cdots & \cdots & \cdots & \cdots & \cdots \\ 1 & t_{1N} & t_{1N}^2 & r_{1N} & t_{1N}r_{1N} & t_{1N}^2 r_{1N} \end{pmatrix} \tag{37}$$

where the subscript "$*_i$" means the GCP index, and "N" is the number of GCPs.

With the estimated compensation coefficients, we can calculate the system phase ϕ_e of each master image pixel according to Equation (34). The compensated phase image can then be acquired by compensating phase ϕ_e; at the same time the influence of azimuth phase error and baseline error can be weakened.

Based on the above discussions, GCPs are used to acquire absolute phase and compensate phase error. Because the GCP data are their three-dimensional coordinates in the geodetic coordinates system, we still need to find the positions of the GCPs in the phase image before the above geolocation processing. First, the GCP coordinates should be transformed from the geodetic coordinates to Earth-centered fixed coordinates. Then, for each GCP, its azimuth time t_p and range distance r_s need to be calculated. These two parameters can determine the position of each GCP in the phase image.

In the calculation of a GCP's t_p and r_s, the corresponding satellite position can be approximated as $\mathbf{S} = \mathbf{S}_0 + \mathbf{v}_0 t_p$, where $\mathbf{S}_0$ and $\mathbf{v}_0$ are the satellite position and velocity at the reference time t_0. Thus, we can calculate the azimuth time t_p as follows:

$$t_p = (-p_b \pm \sqrt{p_b^2 - 4p_a p_c})/(2p_a) \tag{38}$$

$$p_a = |\mathbf{v}_0|^2 - 4|\mathbf{v}_0|^4/(\lambda^2 f_{dc}^2), \quad p_b = -2 \cdot \mathbf{v}_0 \cdot (\mathbf{T}_p - \mathbf{S}_0) + 8|\mathbf{v}_0|^2 \cdot \mathbf{v}_0 \cdot (\mathbf{T}_p - \mathbf{S}_0)/(\lambda^2 f_{dc}^2),$$
$$p_c = |\mathbf{T}_p - \mathbf{S}_0|^2 - 4 \cdot [\mathbf{v}_0 \cdot (\mathbf{T}_p - \mathbf{S}_0)]^2/(\lambda^2 f_{dc}^2) \tag{39}$$

where the sign "$\pm$" is determined by the squint angle of a GCP and $\mathbf{T}_p$ is the coordinates of the GCP.

The approximation "$\mathbf{S} = \mathbf{S}_0 + \mathbf{v}_0 t_p$" does not consider the velocity variation. In order to decrease this influence, we must make a new approximation as $\mathbf{S} = \mathbf{S}_{02} + \mathbf{v}_{02} t_{p2}$, where $\mathbf{S}_{02}$ and $\mathbf{v}_{02}$ are the actual satellite position and velocity at the reference time $t_0 + t_p$. We repeat the calculation as Equations (38) and (39), and a new azimuth time t_{p2} can thus be obtained. With the same method, we can acquire a third new azimuth time t_{p3}. Thus, the final azimuth time "$t_{pf} = t_p + t_{p2} + t_{p3}$", which refers to t_0, can be determined. Range distance r_s at the azimuth time t_{pf} can be calculated with Equation (24). Thus, the GCP can be located in the phase image. From the GCP coordinates, we can obtain the approximate height of the nearest grid point. The above geolocation and compensation can then be carried out.

6. Results and Discussion

The above sections analyzed the coherence of ScanSAR interferometry and studied several problems in Gaofen-3 processing. In this section, we carried out a simulation and practical interferometric processing to explain the analysis and processing methods. For interferometric processing, we used the above-mentioned Gaofen-3 interferometric images over Kunlun Mountain. From the interferometric processing, the iterative filtering method, phase compensation, and DEM geolocation were verified.

6.1. Iterative Filtering Method

In Section 2, we discussed the interferometric performance related to the burst central time difference and burst duration difference. The relationship between the burst central time, burst duration difference, and the coherence is shown in Figure 3.

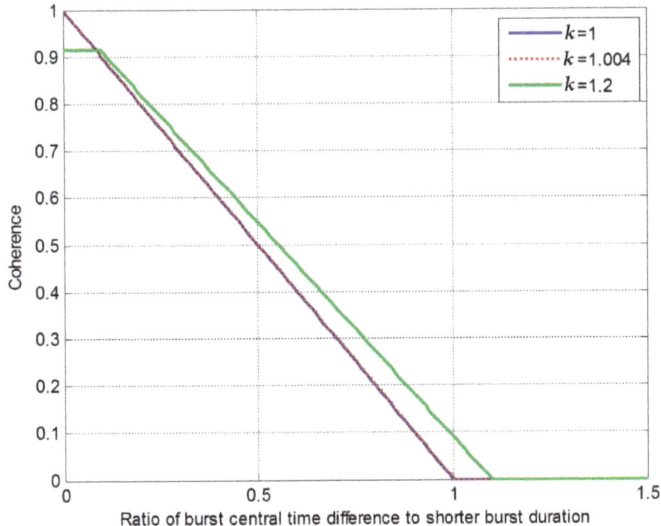

Figure 3. The relationship between the ratio of burst central time difference to shorter burst duration and coherence.

In Figure 3, "k" is a coefficient and $k = T'_{b2}/T_{b1}$, which means the ratio of burst durations. We express the ratio of burst central time difference to shorter burst duration as "k_c". Considering $k = 1$, the coherence is only influenced by the burst central time difference. In this situation, if $k_c = 0$, the coherence is not influenced. With an increase of k_c, the coherence decreases linearly. When k_c exceeds 1—that is to say, when the burst central time difference exceeds the burst duration—the coherence is reduced to 0. The interferometric processing will fail in this decorrelation situation. Considering $k = 1.2$, the coherence will be influenced by burst duration difference. In this situation, when k_c is within 0–0.1, the coherence value is 0.91 and is mostly lower than that in "$k = 1$" situation. When k_c is from 0.1 to 1.1, the coherence decreases linearly, but it is better than that in the $k = 1$ situation. When k_c exceeds 1.1, the coherence reduces to 0. For the images in this paper, k was near 1.004. Thus, the coherence of these images was mainly influenced by the burst central time difference.

In Gaofen-3 interferometric images, it is difficult to maintain a zero burst central time difference. As a consequence, interferometric coherence will be more or less influenced. When the burst central time difference is relatively large, the iterative filtering method described in Section 3 can be used to alleviate the influence.

Two interferometric images, shown in Figure 4, were used to verify the filtering method. These two interferometric images were cut from the Kunlun Mountain images with a relatively big burst central time difference.

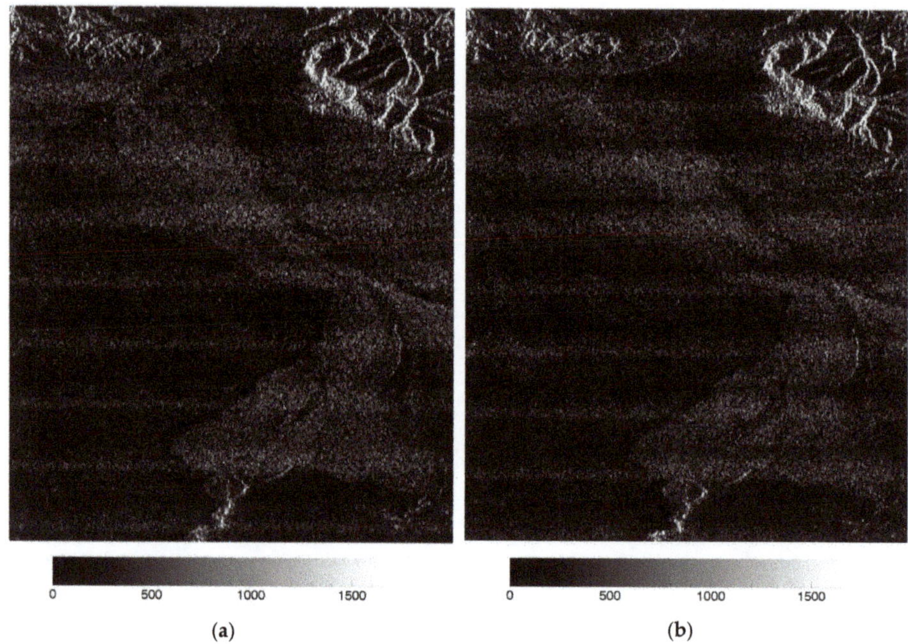

Figure 4. Gaofen-3 SAR images. (**a**) The master image; (**b**) the slave image.

Filtering the two images with different burst central time differences $|\Delta T|$, we found different coherence values after interferometry. This coherence was estimated from the interferometric images. $|\Delta T|$ versus coherence is shown in Figure 5.

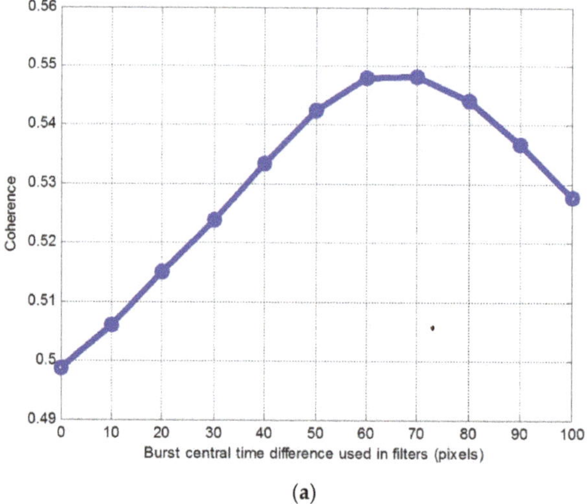

(**a**)

Figure 5. *Cont.*

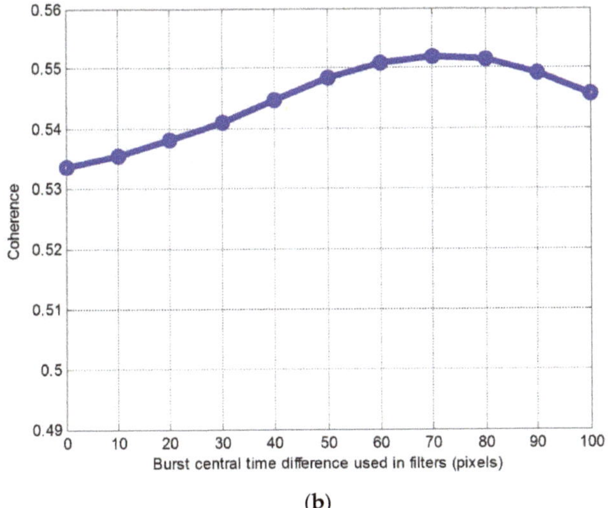

(b)

Figure 5. $|\Delta T|$ versus coherence processed from the two images shown in Figure 4. (**a**) Results when using rectangular filters; (**b**) results when using combined filters.

From Figure 5, when we used the rectangular filters, and $|\Delta T|$ used in the filters reached 70 pixels, the interferometric coherence increased by 0.05. When we used the combined filters, and $|\Delta T|$ used in the filters reached 70 pixels, the interferometric coherence increased by 0.02. With these two kinds of filters, the best $|\Delta T|$ values were all 70 pixels. With the rectangular filters, the decorrelation caused by the burst central difference was $70/582 = 0.12$, and the coherence caused by other factors was 0.55, where 582 was the azimuth band sample. Thus, the coherence increases by $0.12 \times 0.55 = 0.066$ theoretically, and the experiment result of 0.05 was close to the theoretical value. From Figure 5b, when the burst central time difference used in the combined filters was 0 pixels, the coherence was better than that of the value shown in Figure 5a. This is because the Hanning window decreased the amplitude of the unsynchronized signal part. When the burst central time difference used in the combined filters was 70 pixels, the coherence was better than that of the value shown in Figure 5a. This means that the Hanning window increased the coherence. Thus, it was suitable to use combined filters in the iterative filtering method.

6.2. Phase Compensation

In Gaofen-3 ScanSAR interferometry, as discussed in Section 4, a linear phase term and a second-order phase term should be compensated along the azimuth direction. By first applying this compensation method with a linear phase term on a pair of Gaofen-3 interferometric images (Figure 6a,b), we can get a compensated interferometric phase, as shown in Figure 6. These images were also cut from the Kunlun Mountain images.

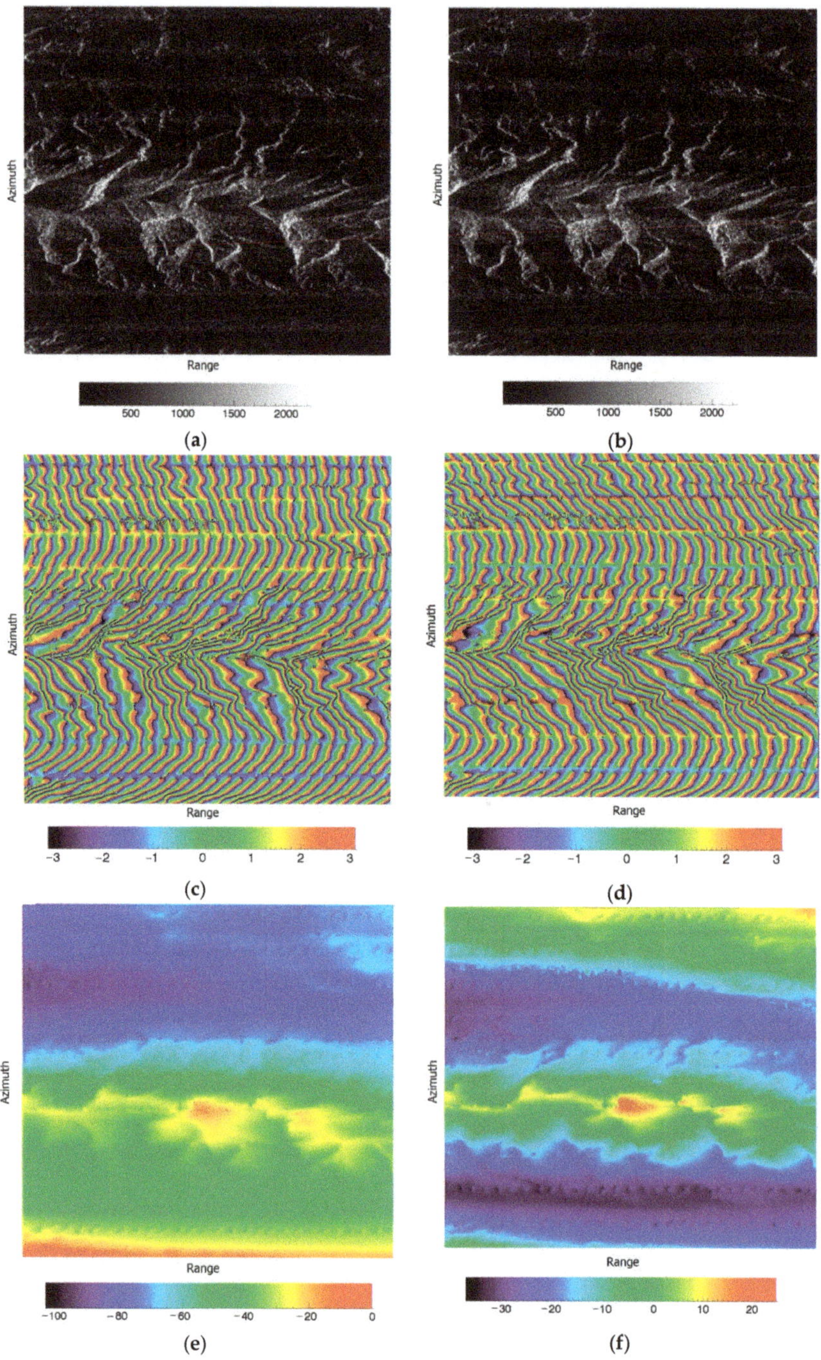

Figure 6. The linear compensation results. (**a**) The master image; (**b**) the slave image; (**c**) the original denoised interferometric phase (rad); (**d**) the compensated denoised interferometric phase (rad); (**e**) the original unwrapped phase after flat Earth removal along the range direction (rad); (**f**) the compensated unwrapped phase after flat Earth removal along the range direction (rad).

After interferometry, the original denoised interferometric phase is shown in Figure 6c. Figure 6d shows the compensated denoised interferometric phase, Figure 6e shows the original unwrapped phase after flat Earth removal along the range direction, and Figure 6f shows the corresponding compensated phase. From these figures, we can see that the compensation solved the phase's linear slope along azimuth direction.

In the above figures, the velocity of the master image was 7.5674 km/s and its PRF was 1185.6 Hz, while the velocity of the slave image was 7.5679 km/s and its PRF was 1190.4 Hz. As discussed in Section 4, these differences resulted in a second-order term along the azimuth direction. Compensating the Gaofen-3 interferometric phase with a second-order term, we obtained the following results.

In these figures, Figure 7a shows the second-order compensated denoised interferometric phase, and Figure 7b shows the second-order compensated unwrapped phase after flat Earth removal along the range direction. From the results, second-order compensation was able to solve the phase curving effect along the azimuth direction.

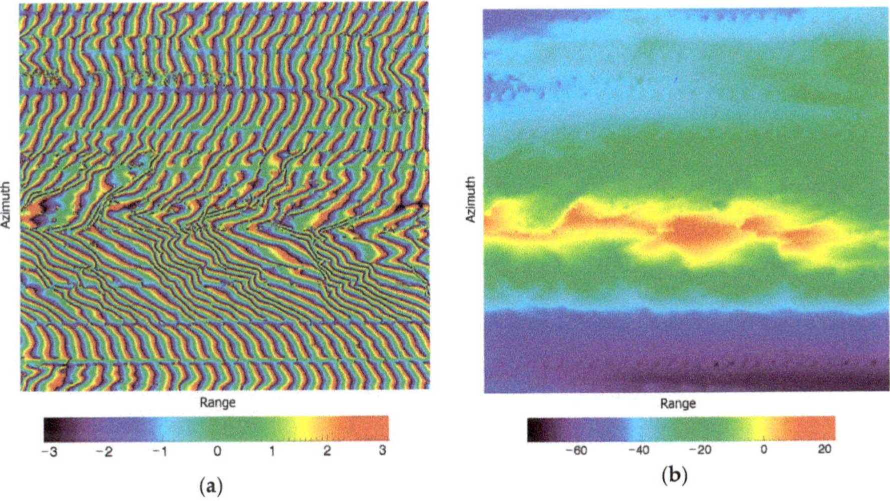

Figure 7. The second-order compensation results. (**a**) The second-order compensated denoised interferometric phase (rad); (**b**) the second-order compensated unwrapped phase after flat Earth removal along the range direction (rad).

In the interferometric phase, we found periodic lines. These lines were located at the areas where different bursts intersected. The burst central time difference in these areas neared the burst cycle time. Thus, based on the discussion in Section 2, the coherence in these areas was 0 and normal interferometric phase stripes could not be formed. This influence can be overcome by bursts aligned between the master and slave images before ScanSAR burst splicing. This aligning method is the best method. However, if we cannot obtain the interferometric images before burst splicing, the interpolation method can be used to fill in the invalid areas.

6.3. DEM Geolocation

From the above processing, a compensated unwrapped interferometric phase image was achieved. Subsequently, the satellite position and velocity during the observing time, as well as the Doppler central frequency and the target range distance were obtained from the Gaofen-3 information file. We then chose several GCPs in the master image. GCP height information can be obtained from a known DEM. According to the method in Section 5, we obtained the DEM of the tested Earth's surface as follows.

In Figure 8, Figure 8a shows the Gaofen-3 DEM, with imaging coordinates covering a 9 km (range) × 20 km (azimuth) area, and Figure 8b shows the top view of the Gaofen-3 DEM. The geographical characteristics of the DEM were coincident with those of the master image. Compared with the SRTM data of the same area (Figure 9), the achieved DEM matched the SRTM DEM (a 30 m × 30 m grid) [30].

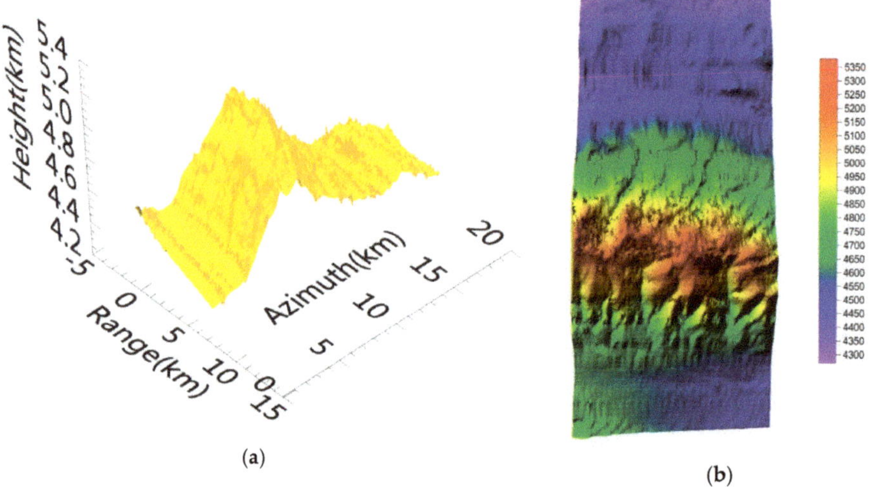

(a)

(b)

Figure 8. Gaofen-3 DEM after geolocation. (**a**) Gaofen-3 DEM in imaging coordinates; (**b**) top view of the Gaofen-3 DEM (m).

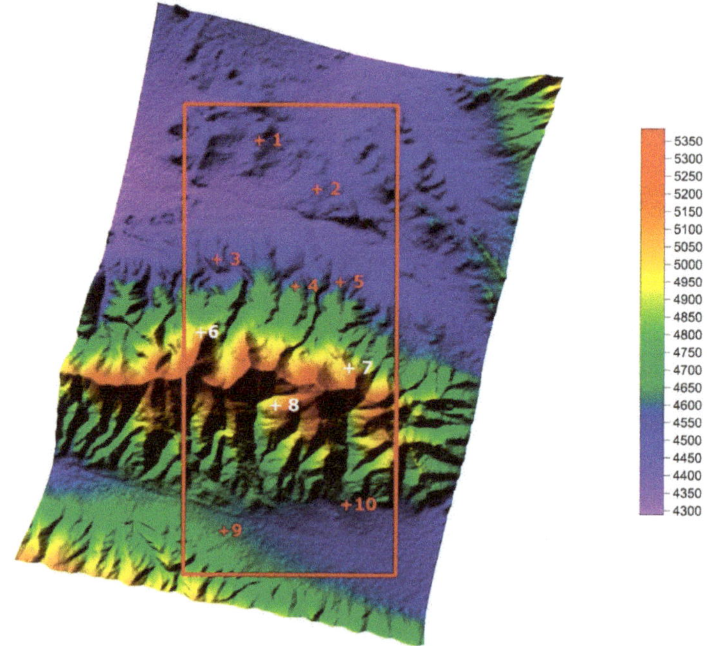

Figure 9. Shuttle Radar Topography mission (SRTM) DEM of the same area (m).

In order to evaluate the Gaofen-3 DEM quantitatively, we chose 10 check points from the SRTM DEM, marked with "+" in Figure 9. The height comparisons of the Gaofen-3 DEM and the SRTM DEM for these check points are listed in Table 2.

Table 2. Height comparisons of the Gaofen-3 DEM and the SRTM DEM for check points.

Index	1	2	3	4	5	6	7	8	9	10
Gaofen-3 DEM (m)	4496	4466	4489	4593	4584	5090	5061	4911	4683	4643
SRTM DEM (m)	4472	4454	4533	4630	4597	5050	5029	4923	4657	4633
Height difference (m)	24	12	−44	−37	−13	40	32	−12	26	10

As seen in Figures 8 and 9, the Gaofen-3 DEM was coarser than the SRTM DEM. As shown in Table 2, the average height precision of the Gaofen-3 DEM was about 25 m, and the maximum height error of the check points reached 44 m (absolute value). Height errors of the SRTM DEM samples were lower than 16 m. These results occurred because of the differences between the Gaofen-3 and SRTM interferometry. The Gaofen-3 DEM was acquired in ScanSAR mode and its grid was about 160 × 160 m, while the SRTM DEM was acquired in stripmap mode, and its grid was about 30 × 30 m; Gaofen-3 has a coarser grid. Gaofen-3 features repeat-pass interferometry and SRTM uses single-pass interferometry, so the coherence of Gaofen-3 should theoretically be lower than that of SRTM. Consequently, the Gaofen-3 DEM's quality was in accord with Gaofen-3's system characteristics. As the geographical characteristics of these DEMs were consistent, the accuracy of the Gaofen-3 DEM was verified.

By applying the above-mentioned interferometric processing to a wide area, we obtained the Gaofen-3 DEM as Figure 10.

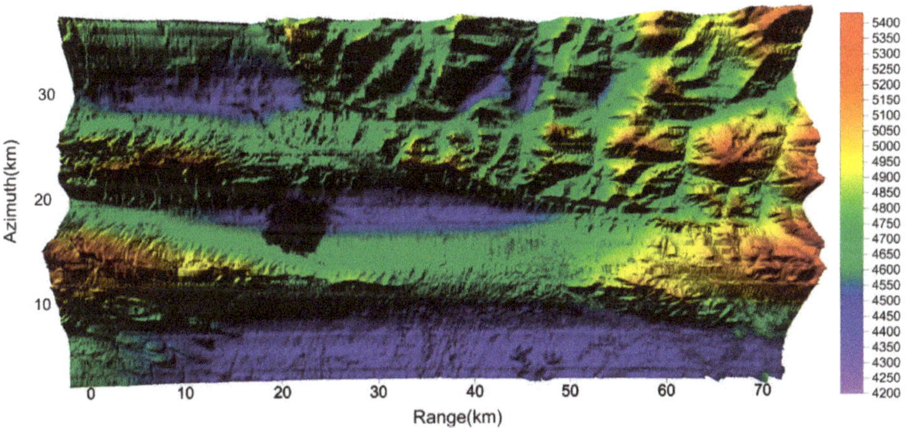

Figure 10. Gaofen-3 DEM covering a 70 km (range) × 35 km (azimuth) area (m).

The produced DEM was also of Kunlun Mountain, covering a 70 (range) × 35 km (azimuth) area. ScanSAR interferometry is suitable for this kind of wide-area mapping. Further, wide-area mapping can be dealt with by block processing and splicing, and the above 70 × 35 km area can be treated as a block.

7. Conclusions

This paper discussed interferometric analyzing and processing methods for Gaofen-3 images in ScanSAR mode. The conditions for ScanSAR interferometry are more rigorous than those of normal stripmap SAR interferometry. We analyzed the coherence in ScanSAR interferometry in detail to determine these conditions. From the analysis, the burst central time difference between the master and slave images was shown the coherence. In order to reduce the influence, we presented an iterative filtering method able to remove the signal parts irrelevant to interferometry, so as to increase the coherence. The analysis and the filtering method can also be influenced by burst duration difference and velocity difference, which should be incorporated in the analysis and filters. In Gaofen-3 ScanSAR interferometry, the phase error along the azimuth direction is severe. We analyzed the cause of the phase error, and correspondingly proposed a linear phase compensation and a second-order phase compensation to determine the right interferometric phase. In the DEM geolocation of Gaofen-3 interferometry, we derived a closed-form solution with GCP information. Without complex iteration in the method, a closed-form solution was able to efficiently retrieve a DEM of the Earth's surface. These methods were applied to Gaofen-3 ScanSAR images and returned good results. These methods could also help to realize ScanSAR interferometry for other similar satellites.

Author Contributions: Methodology, validation and writing—original draft preparation, Z.S.; writing—review and editing, A.Y., Z.D. and H.L.

Funding: This research received no external funding.

Acknowledgments: The authors wish to acknowledge the support of the Gaofen-3 mission and appreciate National Satellite Ocean Application Service for providing the Gaofen-3 data.

Conflicts of Interest: The authors declare no conflict of interest.

References

1. Sun, J.; Yu, W.; Deng, Y. The SAR payload design and performance for the GF-3 mission. *Sensors* **2017**, *17*, 2419. [CrossRef] [PubMed]

2. Wang, H.; Yang, J.; Mouche, A.; Shao, W.; Zhu, J.; Ren, L.; Xie, C. GF-3 SAR ocean wind retrieval: The first view and preliminary assessment. *Remote Sens.* **2017**, *9*, 694. [CrossRef]

3. Wang, H.; Wang, J.; Yang, J.; Ren, L.; Zhu, J.; Yuan, X.; Xie, C. Empirical algorithm for significant wave height retrieval from wave mode data provided by the Chinese satellite Gaofen-3. *Remote Sens.* **2018**, *10*, 363. [CrossRef]

4. Li, X.; Zhang, T.; Huang, B.; Jia, T. Capabilities of Chinese Gaofen-3 synthetic aperture radar in selected topics for coastal and ocean observations. *Remote Sens.* **2018**, *10*, 1929. [CrossRef]

5. Ma, M.; Chen, J.; Liu, W.; Yang, W. Ship classification and detection based on CNN using GF-3 SAR images. *Remote Sens.* **2018**, *10*, 2043. [CrossRef]

6. An, Q.; Pan, Z.; You, H. Ship detection in Gaofen-3 SAR images based on sea clutter distribution analysis and deep convolutional neural network. *Sensors* **2018**, *18*, 334. [CrossRef]

7. Dong, H.; Xu, X.; Wang, L.; Pu, F. Gaofen-3 PolSAR image classification via XGBoost and polarimetric spatial information. *Sensors* **2018**, *18*, 611. [CrossRef]

8. Gao, G.; Ouyang, K.; Luo, Y.; Liang, S.; Zhou, S. Scheme of Parameter estimation for generalized gamma distribution and its application to ship detection in SAR images. *IEEE Trans. Geosci. Remote Sens.* **2017**, *55*, 1812–1832. [CrossRef]

9. Chang, C.Y.; Jin, M.Y.; Lou, Y.L.; Holt, B. First SIR-C ScanSAR results. *IEEE Trans. Geosci. Remote Sens.* **1996**, *34*, 1278–1281. [CrossRef]

10. Guarnieri, A.M.; Prati, C.; Rocca, F. Interferometry with ScanSAR. In Proceedings of the IEEE International Geoscience and Remote Sensing Symposium, Firenze, Italy, 10–14 July 1995; pp. 550–552.

11. Guarnieri, A.M.; Prati, C. ScanSAR focusing and interferometry. *IEEE Trans. Geosci. Remote Sens.* **1996**, *34*, 1029–1038. [CrossRef]

12. Bamler, R.; Geudtner, D.; Schattler, B.; Vachon, P.W.; Steinbrecher, U.; Holzner, J.; Mittermayer, J.; Breit, H.; Moreira, A. RADARSAT ScanSAR interferometry. In Proceedings of the IEEE International Geoscience and Remote Sensing Symposium, Hamburg, Germany, 28 June–2 July 1999; pp. 1517–1521.

13. Holzner, J.; Bamler, R. Burst-mode and ScanSAR interferometry. *IEEE Trans. Geosci. Remote Sens.* **2002**, *40*, 1917–1937. [CrossRef]

14. Hensley, S.; Rosen, P.; Gurrola, E. Topographic map generation from the Shuttle Radar Topography Mission C-band SCANSAR interferometry. In Proceedings of the SPIE, Proceedings of Second International Asia-Pacific Symposium on Remote Sensing of the Atmosphere, Environment, and Space, Sendai, Japan, 21 December 2000; pp. 179–189.

15. Guarnieri, A.M.; Cafforio, C.; Guccione, P.; Pasquali, P.; Desnos, Y.L. ENVISAT ASAR ScanSAR interferometry. In Proceedings of the IEEE International Geoscience and Remote Sensing Symposium, Toulouse, France, 21–25 July 2003; pp. 1124–1126.
16. Guccione, P. Interferometry with ENVISAT wide swath ScanSAR data. *IEEE Geosci. Remote Sens. Lett.* **2006**, *3*, 377–381. [CrossRef]
17. Shimada, M. PALSAR ScanSAR ScanSAR interferometry. In Proceedings of the IEEE International Geoscience and Remote Sensing Symposium, Boston, MA, USA, 7–11 July 2008; pp. 93–96.
18. Liang, C.; Fielding, E.J. Interferometry with ALOS-2 full-aperture ScanSAR data. *IEEE Trans. Geosci. Remote Sens.* **2017**, *55*, 2739–2750. [CrossRef]
19. Hu, Z.; Ge, L.; Li, X. Blind azimuth phase elimination for TerraSAR-X ScanSAR interferometry. In Proceedings of the IEEE International Geoscience and Remote Sensing Symposium, Vancouver, BC, Canada, 24–29 July 2011; pp. 3476–3479.
20. Cumming, I.G.; Wong, F.H. *Digital Processing of Synthetic Aperture Radar Data: Algorithms and Implementation*; Artech House, Inc.: Norwood, MA, USA, 2005.
21. Moreira, A.; Mittermayer, J.; Scheiber, R. Extended chirp scaling algorithm for air—and spaceborne SAR data processing in stripmap and ScanSAR imaging modes. *IEEE Trans. Geosci. Remote Sens.* **1996**, *34*, 1123–1136. [CrossRef]
22. Mittermayer, J.; Moreira, A. A generic formulation of the Extended Chirp Scaling algorithm (ECS) for phase preserving. In Proceedings of the IEEE International Geoscience and Remote Sensing Symposium, Honolulu, HI, USA, 24–28 July 2000; pp. 108–110.
23. Lanari, R.; Hensley, S.; Rosen, P.A. Chirp z-transform based SPECAN approach for phase-preserving ScanSAR image generation. *IEE Proc. Radar Sonar Navig.* **1998**, *145*, 254–261. [CrossRef]
24. Rodirguez, E.; Martin, J.M. Theory and design of interferometric synthetic aperture radars. *IEE Proc. F* **1992**, *139*, 147–159. [CrossRef]
25. Bamler, R.; Hartl, P. Synthetic aperture radar interferometry. *Inverse Probl.* **1998**, *14*, R1–R54. [CrossRef]
26. Han, B.; Ding, C.; Zhong, L.; Liu, J.; Qiu, X.; Hu, Y.; Lei, B. The GF-3 SAR data processor. *Sensors* **2018**, *18*, 835. [CrossRef]
27. Rosen, P.A.; Hensley, S.; Joughin, I.R.; Li, F.K.; Madsen, S.N.; Rodriguez, E.; Goldstein, R.M. Synthetic aperture radar interferometry. *Proc. IEEE* **2000**, *88*, 333–382. [CrossRef]
28. Nico, G. Exact closed-form geolocation for SAR interferometry. *IEEE Trans. Geosci. Remote Sens.* **2002**, *40*, 220–222. [CrossRef]
29. Sansosti, E.A. Simple and exact solution for the interferometric and stereo SAR geolocation problem. *IEEE Trans. Geosci. Remote Sens.* **2004**, *15*, 1625–1634. [CrossRef]
30. Global Data Explorer. Available online: https://gdex.cr.usgs.gov/gdex (accessed on 2 March 2018).

Article

A Highly Efficient Heterogeneous Processor for SAR Imaging

Shiyu Wang *, Shengbing Zhang *, Xiaoping Huang, Jianfeng An and Libo Chang

School of Computer Science and Engineering, Northwestern Polytechnical University, Xi'an 710072, China
* Correspondence: onion0709@mail.nwpu.edu.cn (S.W.); zhangsb@nwpu.edu.cn (S.Z.);
Tel.: +86-1331-927-0830 (S.W.)

Received: 4 June 2019; Accepted: 1 August 2019; Published: 3 August 2019

Abstract: The expansion and improvement of synthetic aperture radar (SAR) technology have greatly enhanced its practicality. SAR imaging requires real-time processing with limited power consumption for large input images. Designing a specific heterogeneous array processor is an effective approach to meet the power consumption constraints and real-time processing requirements of an application system. In this paper, taking a commonly used algorithm for SAR imaging—the chirp scaling algorithm (CSA)—as an example, the characteristics of each calculation stage in the SAR imaging process is analyzed, and the data flow model of SAR imaging is extracted. A heterogeneous array architecture for SAR imaging that effectively supports Fast Fourier Transformation/Inverse Fast Fourier Transform (FFT/IFFT) and phase compensation operations is proposed. First, a heterogeneous array architecture consisting of fixed-point PE units and floating-point FPE units, which are respectively proposed for the FFT/IFFT and phase compensation operations, increasing energy efficiency by 50% compared with the architecture using floating-point units. Second, data cross-placement and simultaneous access strategies are proposed to support the intra-block parallel processing of SAR block imaging, achieving up to 115.2 GOPS throughput. Third, a resource management strategy for heterogeneous computing arrays is designed, which supports the pipeline processing of FFT/IFFT and phase compensation operation, improving PE utilization by a factor of 1.82 and increasing energy efficiency by a factor of 1.5. Implemented in 65-nm technology, the experimental results show that the processor can achieve energy efficiency of up to 254 GOPS/W. The imaging fidelity and accuracy of the proposed processor were verified by evaluating the image quality of the actual scene.

Keywords: heterogeneous array; SAR imaging; data cross-placement; computing resource management

1. Introduction

Aerospace synthetic aperture radar (SAR) can be all-time and all-weather to obtain high-precision microwave images and other value-added products over large areas, and it has an extensive range of applications in remote sensing, environmental monitoring, geographical mapping, war zone surveillance, precision guidance, and reconnaissance [1–4].

Extensions and modifications of the SAR technology have significantly increased its practicality and applications. The demand for high-resolution and wide-swath (HRWS) SAR imaging is growing, especially in the areas of ocean observation, geological survey, and environmental protection. In 1978, the United States launched the first spaceborne SAR named Seasat-1. It is a satellite specifically designed for telemetry of the Earth's oceans, and is aimed at realizing the possibility of global satellite monitoring of the oceans and determining the system requirements for marine remote sensing satellites. RADARSAT-1 was successfully launched in Canada in 1995 [5]. It not only provided Canada with a large amount of all-weather and all-time SAR data, but also provided useful information for commercial and scientific users in disaster management, agriculture, mapping, hydrology, forestry, oceanography, ice research, and coastal monitoring. In January 2006, Japan launched the Advanced Land Observing Satellite (ALOS) [6]. The Phased Array type L-band Synthetic Aperture Radar (PALSAR) that it carried was an L-band SAR sensor that is not affected by atmospheric conditions, cloud cover, and other related conditions, so it can be used for ground observations around the clock. In June 2007, the Terra SAR-X was launched by the German National Space Center. Its X-band SAR radar reliably provided high-resolution weather conditions and wide-area radar images with superior geometric accuracy over any other spaceborne SAR sensor [7]. Moreover, for both civil and military applications, it is desired to monitor moving targets, including ground moving target indication/ground moving target imaging (GMTI/GMTIm) [8,9].

For SAR processing systems, SAR imaging time accounts for most of the processing time, and directly brings a significant impact on system throughput and rapid response capability. The imaging delay of SAR will seriously affect the subsequent image processing, such as content analysis, risk diagnosis, and feature extraction. SAR imaging efficiency plays a very important role in the SAR system platform, and it can directly affect the throughput and rapid response capability of the entire platform.

Spaceborne and airborne real-time SAR imaging is the most direct and effective real-time imaging implementation approach, which can quickly provide SAR image data for SAR applications while significantly reducing the communication burden of air-to-ground data links [10]. At the same time, the working environment of spaceborne and airborne imaging systems is harsh, and the power consumption of the processor is also severely limited. Therefore, real-time and low power consumption are two essential items that must be met by spaceborne/airborne SAR imaging processors.

Since SAR imaging requires a large amount of two-dimensional parallel computing, it is difficult for a single multi-core central processing unit (CPU) to meet its real-time requirements. The SAR imaging scheme with multiple CPU nodes has high power consumption and low processing efficiency, and cannot be applied in spaceborne/airborne SAR processing. Generally, heterogeneous schemes such as CPU + GPU, CPU + DSP (s), and CPU + FPGA (one or more) can meet the performance requirements of real-time processing, but their power consumption is above 10 W, or even more than 100 W. A dedicated chip that fully implements the imaging algorithm can achieve better results in real-time, and low power consumption and is suitable for applications with strict power constraints, but the scheme hardens the algorithm, resulting in poor flexibility. ASIP (Application Specific Instruction Set Processor) is a dedicated processor solution between a general-purpose processor and an application specific integrated circuit (ASIC). This processor combines the flexibility of a general-purpose processor and the efficiency of an ASIC. In order to achieve a good trade-off between flexibility and processing efficiency, the development of a dedicated processor that is capable of fully implementing the SAR imaging process is an effective solution to meet its power consumption and real-time requirements for spaceborne/airborne SAR processing.

The chirp scaling algorithm (CSA) is one of the most commonly used algorithms for SAR imaging [11]. Its calculations mainly include Fast Fourier Transformation/Inverse Fast Fourier Transform (FFT/IFFT), phase multiplication, interpolation, etc., especially FFT/IFFT operations account for the highest proportion. The accuracy requirements and computing flow of these operations are different. Therefore, how to design an array structure and storage structure suitable for such processing is a key issue to be solved.

With the progress of integrated circuit (IC) technology, more processing units and memory blocks can be integrated on a single chip. Based on the abundant computational and memory resources on the chip, to make full use of bandwidth resources, this paper proposes a heterogeneous array structure that efficiently supports CSA imaging processing by combining block parallelism and pipeline processing while buffering the intermediate results on-chip.

It can support the parallel and pipeline processing and increases the maximum utilization of computing units. Moreover, we have designed an on-chip multi-level data buffer structure matching the heterogeneous array structure to ensure data supply for pipeline processing. This solution can reduce the complexity of the system while improving real-time performance.

The paper is organized as follows. Section 2 outlines related work and background. Section 3 analyzes the characteristics of CSA and proposes the design of the processor. Section 4 presents the heterogeneous architecture implementations. We present the evaluation of experimental results in Section 5, and the conclusions in Section 6.

2. Related Work

Digital signal processors (DSPs), CPUs, and graphics processing units (GPUs) have respective advantages in real-time SAR processing. As the system adopts CPU, it has good flexibility and portability [12]. However, their power efficiency for computing is quite low, which is a bottleneck in real-time SAR applications. Due to GPU's powerful parallel computation capability and programmability, the new method makes full use of GPU's powerful computation ability, which effectively improves the real-time quality of SAR scene generation [13–16]. At present, the GPU + CPU method can effectively combine the advantages of the two processors to improve imaging efficiency [17,18]. However, the average power consumption which is up to 150 W, limits the application of GPU in micro air vehicles.

Nowadays, high capability DSPs easily realize many complex theories and algorithms on hardware, and promote the development of SAR technology [19–21]. In 2003, Hanover University implemented a SAR real-time processing system using a multi-DSP architecture. This system uses highly parallel digital signal processor technology (HiPAR-DSP) for SAR signal processing [22]. The Indian Space Research Organization (IRSO) developed the SAR Specialized Processor (NRTP) based on Analog Devices' DSP multiprocessor, which approximates the real-time imaging of SAR [23]. However, for some applications with strictly constrained power, DSP has lower energy efficiency, resulting in lower imaging efficiency.

The rapid development of field-programmable gate array (FPGA) has been one of the most important technologies of realizing digital signal processing. With its rich on-chip memory and computational resources, FPGA can be configured as a SAR imaging platform to meet the high throughput rate SAR signal processing requirements [24–26]. An FPGA based on fault-tolerant architecture (Xilinx Virtex-II Pro) is applied to SAR processing systems [27,28]. In 2006, the University of Florida developed a high-performance heterogeneous spatial computing framework based on hardware/software interfaces. In this architecture, the CPU is responsible for scheduling and task management, and the FPGA acts as a coprocessor for computational acceleration [29]. With the rapid development of storage capacity and computing power of commercial FPGAs, SAR real-time imaging systems can all be built by FPGA (Xilinx Virtex-6) [30]. However, for highly complex algorithms, the development cycle of FPGA is relatively long.

For the real-time requirements and physical implementation limitations of SAR imaging, ASIC implementation is generally employed [31,32]. The Massachusetts Institute of Technology (MIT) Lincoln Laboratory uses bit-level systolic-array technology to design a SAR signal processor with high throughput and low power consumption [33]. The jet propulsion laboratory has also developed an airborne SAR processing system using a VLSI+SOC (very large scale integration+system on chip) hardware solution [10]. The processor's low power consumption and small size make it suitable for small SAR imaging systems.

In general, the DSP solution is used to implement SAR imaging through software programming. Since the DSP is designed for general purposes, this implementation has high flexibility and a short design cycle. It is more suitable for real-time SAR imaging than a CPU, but for low power applications, it is still not the most suitable choice. The ASIC solution for SAR imaging has the optimal power and performance for a single computational process. However, SAR imaging is a combination of multiple calculations on one device, which causes the design cost and power consumption of SAR imaging to soar, the design cycle to become longer, and poor flexibility. ASIP makes a good trade-off between the high flexibility of a general purpose processor and the high processing efficiency of an ASIC, and can be tailored and optimized for a certain type of algorithm or domain application to meet constraints such as performance, area, and power consumption. Moreover, it can effectively reduce design cycles and the design risk. Thus, many advantages of ASIP make it a very important implementation method in the field of signal processing.

Making trade-offs between speed, cost, power consumption, and flexibility, ASIP design methodology in the design of SAR real-time signal processing system can not only satisfy the real-time and performance requirements of aerospace systems, but also shorten the lead time of the processors. ASIP, when designed with a specific architecture with higher parallelism and higher complexity, also has good scalability. Therefore, we have designed a dedicated processor that can fully implement the SAR imaging process to meet the power consumption and real-time requirements of the application environment.

3. Processor Architecture Design

The CSA is one of the most commonly used algorithms for SAR imaging [11]. Compared with other algorithms, the CSA has the advantages of a simple operation process, low computational complexity, and high imaging efficiency. On the other hand, the CSA improves the fidelity of the image, especially the preservation of the phase information. Moreover, the CSA can adapt to different radar scanning modes, for example, spotlight, strip-map, scan SAR, sliding spotlight, Tops, and Mosaic modes [34,35].

3.1. CSA Flow Analysis

The imaging principle of the CSA is shown in Figure 1. The CSA can be divided into three modules according to functions, or divided into seven steps according to the operation sequence. The algorithm is executed step by step, and in the algorithm process, we perform the alternating operation of FFT/IFFT and phase compensation. To perform a SAR imaging, four Fourier transform and three-phase multiplication are needed.

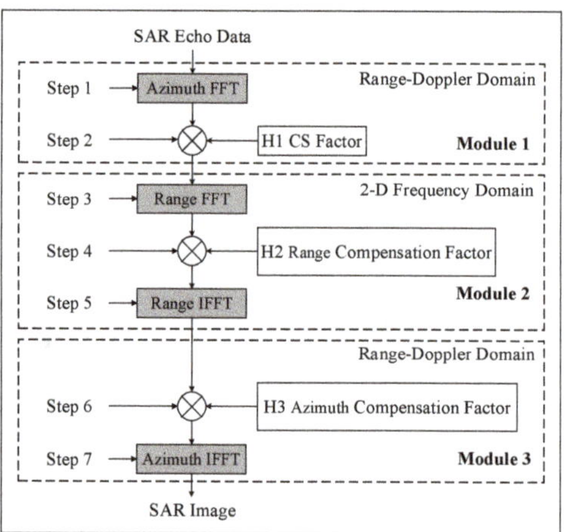

Figure 1. Chirp scaling algorithm (CSA) flow chart. SAR: synthetic aperture radar.

The Q-point FFT/IFFT can be decomposed into $2Q \log_2 Q$ real multiplications and $3Q \log_2 Q$ real additions [36]. Table 1 lists the computation quantity of the seven-step operation.

From Table 1, we can see that the proportion of FFT(IFFT) in all operations is:

$$W = \frac{(2+2+3+3)NM \log_2 M + (2+2+3+3)NM \log_2 N}{10NM \log_2 M + 10NM \log_2 N + 18NM} \tag{1}$$

For different imaging matrix sizes, the proportion W of FFT(IFFT) is slightly different, as shown in Table 2. It can be shown from Table 2 that the W values are basically above 90% and can reach up to 95% as the matrix size becomes larger. Therefore, accelerating the FFT/IFFT operation will inevitably reduce the imaging time and optimize the imaging efficiency.

Table 1. Computational statistics of CSA.

Calculation Content	Step	Real-Multi	Real-Add	Total
Azimuth-FFT [1]	1	$2NM \log_2 M$ [2]	$3NM \log_2 M$	$5NM \log_2 M$
CS Factor-Multi [3]	2	$4NM$	$2NM$	$6NM$
Range-FFT	3	$2NM \log_2 N$	$3NM \log_2 N$	$5NM \log_2 N$
RC Factor [4]-Multi	4	$4NM$	$2NM$	$6NM$
Range-IFFT [5]	5	$2NM \log_2 N$	$3NM \log_2 N$	$5NM \log_2 N$
AC Factor [6]-Multi	6	$4NM$	$2NM$	$6NM$
Azimuth-IFFT	7	$2NM \log_2 M$	$3NM \log_2 M$	$5NM \log_2 M$
Total	-	$4NM \log_2 M + 4NM\log_2 N + 12NM$	$6NM \log_2 M + 6NM \log_2 N + 6NM$	$10NM \log_2 M + 10NM \log_2 N + 18NM$

[1] **FFT**: Fast Fourier Transformation; [2] *M*: Azimuth direction sample numbers; *N*: Range direction sample numbers; [3] CS Factor: Chirp Scaling Factor; [4] RC Factor: Range Compensation Factor; [5] IFFT: Inverse Fast Fourier Transform. [6] AC Factor: Azimuth Compensation Factor.

Table 2. Computational load statistics.

Image Size	256×256	1024×1024	4096×4096	$16,384 \times 16,384$	$65,536 \times 65,536$
FFT Computational Load	10^7	2.1×10^8	4.1×10^9	7.6×10^{10}	1.2×10^{12}
Phase Compensation Computational Load	10^6	1.8×10^7	3×10^8	4.8×10^9	1.2×10^{10}
W-Value	89.8%	91.7%	93%	94%	94.7%

3.2. Computation Flow Strategy

In the imaging process, we take the block imaging method and perform parallel processing between blocks. In the algorithm process, four FFT/IFFT and three phase operations are pipelined according to the algorithm flow, while each multi-range (multi-azimuth) FFT/IFFT and phase operation can be parallel processing individually. To organize the pipeline processing of two types of operations in SAR imaging, we designed a calculation process based on space–time flow (ST-Flow), as shown in Figure 2. At a time, in space, multi-line FFT/IFFT can be performed in parallel, and phase compensation operation can be calculated simultaneously at multiple points, so no calculation unit is idle. On the timeline, data is continuously fed into the processing unit, and the calculation unit does not have a stall due to waiting for data. With this ST-Flow, SAR imaging can be done in a continuous process.

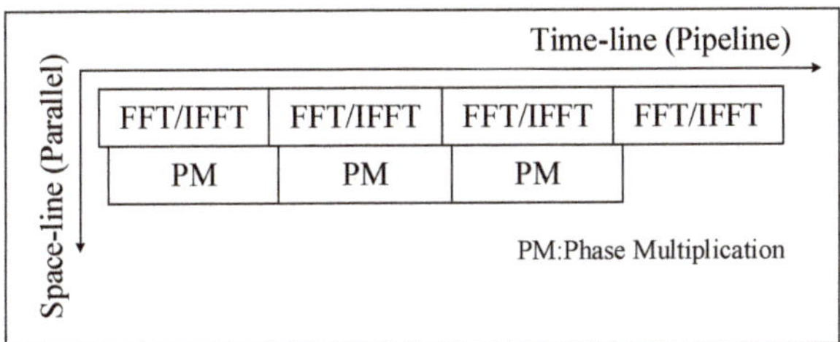

Figure 2. SAR imaging flow.

3.3. Heterogeneous Arrays

CSA includes scalar operation for phase multiplication and vector operation FFT (IFFT). As Table 2 shown, FFT/IFFT operations account for up to 95% of SAR imaging, so accelerating the FFT/IFFT operation efficiently is the most important approach for imaging processors.

The fixed-point FFT/IFFT operation with lower accuracy has a small loss of imaging accuracy, and can significantly improve the processing throughput. In [37], the quantization error power of the fixed-point processing CSA was evaluated in detail. The analysis results showed that as the word length increases from 12 to 16, the quantization error power remains essentially unchanged, and the imaging quality with a 15 or 16-bit word length is very close to that of a single precision floating-point. Therefore, we design PE arrays to support 12-bit, 14-bit, and 16-bit fixed-point FFT/IFFT. For applications with lower accuracy requirements, low-bit width operation can be selected.

However, the phase compensation operation requires high precision and must use floating-point arithmetic operations. Based on the earlier description and discussion, a heterogeneous array is designed, which includes two types of computing units named PE and FPE. PE is used for FFT/IFFT operation and FPE is used for phase compensation operation.

Since the operation ratio of FFT/IFFT against phase compensation is approximately 9:1, the configuration of PE and FPE should also follow this proportional relationship. For smaller matrix sizes, the ratio is near 90%; to meet the different matrix sizes, we design the processing array, in which the ratio of PE and FPE is 8:1, as shown in Figure 3.

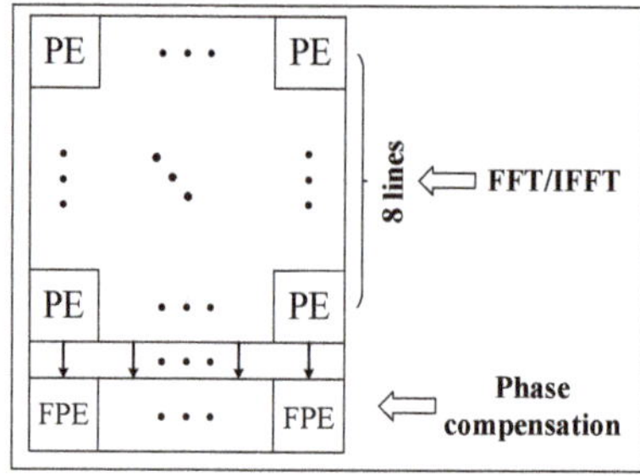

Figure 3. PE array structure diagram. PE: computing unit for FFT/IFFT operation in a heterogenous array.

In CSA flow, each range/azimuth FFT/IFFT operation is relatively independent, and there is no data dependency between range/azimuth, so each range/azimuth FFT/IFFT operation can be performed in parallel. Moreover, in the FFT/IFFT operation, each butterfly operation is relatively independent, and multiple butterfly operations can be performed in parallel. The phase compensation process performs independent operations at a single point so that multiple independent operations can be performed in parallel.

In CSA flow, four FFT/IFFT and three phase operations are data dependent; they are processed in the pipeline. As shown in Figure 4, to establish a pipeline between the FFT and the phase operation, the parallel FFT/IFFT differ by 1/8 computation cycles.

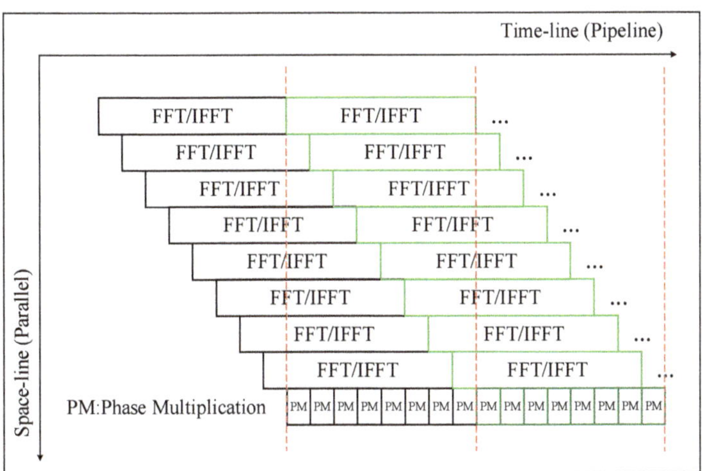

Figure 4. Pipeline between FFT/IFFT and phase operation.

3.4. Data Placement and Simultaneous Access

In the FFT/IFFT process, the data transfer has a bit-reverse address sequence. To support this data access pattern, we use a multi-bank distributed data placement strategy, as shown in Table 3. According to the calculation requirements, one row of PE parallel performs 16 butterfly operations, and needs to provide 32 data at the same time. Therefore, data access is performed in parallel. As shown in Figure 5, 32 data are simultaneously accessed from Bank 0 and Bank 1 in the first cycle. In the second cycle, data are read simultaneously from Bank 2 and Bank 3. Bank selection and the address in a bank are generated to follow each step in the FFT/IFFT processing flow. Although each PE performs a different FFT/IFFT operation, they use similar data placement and access strategies.

Table 3. 4096 points input data storage in four banks.

Bank_NO.	Input Data Storage Status in Bank			
Bank_0	0–15	64–79	-	-
Bank_1	16–31	80–95	-	-
Bank_2	32–47	96–111	-	-
Bank_3	48–63	112–127	-	4080–4095

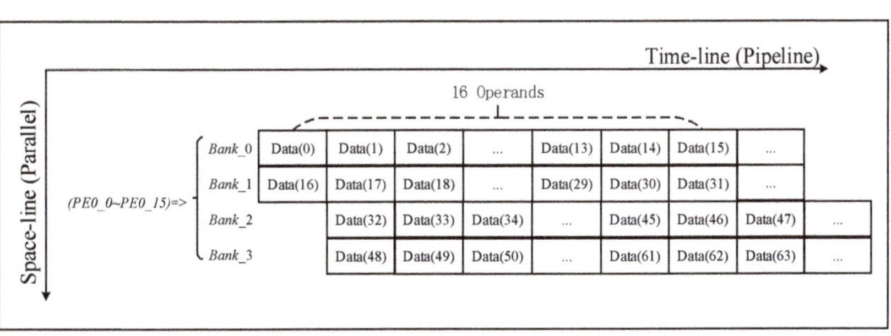

Figure 5. 16 PEs perform base-4 butterfly operation timing diagram for four banks.

There is no special requirement for the sequence of data in the phase compensation calculation process; therefore, as shown in Figure 6, the calculation process only needs to access the data in parallel.

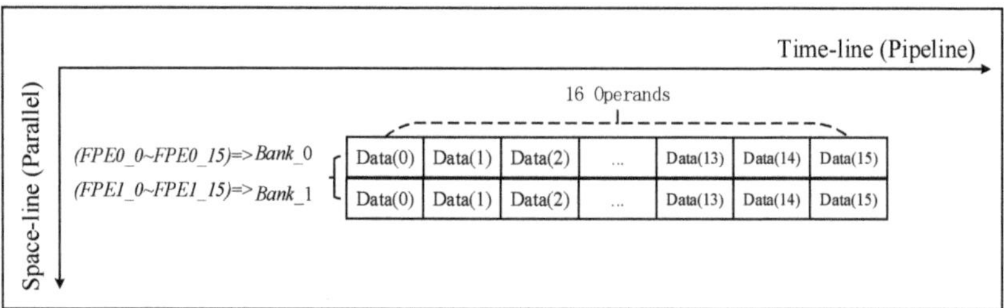

Figure 6. 32 FPEs perform phase operation for two banks. FPE: computing unit in a heterogeneous array used for phase compensation operation.

4. Architectural Implementations

4.1. Overall Architecture

A highly efficient heterogeneous processor for SAR imaging is designed. Figure 7 shows the top-level architecture of the proposed SAR imaging processor. This section describes the overall hardware block diagram and functional modules. Essentially, the architecture consists of three major components: a hybrid–PE array, an on-chip buffer module, and a data systolic engine.

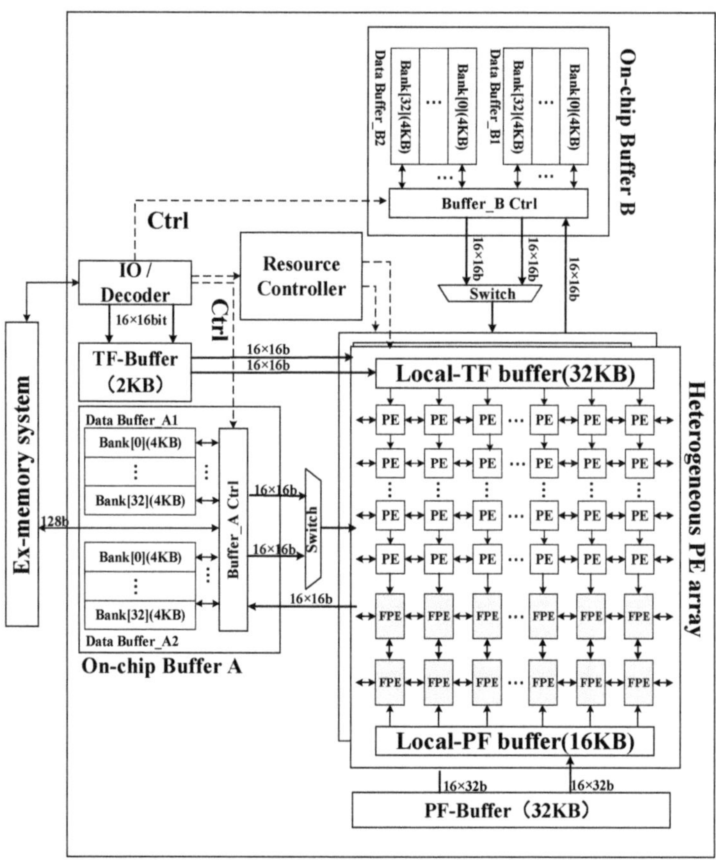

Figure 7. Top-level architecture of processor.

To meet the throughput requirement of SAR imaging, two identical sets of heterogeneous arrays are implemented, which can perform different block imaging processing computations in parallel. Each of the heterogeneous arrays contains 16×16 PEs and 2×16 FPEs. The number ratio of PE and FPE satisfies the proportional relationship of 8:1.

To feed the processing array with adequate data supply, three types of buffers are implemented on chip. In a processing array, all the data banks for 16-line PEs and two-line FPEs are organized as a 264-KB data buffer with two sub-buffers, each of which contains 32 banks for PEs and one bank for an FPE. A 32-KB twiddle factor dedicated local buffer (Local-TF buffer) and a 16-KB phase factor dedicated local buffer (Local-PF buffer) for the phase compensation operation is also implemented inside a processing array.

To organize the data transfer between off-chip RAM and on-chip buffers, a data systolic engine is implemented. With this data systolic engine, the input raw image echo can be read and the imaging output can be written back following the processing flow.

4.2. Heterogeneous PE Arrays

Each PE pipelined performs a four-point butterfly operation in six cycles, and all of the PE in a row parallel perform butterfly operations in a block. During the FFT/IFFT operation, all 64 input data are sent to one row of PEs in two cycles from the data buffer, and the 64 output data are written back to the data buffer in two cycles.

In a heterogeneous array, as shown in Figure 7, PEs are interconnected to pass a twiddle factor, the Local-TF buffer distributes the twiddle factor to the PE from top to bottom. The twiddle factor passes two rows down each cycle, and the required twiddle factors are assigned to 16 rows of PEs in eight cycles. Besides, each PE supports zero-padding to expand the raw data to an integer power of two.

During the phase compensation operation, the two input data banks send 32 input data to two rows of FPEs (32 FPEs) in parallel. The Local-PF buffer passes and distributes the phase compensation factor from bottom to top.

4.3. Alternate Systolic-Memory and On-Chip Buffer Organization

Since on-chip memory space is limited, all of the radar echo data is stored in the external memory first. As shown in Figure 8, the data systolic engine (DSE) fetches the data from dynamic random access memory (DRAM) and pushes the data into on-chip memory. To hide the communication latency of data transfer between DSEs and arithmetic components, we employ the alternate systolic technique. In order to avoid DSE competition in hardware resources, we use two alternate systolic memory modules for each of the input/output interfaces for the whole system. At the same time, we adopt two DSE channels for input data and weight at the input end. The proposed memory architecture can provide 4 GB/s of read/write memory bandwidth at 250-MHz frequency to satisfy the data requirements of the processor.

As shown in Figure 8, our storage architecture consists of three layers: DRAM, a data transfer engine system, and an on-chip buffer. Since the on-chip storage resources are limited in size, all the pending radar echo data is first stored in off-chip memory (DRAM). During data processing, the data is first cached by the data transfer engine system into the on-chip buffer, and then sent to the PE array for processing by the on-chip buffer. As shown in Figure 8, in order to hide the communication latency between the off-chip memory and the on-chip buffer, we use the double-buffered data alternate transmission method.

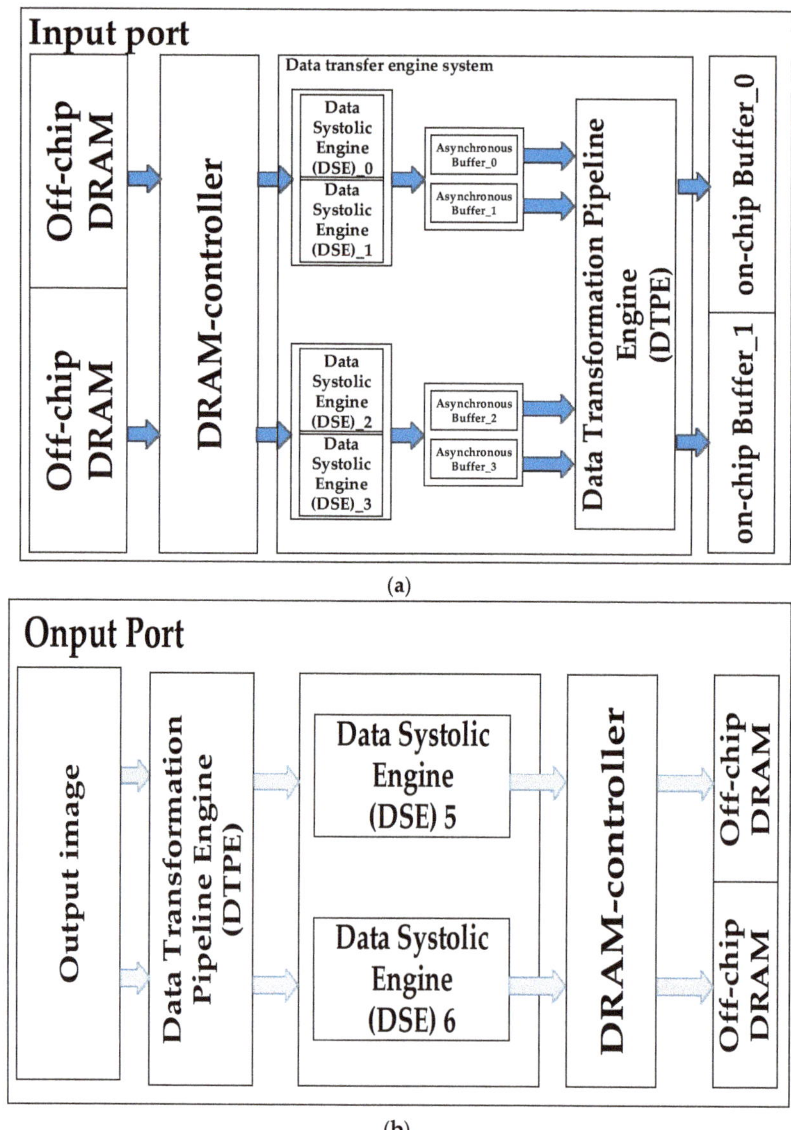

Figure 8. Memory hierarchy architecture. (**a**): Input Port; (**b**): Output Port.

4.4. Resource Controller

The resource controller is responsible for allocating the execution unit and arranging the access flow of the on-chip buffer.

Two imaging blocks are respectively assigned to two arrays for parallel processing. The FFT/IFFT and phase compensation operations are involved in the intra-block processing, so the PE is assigned to the FFT/IFFT during the calculation and the FPE is assigned to the phase compensation operation.

According to the designed data mapping and access strategy, in order to support the parallel access of data, the resource controller allocates bank and bank addresses for each range of data.

When performing range FFT/IFFT, each row of data is stored in four banks according to a distributed storage strategy.

As shown in Figure 9, we take a row of 1024 points as an example ($r = 1024$). When performing FFT/IFFT, 1024 points are segmented and stored in four banks according to the distributed storage strategy. A total of 16 consecutive points are used as a segment, in which approximately 0 to 15 are placed in Bank_0, 16 to 31 are placed in Bank_1, 32 to 47 are placed in Bank_2, and 48 to 63 are placed in Bank_3; the above operation is repeated until all data of 256 segments are stored. A base-4 FFT/IFFT operation at 1024 points requires a total of five levels of operation. The calculation process uses multi-bank parallel data access. Taking the first stage as an example, data 0 to 31 is read from Bank_0 and Bank_1 in the first cycle, and data 992 to 1023 is read from Bank_2 and Bank_3 in the second cycle. The latter four levels of the operational data access process are similar to the first level.

Similarly, when performing azimuth FFT/IFFT, each azimuth of data is stored in four banks according to the storage strategy (taking 1024 points as an example, a = 1024). The data access process is similar to the FFT/IFFT range.

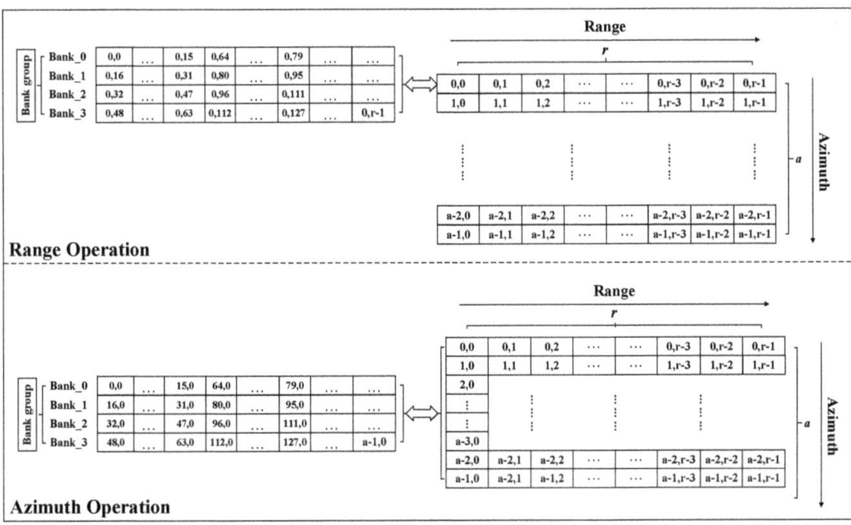

Figure 9. Data access pattern.

SAR imaging is a continuous process with huge differences in operational density between FFT/IFFT and the phase compensation operation. For the characteristics of the computational process, we have designed a way to organize the processing of SAR imaging in space and time flow (ST-Flow), as shown in Figure 10.

Taking 1024 points FFT/IFFT as an example, each FPE performs a one-point phase compensation operation in one cycle, and all the FPE in a row parallel perform phase compensation operations. During the phase compensation operation, all 16-input data are sent to one row of FPEs in one cycle from the data buffer, and 16 output data are written back to the data buffer in one cycle. It can be seen that the 1024-point phase compensation operation requires 64 cycles. In order to satisfy the task saturation and parallelism of the parallel pipeline between phase compensation and FFT/IFFT, the resource controller sets the start time for each row of PE to be delayed by 64 cycles from the previous row. Considering the different matrix sizes, the ratio of PEs to FPEs is configured to be 8:1, so for larger matrices, the FPE will be idle. During the processing of the FPE, it is necessary to wait for the PE to complete the FFT operation before starting the processing of the next frame.

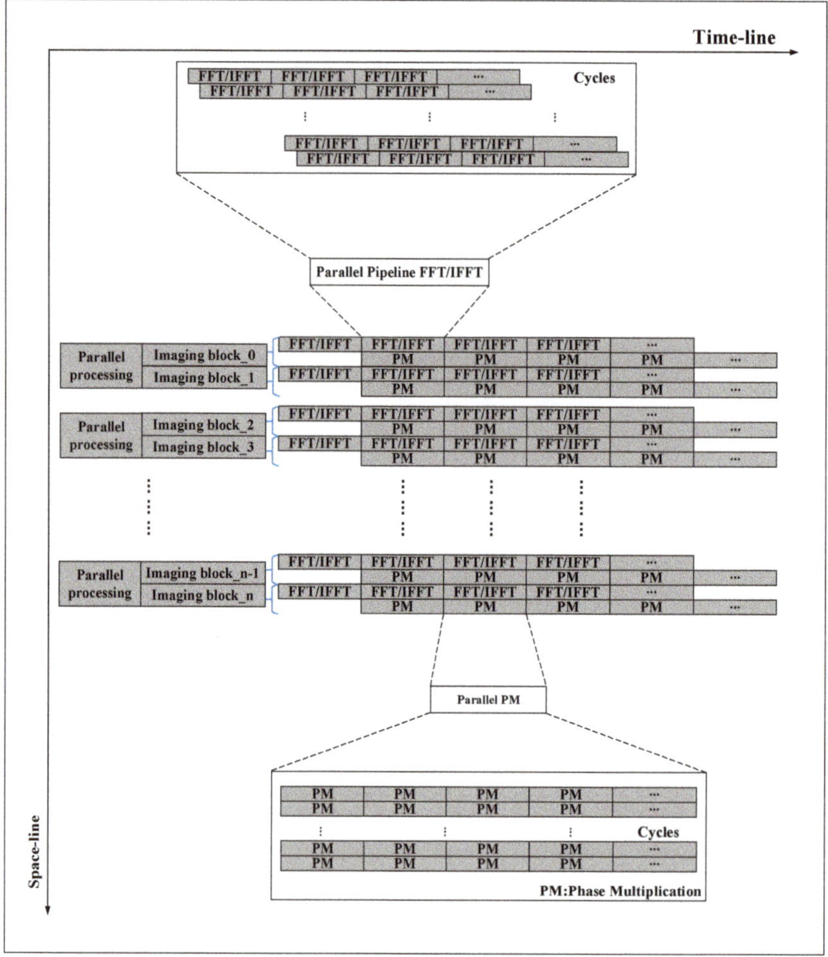

Figure 10. Space–time flow (ST-Flow) of imaging processing.

5. Processor Performance Evaluation

We implemented the SAR imaging processor at 65-nm CMOS (complementary metal oxide semiconductor) technology with 1.2 V of supply voltage using Synopsys tools. Figure 11 shows the die photograph of the chip. In the evaluation, the CS imaging algorithm is selected as the benchmark.

5.1. Performance Analysis

In this section, we configure the processor with fixed-point PE and single-precision floating-point FPE. We evaluate the processor performance at 200 MHz with different fixed-point lengths. The test echo data matrix size is $16,384 \times 16,384$. We perform two operations in parallel on the heterogeneous PE, which can take advantage of the computing power and increase the throughput. When the CSA is processed in heterogeneous PE mode, the throughput is achieved to 115.2 Giga operations per second (GOPS), with 463 mW of power consumption. As shown in Table 4, when all the imaging processes use single-precision floating-point units, the power consumption of the processor is up to 713 mW, and its energy efficiency is only 67% of the fixed/floating point heterogeneous imaging mode. Also, the processor can reduce a small amount of power consumption when selecting low-bit fixed-point

FFT/IFFT operation. The processor consumes 463 mW for 16-bit fixed-point FFT/IFFT and reduces to 454 mW for 12-bit fixed-point FFT/IFFT, as shown in Table 4.

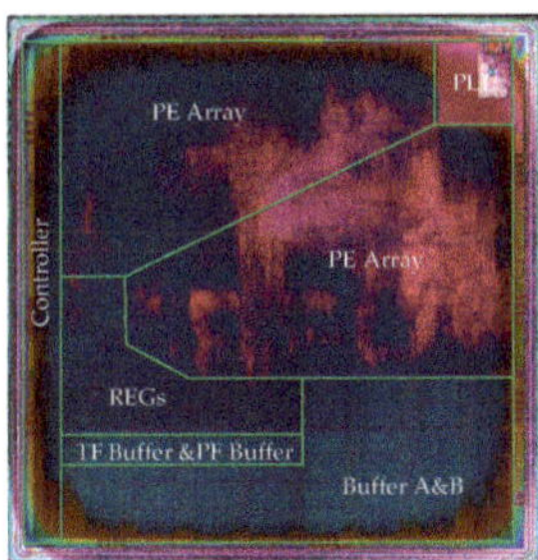

Figure 11. Die photograph of the chip.

Table 4. System performance assessment with different fixed-point length FFT.[1]

PE Bit-Width (Bits)	12	14	16	Single-Precision Floating
Throughput (GOPS)	115.2	115.2	115.2	115.2
Power (mW)	454	459	463	713
Energy efficiency (GOPS/mW)	0.254	0.250	0.240	0.16

[1] Table 4 provides statistics on throughput, power consumption, and energy efficiency for the entire heterogeneous processor.

5.2. Array Utilization Analysis

As shown in Table 5, we can see that in the algorithm processing, the ST-flow two-dimensional parallel pipeline achieves better array utilization than one-dimensional time-based computational flow (TI-flow). The high utilization of the array can increase the throughput of the system. The time-based computational flow (TI-flow) that is employed in existing processors is inefficient for SAR imaging processing. As shown in Table 5, in the ST-Flow mode, the FFT operation and the phase mean (PM) operation are pipelined, the throughput reaches 115.2 GOPS, the resource utilization rate can reach 98.8%, and the energy efficiency is 0.24 GOP/mW. In TI-Flow mode, the FFT operation and the PM operation are executed sequentially, the throughput is only 62.6 GOPS, the resource utilization rate is 54.3%, and the energy efficiency is only 0.16 GOP/mW. Compared with the TI-Flow mode, the resource utilization in ST-Flow mode significantly increases, the throughput increases by 84.5%, and the average power consumption only increases by 21.2%.

Table 5. Array utilization with ST-flow and time-based computational flow (TI-flow). GOPS: Giga operations per second.

-	ST-Flow	TI-Flow		
		FFT	Phase Compensation	Overall
Array utilizations	98.8%	88.9%	11.1%	54.3%
Throughput (GOPS)	115.2	102.4	12.8	62.6
Power (mW)	463	435	317	382
Energy efficiency (GOP/mW)	0.24	-	-	0.16

5.3. Analyzes of Array Scalability

We analyze the performance of a single heterogeneous array, as shown in Figures 12 and 13. On the horizontal (X) axis, the numbers 5, 9, 18, and 36 represent the array scales of 5×4, 9×8, 18×16, and 36×32, respectively.

As the size of the array increases, the throughput and imaging efficiency of the system increase significantly, but the power consumption of the processor also rises sharply. In general, the power-delay product and energy efficiency of large PE arrays are better than those of small PE arrays. On the other hand, the array size must be closely matched to the buffer size; an oversized or undersized array configuration will result in wasted PE resources or low memory bandwidth utilization. Therefore, the size of a single heterogeneous processing array is designed to be 18×16 after a trade-off between the chip implementation complexity and processing performance.

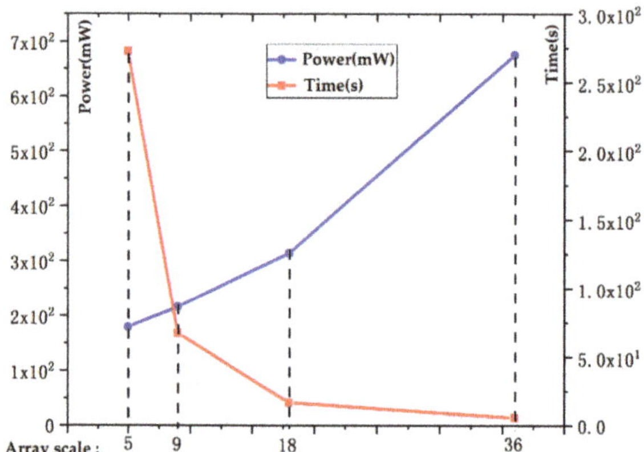

Figure 12. Power consumption and imaging time with different array sizes.

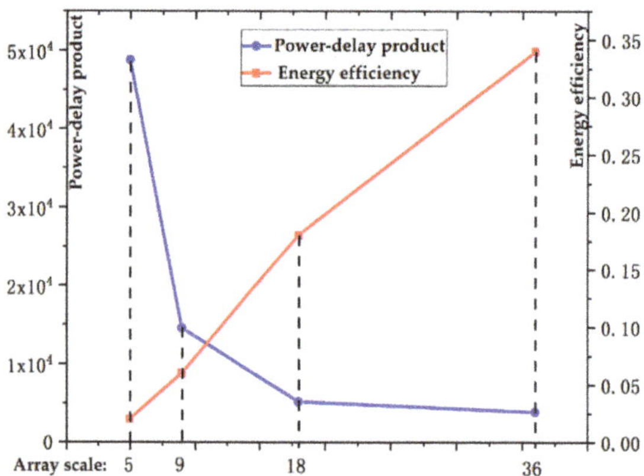

Figure 13. Energy efficiency and power-delay product with different array sizes.

5.4. Comparison with Other Schemes

Table 6 lists the SAR imaging time for different sizes of input. For the ordinary SAR radar (for instance, the Chinese Gaofen-3 satellite, pulse repetition frequency: 2000 Hz), the real-time processing time of $16,384 \times 16,384$ SAR raw data requires 8 s. The proposed scheme can meet the real-time requirements.

The power consumption and SAR imaging time for other studies are also listed in Table 6. As can be seen from Table 6, the power consumption of the proposed scheme is the smallest, because the proposed scheme can completely realize the entire SAR imaging process without additional microcontroller unit (MCU) or CPU. Similar to [15], the Mobile-GPU architecture uses a lower power cost (5 W) to achieve better real-time performance. Compared with [15], the proposed architecture is better in performance-to-power ratio and improves by a factor of 230.4. From the real-time performance perspective, the CPU + GPU scheme is the best, but its power consumption exceeds 300 W. The real-time performance of the proposed scheme is only 8.6% of [17], but the performance-to-power ratio improves by a factor of 63.4. Table 7 shows the comparison of the proposed scheme and related research in real-time performance. As can be seen from Table 7, compared with [15], the speedup ratio reached 21.33.

Table 6. Comparison with previous works.

Architectural Model	Operating Frequency	Power Consumption	SAR Imaging Algorithms	Frame Size	SAR Signal Processing Time (s)
Proposed solution	200 MHZ	463 mW	CS	1024×1024	0.04
				2048×2048	0.15
				6472×3328	0.68
				$16,384 \times 16,384$	8.2
				$30,000 \times 6000$	5.54
				$32,768 \times 32,768$	32.9
GPGPU [14]	-	>500 W	Omega-k	$30,000 \times 6000$	8.5
CPU + GPU [18]	-	345 W	CS	$32,768 \times 32,768$	2.8
Mobile-GPU [16]	2.3 GHZ	5 W	CS	2048×2048	3.2
Microprocessor + FPGA [15]	-	68 W	CS	6472×3328	8
CPU + ASIC [28]	100 MHZ	10 W	-	1024×1024	-

Table 7. Speed-up ratio to previous works.

Architectural	Imaging Time	Imaging Time in Proposed Solution	Speed-Up Ratio	Frame Size
CPU + ASIC [28]	-	0.04 s	-	1024×1024
Mobile-GPU [16]	3.2 s	0.15 s	21.33	2048×2048
Microprocessor + FPGA [15]	8 s	0.68 s	11.76	6472×3328
GPGPU [14]	8.5 s	5.54 s	1.54	$30,000 \times 6000$
CPU + GPU [18]	2.8 s	32.9 s	0.086	$32,768 \times 32,768$

5.5. SAR Imaging Quality Evaluation

We compared the scene SAR imaging results of different fixed-point length FFT. Radar data were obtained from RADARSAT-1 of Canada (width: 50 km; resolution: 6 m) [38]. The imaging effect is shown in Figure 14.

For the actual scenes, the mean square error (*MSE*), peak signal-to-noise ratio (*PSNR*) [39], structural similarity index (*SSIM*) [40], and radiometric resolution (*RL*) [41] are commonly adopted to evaluate SAR imaging quality.

Sufficient imaging accuracy can be achieved with single-precision floating-point imaging. Fixed-point processing methods will cause a certain loss of precision. We take the single-precision floating-point imaging as the test reference to evaluate the fixed-point FFT SAR image quality.

The *MSE* is adopted to calculate the squared intensity difference between the pixels of the partial fixed-point image and the pixels of the full single-precision floating-point image. The *PSNR* is essentially the same as the *MSE*, but it is associated with the quantized gray level of the SAR image. The *MSE* and *PSNR* are calculated as shown in Formulas (2) and (3):

$$MSE = \frac{1}{M \times N} \sum_{i=1}^{M} \sum_{j=1}^{N} (f'(i,j) - f(i,j))^2 \tag{2}$$

$$PSNR = 10 \log_{10} \frac{Q^2 \times M \times N}{\sum_{i=1}^{M} \sum_{j=1}^{N} \left(f'(i,j) - f(i,j)\right)^2} \tag{3}$$

where $f'(i,j)$ and $f(i,j)$ represent the image pixels to be evaluated and the reference image pixels, respectively; M, N represent the length and width of the image, respectively. Q represents the gray level of the image ($Q = 255$).

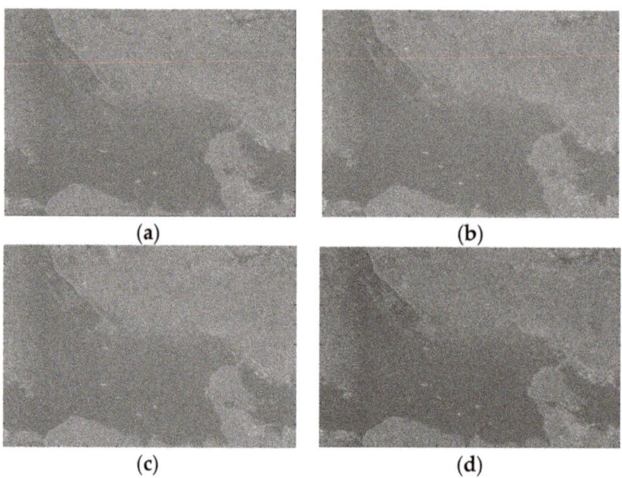

Figure 14. The scene SAR imaging results for different fixed-point length FFT. (**a**) 12-bit fixed-point FFT; (**b**) 14-bit fixed-point FFT; (**c**) 16-bit fixed-point FFT; (**d**) single-precision float-point FFT.

PSNR and *MSE* are simple and straightforward SAR image quality assessments based on the visibility of errors. Due to the *PSNR* index not being exactly the same as the visual quality seen by the human eye, the evaluation requirements of the human visual system (HVS) cannot be met [40]. Therefore, we also adopt *SSIM* (the Structural Similarity Index) to evaluate the SAR images. As shown in Formula (4):

$$SSIM(x, y) = \frac{\left(2\varphi_x\varphi_y + \varepsilon_1\right)\left(2\delta_{xy} + \varepsilon_2\right)}{\left(\varphi_x^2 + \varphi_y^2 + \varepsilon_1\right)\left(\delta_x^2 + \delta_y^2 + \varepsilon_2\right)} \tag{4}$$

where δ_x^2 represents the fixed-point image variance, and δ_y^2 represents the single-precision floating-point image variance; φ_x represents the mean value of the fixed-point image, and φ_y represents the mean value of the single-precision floating-point image. The *SSIM* value range is [0, 1], and the larger the SSIM value, the smaller image distortion.

RL is also a very important evaluation indicator. *RL* is adopted to evaluate the minimum variation of target reflection that radar sensors can distinguish. As shown in Formula (5):

$$RL = 10 \log_{10}\left(\frac{\alpha}{\beta} + 1\right) \tag{5}$$

where α represents the standard deviation of the image, and β represents the mean value of the image.

Table 8 lists the loss of precision due to the different data widths. As can be seen from Table 8, the *PSNR* value of a partial 16-bit fixed-point image can reach 29.1 dB, the results show that the partial 16-bit fixed-point image and the single-precision floating-point image differ only by 0.02 and 0.05 dB on the two indexes of *SSIM* and *RL*, respectively. For the actual scene SAR imaging, compared with a single-precision floating-point image, the accuracy loss of a partial 16-bit fixed-point image is within 2%.

Table 8. Quantitative evaluation of actual scene SAR imaging.

FFT Pro-Acc [1]	*PSNR* [2] (dB)	*MSE* [3] (dB)	*SSIM* [4] (dB)	*RL* [5] (dB)
Single-precision float-point	∞	0	1	4.99
12-bit fixed-point	13.7	2765.2	0.23	4.11
14-bit fixed-point	22.4	377.4	0.77	4.71
16-bit fixed-point	29.1	81.8	0.98	4.94

[1] FFT pro-acc: FFT processing accuracy; [2] PSNR: peak signal-to-noise ratio; [3] MSE: mean square error; [4] SSIM: Structural Similarity Index; [5] RL: Radiometric Resolution.

Phase is also important information for a SAR image. The phase mean (*PM*) and phase deviations (*PD*) are estimated by the method proposed in [42]. Table 9 lists the phase precision with different fixed-point SAR imaging. As can be seen from Table 9, the loss of phase precision with partial 16-bit fixed-point imaging is less than 3%.

Table 9. Phase information evaluation of actual scene SAR imaging.

FFT Pro-Acc [1]	*PM* [2]	*PD* [3]
Single-precision float-point	0.00244°	3.3026°
12-bit fixed-point	0.00916°	3.3054°
14-bit fixed-point	0.00398°	3.3032°
16-bit fixed-point	0.00252°	3.3026°

[1] FFT pro-acc: FFT processing accuracy; [2] PM: phase mean; [3] PD: phase deviation.

For the point target imaging quality evaluation, we adopted the point target simulation echo data. We compared the point target SAR imaging results for FFT with different fixed-point lengths, as shown in Figure 15. For the point target image, spatial resolution (RES), peak side lobe ratio (PSLR) and integrated side lobe ratio (ISLR) are commonly adopted to assess imaging quality [38,43]. Table 10 shows the results of the point targets imaging quality assessment and comparison.

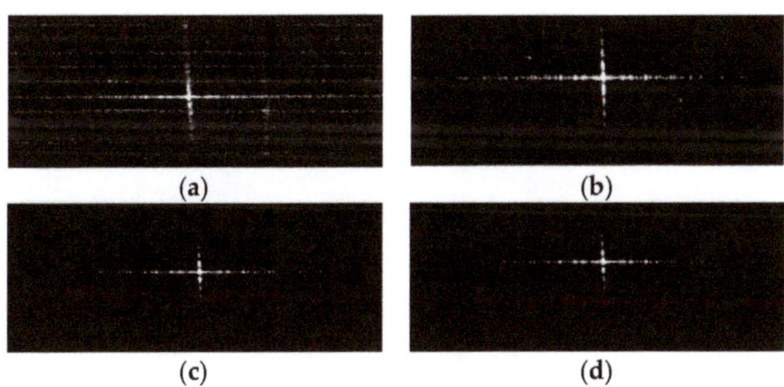

Figure 15. The point target SAR imaging results for different fixed-point length FFT. (**a**) 12-bit fixed-point FFT; (**b**) 14-bit fixed-point FFT; (**c**) 16-bit fixed-point FFT; (**d**) single-precision float-point FFT.

Table 10. Quantitative evaluation of point target SAR imaging.

FFT Pro-Acc [1]	Azimuth Direction			Range Direction		
	RES [2] (m)	PSLR [3] (dB)	ISLR [4] (dB)	RES (m)	PSLR (dB)	ISLR (dB)
Single-precision float-point	4.74	−12.91	−9.64	2.58	−13.31	−9.96
12-bit fixed-point	5.43	−5.68	−2.99	3.71	−5.88	−3.22
14-bit fixed-point	4.81	−11.85	−8.22	2.83	−11.55	−9.09
16-bit fixed-point	4.77	−12.86	−9.53	2.61(m)	−13.28	−9.93

[1] FFT pro-acc: FFT processing accuracy; [2] RES: spatial resolution; [3] PSLR: peak side lobe ratio; [4] ISLR: integrated side lobe ratio.

For partial 16-bit fixed-point imaging, in the azimuth direction, the PSLR and ISLR precision loss of the image are 0.3% and 0.8%, respectively; the RES precision loss is 0.2%. In the range direction, the PSLR and ISLR precision losses of the image are 0.2% and 0.2%, respectively; the RES precision loss is 0.7%.

According to the actual scene and the point target image quantization analysis, as shown in Tables 8–10, the partial 16-bit fixed-point imaging accuracy is close to the single-precision floating-point imaging accuracy, which meets the requirements of on-orbit SAR imaging applications.

6. Conclusions

This paper proposes a heterogeneous imaging processor using fixed-floating point heterogeneous parallel acceleration technology to perform SAR imaging in the aerospace field. The processor consists of two 18×16 heterogeneous arrays that provide 115.2 GOPS throughput. To improve energy efficiency, each array supports fixed-floating hybrid calculations to take full advantage of computing resources, which can increase the throughput of imaging processing by 1.82 times. At the same time, the PE array can be partitioned by rows through a sensible algorithm-to-hardware architecture mapping, process the imaging process in parallel, provide high-utilization hardware resources, and improve the efficiency by a factor of 1.5. A single processor requires 8 s and consumes 463 mW to process SAR raw data with a granularity of $16,384 \times 16,384$, which meets the limits real-time and power consumption of the on-orbit SAR imaging platform. The proposed solution also has good scalability, by extending the size of the processor array, the real-time requirements of larger-scale SAR imaging can be met.

Author Contributions: Investigation, J.A.; Methodology, S.W.; Project administration, S.W. and S.Z.; Software, X.H. and J.A.; Writing—original draft, S.W.; Writing—review & editing, L.C.

Funding: This research received no external funding.

Conflicts of Interest: The authors declare no conflict of interest.

References

1. Percivall, G.S.; Alameh, N.S.; Caumont, H.; Moe, K.L.; Evans, J.D. Improving Disaster Management Using Earth Observations—GEOSS and CEOS Activities. *IEEE J. Sel. Top. Appl. Earth Obs. Remote Sens.* **2013**, *6*, 1368–1375. [CrossRef]
2. Joyce, K.E.; Belliss, S.E.; Samsonov, S.V.; McNeill, S.J.; Glassey, P.J. A review of the status of satellite remote sensing and image processing techniques for mapping natural hazards and disasters. *Prog. Phys. Geog.* **2009**, *33*, 183–207. [CrossRef]
3. Tralli, D.M.; Blom, R.G.; Zlotnicki, V.; Donnellan, A.; Evans, D.L. Satellite remote sensing of earthquake, volcano, flood, landslide and coastal inundation hazards. *ISPRS J. Photogramm. Remote Sens.* **2005**, *59*, 185–198. [CrossRef]
4. Gierull, C.H.; Vachon, P.W. Foreword to the Special Issue on Multichannel Space-Based SAR. *IEEE J. Sel. Top. Appl. Earth Obs. Remote Sens.* **2015**, *8*, 4995–4997. [CrossRef]
5. Radarsat Constellation Mission. Available online: http://directory.eoportal.org/web/eoportal/satellite-missions/r (accessed on 23 March 2017).
6. Advanced Land Observing Satellite-2. Available online: https://directory.eoportal.org/web/eoportal/satellite-missions (accessed on 23 March 2017).
7. TerraSAR-X Add-on for Digital Elevation Measurement. Available online: https://directory.eoportal.org/web/eoportal/satellite-missions/t/tandem-x (accessed on 23 march 2017).
8. Yang, L.; Zhao, L.; Zhou, S.; Guo, B. Sparsity-Driven SAR Imaging for Highly Maneuvering Ground Target by the Combination of Time-Frequency Analysis and Parametric Bayesian Learning. *IEEE J. Sel. Top. Appl. Earth Obs. Remote Sens.* **2017**, *4*, 1443–1455. [CrossRef]
9. Zhu, S.; Liao, G.; Tao, H.; Yang, Z. Estimating Ambiguity-Free Motion Parameters of Ground Moving Targets from Dual-Channel SAR Sensors. *IEEE J. Sel. Top. Appl. Earth Obs. Remote Sens.* **2014**, *7*, 3328–3349. [CrossRef]

10. Wai-Chi, F.; Jin, M.Y. On board processor development for NASA's spaceborne imaging radar with VLSI system-on-chip technology. In Proceedings of the 2004 IEEE International Symposium on Circuits and Systems (IEEE Cat. No. 04CH37512), Vancouver, BC, Canada, 23–26 May 2004; p; Volume 2, p. II-901-4.

11. Raney, R.K.; Runge, H.; Bamler, R.; Cumming, I.G.; Wong, F.H. Precision SAR processing using chirp scaling. *IEEE Trans. Geosci. Remote Sens.* **1994**, *32*, 786–799. [CrossRef]

12. Li, G.; Zhang, F.; Ma, L.; Hu, W.; Li, W. Accelerating SAR imaging using vector extension on multi-core SIMD CPU. In Proceedings of the 2015 IEEE International Geoscience and Remote Sensing Symposium (IGARSS), Milan, Italy, 26–1 July 2015.

13. Peternier, A.; Boncori, J.P.M.; Pasquali, P. Near-real-time focusing of ENVISAT ASAR Stripmap and Sentinel-1 TOPS imagery exploiting OpenCL GPGPU technology. *Remote Sens. Environ.* **2017**, *202*, 45–53. [CrossRef]

14. Lou, Y.; Clark, D.; Marks, P.; Muellerschoen, R.J.; Wang, C.C. Onboard Radar Processor Development for Rapid Response to Natural Hazards. *IEEE J. Sel. Top. Appl. Earth Obs. Remote Sens.* **2016**, *9*, 2770–2776. [CrossRef]

15. Tang, H.; Li, G.; Zhang, F.; Hu, W.; Li, W. A Spaceborne SAR on-board processing simulator using mobile GPU. In Proceedings of the 2016 IEEE International Geoscience and Remote Sensing Symposium (IGARSS), Beijing, China, 10–15 July 2016.

16. Ge, B.; Chen, L.; An, D.; Zhou, Z. GPU-based FFBP algorithm for high-resolution spotlight SAR imaging. In Proceedings of the 2017 IEEE International Conference on Signal Processing, Communications and Computing (ICSPCC), Xiamen, China, 22–25 October 2017; pp. 1–5.

17. Zhang, F.; Li, G.; Li, W.; Hu, W.; Hu, Y. Accelerating Spaceborne SAR Imaging Using Multiple CPU/GPU Deep Collaborative Computing. *Sensors* **2016**, *16*, 494. [CrossRef] [PubMed]

18. Wu, Z.; Liu, Y.; Zhang, L.; Li, N.; Du, K.; Balz, T. Highly efficient synthetic aperture radar processing system for airborne sensors using CPU+GPU architecture. *J. Appl. Remote Sens.* **2015**, *9*, 097293. [CrossRef]

19. Ye, J.; Shanqing, H.; Jiayun, Z.; Teng, L. Virtual single-node processing for SAR imaging based on multi-DSP. In Proceedings of the 2016 IEEE International Conference on Signal Processing, Communications and Computing (ICSPCC), Hong Kong, China, 5–8 August 2016.

20. Zhijun, Y.; Xiangfei, N.; Wenyi, X.; Xiaowei, N.; Weiming, T. Real time imaging processing of ground-based SAR based on multicore DSP. In Proceedings of the 2017 IEEE International Conference on Imaging Systems and Techniques (IST), Beijing, China; 2017; pp. 1–5.

21. Xingyu, X.; Abou-Khousa, M.A.; Al-Wahedi, K. Embedded Synthetic Aperture Radar Imaging System on Compact DSP Platform. In Proceedings of the 2017 International Conference on Electrical and Computing Technologies and Applications, Ras Al Khaimah, UAE, 21–23 November 2017.

22. Langemeyer, S.; Kloos, H.; Simon-Klar, C.; Friebe, L.; Hinrichs, W.; Lieske, H.; Pirsch, P. A compact and flexible multi-DSP system for real-time SAR applications. In Proceedings of the IEEE International Geoscience & Remote Sensing Symposium, Anchorage, AL, USA, 20–24 September 2004.

23. Desai, N.M.; Kumar, B.S.; Sharma, R.K.; Kunal, A.; Gameti, R.B.; Gujraty, V.R. Near Real Time SAR Processors for ISRO's Multi-Mode RISAT-I and DMSAR. In Proceedings of the 7th European Conference on Synthetic Aperture Radar, Friedrichshafen, Germany, 2–5 June 2008; pp. 1–4.

24. Di, W.; Chen, C.; Liu, Y. FPGA-Based Multi-core Reconfigurable System for SAR Imaging. In Proceedings of the IGARSS 2018-2018 IEEE International Geoscience and Remote Sensing Symposium, Valencia, Spain, 2018, 22–27 July; pp. 8921–8924.

25. Li, B.; Li, C.; Xie, Y.; Chen, L.; Shi, H.; Deng, Y. A SoPC based Fixed Point System for Spaceborne SAR Real-Time Imaging Processing. In Proceedings of the 2018 IEEE High Performance Extreme Computing Conference (HPEC), Waltham, MA, USA; 2018; pp. 1–6.

26. Bierens, L.; Vollmuller, B.J. On-board Payload Data Processor (OPDP) and its application in advanced multi-mode, multi-spectral and interferometric satellite SAR instruments. In Proceedings of the EUSAR 2012, 9th European Conference on Synthetic Aperture Radar, Nuremberg, Germany, 23–26 April 2012.

27. Le, C.; Chan, S.; Cheng, F.; Fang, W.; Fischman, M.; Hensley, S.; Johnson, R.; Jourdan, M.; Marina, M.; Parham, B.; et al. Onboard FPGA-based SAR processing for future spaceborne systems. In Proceedings of the IEEE 2004 Radar Conference, Philadelphia, PA, USA, 29–29 April 2004; pp. 15–20.

28. Fang, W.C.; Le, C.; Taft, S. On-board fault-tolerant SAR processor for spaceborne imaging radar systems. In Proceedings of the 2005 IEEE International Symposium on Circuits and Systems, Kobe, Japan, 23–26 May 2005; pp. 420–423.

29. Greco, J.; Cieslewski, G.; Jacobs, A.; Troxel, I.A.; George, A.D. Hardware/software Interface for High-performance Space Computing with FPGA Coprocessors. In Proceedings of the 2006 IEEE Aerospace Conference, Big Sky, MT, USA, 4–11 March 2006; pp. 1–10.

30. Pfitzner, M.; Cholewa, F.; Pirsch, P.; Blume, H. FPGA based Architecture for real-time SAR processing with integrated Motion Compensation. In Proceedings of the 2013 Asia-Pacific Conference on Synthetic Aperture Radar (Apsar), Tsukuba, Japan, 23–27 September 2013; pp. 521–524.

31. Bi, D.; Xie, Y.; Li, X.; Zheng, Y.R. Efficient 2-D synthetic aperture radar image reconstruction from compressed sampling using a parallel operator splitting structure. *Digit. Signal Process.* **2016**, *50*, 171–179. [CrossRef]

32. Long, T.; Yang, Z.; Li, B.; Chen, L.; Ding, Z.; Chen, H.; Xie, Y. A Multi-mode SAR Imaging Chip based on a Dynamically Reconfigurable SoC Architecture Consisting of Dual-operation Engines and Multilayer Switching Network. *Preprints* **2018**. [CrossRef]

33. Song, W.S.; Baranoski, E.J.; Martinez, D.R. One trillion operations per second on-board VLSI signal processor for Discoverer II space based radar. In Proceedings of the Aerospace Conference Proceedings, Big Sky, MT, USA, 25–25 March 2000.

34. Chen, Q.; Yu, A.; Sun, Z.; Huang, H. A multi-mode space-borne SAR simulator based on SBRAS. In Proceedings of the 2012 IEEE International Geoscience and Remote Sensing Symposium, Munich, Germany, 22–27 July 2012; pp. 4567–4570.

35. Stangl, M.; Werninghaus, R.; Schweizer, B.; Fischer, C.; Brandfass, M.; Mittermayer, J.; Breit, H. TerraSAR-X technologies and first results. *IEE Proc. Sonar Navig.* **2006**, *153*, 86–95. [CrossRef]

36. Cooley, J.W.; Tukey, J.W. An algorithm for the machine calculation of complex Fourier series. *Math. Comput.* **1965**, *19*, 297–301. [CrossRef]

37. Yang, C.; Li, B.; Chen, L.; Wei, C.; Xie, Y.; Chen, H.; Yu, W. A Spaceborne Synthetic Aperture Radar Partial Fixed-Point Imaging System Using a Field-Programmable Gate Array—Application-Specific Integrated Circuit Hybrid Heterogeneous Parallel Acceleration Technique. *Sensors* **2017**, *17*, 1493. [CrossRef] [PubMed]

38. Cumming, I.G.; Wong, F.H. *Digital Processing of Synthetic Aperture Radar Data: Algorithms and Implementation*; Artech house: Norwood, MA, USA, 2005.

39. Hu, A.; Zhang, R.; Yin, D.; Chen, Y. Perceptual quality assessment of SAR image compression. *Int. J. Remote Sens.* **2013**, *34*, 8764–8788. [CrossRef]

40. Wang, Z.; Bovik, A.C.; Sheikh, H.R.; Simoncelli, E.P. Image quality assessment: From error visibility to structural similarity. *IEEE Trans. Image Process.* **2004**, *13*, 600–612. [CrossRef] [PubMed]

41. Márquez-Martínez, J.; Mittermayer, J.; Rodríguez-Cassolà, M. Radiometric resolution optimization for future SAR systems. In Proceedings of the IEEE International Geoscience & Remote Sensing Symposium, Anchorage, AL, USA, 20–24 September 2004.

42. Tell, B.R.; Laur, H. Phase preservation in SAR processing: The interferometric offset test. In Proceedings of the 'Remote Sensing for a Sustainable Future', International Geoscience and Remote Sensing Symposium, Lincoln, NE, USA, 27–31 May 1996; Volume 1, pp. 477–480.

43. Zhang, H.; Li, Y.; Su, Y. SAR image quality assessment using coherent correlation function. In Proceedings of the 2012 5th International Congress on Image and Signal Processing, Chongqin, China, 16–18 October 2012.

MDPI

St. Alban-Anlage 66

4052 Basel

Switzerland

Tel. +41 61 683 77 34

Fax +41 61 302 89 18

www.mdpi.com

Sensors Editorial Office

E-mail: sensors@mdpi.com

www.mdpi.com/journal/sensors